FUNDAMENTALS OF FARMING AND FARM LIFE

FIG. 1a. The native Texas Longhorn.
Courtesy Mr. L. A. Schreiner.

FIG. 1b. The result of scientific breeding, selection, and feeding. A grand champion Aberdeen-Angus steer.
Courtesy Agricultural and Mechanical College of Texas.

FUNDAMENTALS OF FARMING AND FARM LIFE

Edwin Jackson Kyle
Alexander Caswell Ellis

1999

BIOTECH BOOKS
Delhi-110035

FIRST INDIAN EDITION 1999

KYLE, EDWIN JACKSON 1876–

ISBN 81 - 7622 - 032 - 9

Published by : **BIOTECH BOOKS**
1123/74, Tri Nagar,
Delhi-110035
Phone : 7109765

Printed at : **Chawla Offset Printers**
Delhi-110052

PREFACE

WHILE there have been great strides in recent years in the development of textbooks on elementary agriculture, the best texts are still unsatisfactory in many ways. The defects complained of arise largely out of the fact that these texts are usually prepared by busy scientists who, though they may know agriculture, have made little study of the mental processes of children, have had little or no experience in teaching children, and are not very familiar with the exact conditions in the schools in which the texts are to be used. The result is that the subject-matter has not always been wisely chosen in the light of children's natural interests and needs, nor has the subject-matter that was chosen been organized in a careful, pedagogical manner, by starting always with the known, leading to the unknown through the known, and making the study a closely knit, coherent, living whole. Indeed, several of the texts are mere collections of articles on a dozen or more varied topics written by specialists in different institutions and without any common plan of organization or method of presentation. The children are not introduced gradually to the new technical terms, nor are proper apperceptive bases first laid before attempting to present the more complex processes of plant or animal growth.

It was thought that by combining the efforts of the agricultural professor with those of the pedagogue it would be possible to provide a textbook on elementary agriculture that would have the advantage of the accuracy and scholarship of the specialist in agriculture, and have also the organization and pedagogical manner of presentation given to it by the specialist in education familiar with the conditions in the rural schools. All of the chapters were therefore first prepared on a uniform general plan by experts in the several fields and turned in to the pedagogical editor, who attempted to put each chapter into more teachable form and to organize the whole. It is believed that because of the carefully graded steps in the subject-matter, the preparation that is always provided for each new subject, and the simple language used the pupil can in this text cover with little difficulty much more ground and learn more complicated processes than he can with the usual text.

It was also the purpose of the authors to provide a book that did not stop with giving mere advice about farm practices to be memorized by the pupils, but taught first of all the fundamental principles of plant and animal growth and reproduction and of soil management. In this way the pupils are made able to understand the reasons for farm practices and to criticise any old or new practice independently. By emphasizing these far-reaching principles first, the thousand and one details of practice in the study of farm crops and animal husbandry which are later given become a matter for simple reasoning instead of a mere dead weight to be carried in memory. In order still further to develop the powers of observation, exercise the reason, and connect the lessons with the daily life and home needs of the pupils, lists of suggestive questions, exercises, and problems are provided for each subject.

We have in general tried to broaden the conception of the course in elementary agriculture from that of a mere treatise on raising plants and animals for sale to that of a means of preparation for living intelligently and happily, as well as profitably, on the farm.

It is a great pleasure to express here our appreciation of the assistance generously given by our colleagues in preparing this book.

Members of the faculty of the Agricultural and Mechanical College of Texas prepared material for chapters as follows: Professor J. C. Burns, Professor of Animal Husbandry, the chapters on animal husbandry, cattle, horses, sheep, hogs, and the care and feeding of animals; Professor

J. O. Morgan, Professor of Agronomy, the chapters on farm crops and manures, fertilizers and rotation; Professor J. W. Ridgway, Professor of Dairy Husbandry, the chapter on the care of milk and its products; Professor C. M. Evans, Superintendent of Agricultural Extension, the chapter on silos; Professor R. J. Potts, Professor of Highway Engineering, the chapter on roads; Instructor E. F. Ferrin, the chapter on poultry. In every case the limits of an elementary text necessitated cutting down and modifying the valuable material and illustrations furnished. Assistance in securing illustrations or criticism and suggestions were also received from Professors G. S. Fraps, State Chemist; Wilmon Newell, State Entomologist; G. S. Templeton, S. A. McMillan, and Harper Dean.

From the Texas Agricultural Experiment Station Professor H. Ness gave assistance through criticism and suggestion, and Director B. Youngblood and Mr. J. M. Johnson, Expert in Farm Management of the United States Department of Agriculture, stationed in Texas, prepared the material for the chapter on farm planning and accounting.

Of the faculty of the University of Texas we are indebted for valuable criticism and suggestions on the first twenty-eight pages to Professor F. DeF. Heald; on plant growth, reproduction, and soils to Doctor I. McK. Lewis; on plant growth, farm crops, the garden, orchard, and shade trees, and plant enemies to Instructor C. H. Winkler.

We are also indebted to Professor O. S. Morgan, of Columbia University, for very helpful criticisms and suggestions on many of the chapters.

While the authors would make every acknowledgment to their generous colleagues for the excellence of the material in the several chapters, we wish to accept entire responsibility for any inadequacy or inaccuracy that may have resulted from the revision and reduction of the material to its present form.

The many acknowledgments due for courtesies in the use

of illustrative material are made in the body of the text. To Mr. D. T. Stephens we are indebted for the preparation of a large part of the drawings in the first four chapters, and to Miss Florence Rhine for several drawings in later chapters. We wish to express also our appreciation of the great assistance given by both the editorial and art departments of the publishers.

E. J. Kyle.
A. Caswell Ellis.

PREFACE TO THE THIRD REVISED EDITION

TWENTY-FIVE years of successful use of this text in the classroom have proved the soundness of the plan upon which it is written and its teachability. It actually arouses the interest not only of the pupils but of the teachers, and even the parents at home, who also read it.

However, the rapid advances in the science of agriculture and in the possibilities of farm living demand that a third revised edition be prepared. The publishers have therefore agreed to publish this very fully revised edition, which, in such fields as soil conservation, farm crops, soils and fertilizers, insects and diseases, wild game, vocational agriculture, agricultural clubs, and in parts of animal husbandry, contains new materials that have never before been available in an elementary text.

The authors wish to acknowledge their great indebtedness for valuable help in their special fields to the following members of the staff of the Agricultural and Mechanical College of Texas: Doctor G. W. Adriance, Head of the Department of Horticulture; Doctor S. W. Bilsing, Head of the Department of Entomology; Professor F. R. Brison, Department of Horticulture; Mrs. Minnie Fisher Cunningham, Extension Editor, for Appendix on 4-H Clubs; Professor A. L. Darnell, Department of Dairy Husbandry; Doctor G. S. Fraps, State Chemist; Doctor R. L. Hunt, Department of Agricultural Economics; Professor F. R. Jones, Department of Agricultural Engineering; Doctor L. G. Jones, Department of Agronomy; Professor J. S. Mogford, Department of Agronomy; Professor D. H. Reid, Head of the Department of Poultry Husbandry; Professor C. N. Shepardson, Head of the Department of Dairy

Husbandry; Doctor J. J. Taubenhaus, Plant Pathologist of the Agricultural Experiment Station; Doctor W. P. Taylor, Chief of the Division of Wild Life, Agricultural Experiment Station; Doctor F. L. Thomas, State Entomologist; Professor L. M. Thompson, Department of Agronomy; Mr. P. H. Walser, Senior Soil Conservationist; Professor D. W. Williams, Head of the Department of Animal Husbandry. We are also greatly indebted to Mr. J. B. Rutland, State Supervisor of Agricultural Education for the State Department of Education, for valuable suggestions in the planning of this revision; and to Miss Ruth Peskin for efficient help in preparing the manuscript, correcting the proof, and preparing the Glossary and Index.

Numerous other indebtednesses are acknowledged in the text.

Again, we are further indebted to our always co-operative and efficient publishers.

E. J. KYLE.
A. CASWELL ELLIS.

College Station, Texas.

SUGGESTIONS FOR THE TEACHER

THE teacher should encourage the natural interest which both country boys and girls, and town boys and girls, have in growing plants and animals; should show them how agriculture is receiving the best thought of many of the most intelligent men and women in the world; how it offers to them not merely a happy and useful life, but as great a field for the exercise of intelligence and character and the application of scientific methods as do commerce, law, medicine, or any other field of effort. If the subject is fairly presented in both town and country schools it will attract not only such farmers' sons and daughters as are suited to farm life, but other children whose talents lie in this direction. Whether one ever engages in farming or not, his sympathies and outlook in life and his usefulness as a citizen will be vastly broadened by a well-planned study of the problems of agriculture.

Before one begins to teach agriculture or any other subject, he should get a clear idea of the ends to be attained in teaching that subject. Is the aim of the school course in agriculture to teach children certain facts and principles concerning farming and to have them gain a certain amount of skill in farm operations? Undoubtedly it is; but this is not all. Those children who will be farmers will be something else besides; they will be human beings with rich and varied mental capacities that need exercise and opportunity to develop. In a word, the farmer is a man, and his course in agriculture should not only teach him about farming operations, but such subject-matter should be selected, and such methods employed in teaching, that, while he learns his agriculture, he learns also to use effectively his mental and moral powers, and to understand better his relation to his neighbors, to nature, and to the Author of nature.

Teaching the child the bare fact that a certain fertilizer is good for cotton, and a certain food good for hogs, has no special cultural value, and may have very little value for his future farming; but helping him to learn how plants and animals grow and are nourished, and to apply these principles to cotton or pig culture, gives him a kind of mental exercise that will help him in all similar problems in life. Moreover, it gives him certain general principles of feeding and of growth that will make his thinking clearer when later he tries to care intelligently for his own body or provide for his family or his nation. It is not the subject you are teaching that determines whether you are cultivating the minds of your pupils, but the character of the subject-matter selected within that subject and the methods employed in teaching it. For this reason great care has been exercised in this book to select those great but simple, far-reaching principles that underlie successful farming and farm life, and to show by concrete observations how these principles apply, instead of teaching merely unrelated facts and concrete methods of doing farm work. Don't require pupils to memorize all the varieties of cane or milo, or all the sections in which each crop is raised, or the exact number of hogs in Iowa, or the exact limits of weights for Tamworths, or other such details. The exact figures and other details are given in the text to use for reference when needed and to round out the picture for the child when studying. It is worse than a waste of the child's time to have him memorize from day to day masses of details which he forgets in less than a week. The needed details regarding a few crops and animals especially important for the neighborhood may be thoroughly mastered and held in mind by reviews.

We would urge the teacher not to be satisfied with merely getting the children to understand the explanations and our illustrations, but always to stimulate them to exercise their own reasoning powers by finding other applications of these principles in their own experience or observation. To help the teacher, we have provided all through the book suggestive questions and problems for the pupils. Insist on the pupils thinking these answers out, or working them out for themselves. Don't tell them the answers, but help them to think for themselves. We have provided also definite observation problems and tests and experiments, so that pupils may get practice in reading and following directions, in observing carefully, in recording accurately their observations, and in drawing rational conclusions. We have carefully selected such problems and experiments as are not beyond the powers of the sixth or seventh grade boy or girl. While it is difficult for children at the beginning to carry out directions carefully, record observations accurately, and reason clearly, most of them soon learn to do these things if required to do so and if properly encouraged. The exercise of their new power and the recognition of their own new strength give them great pleasure and increase their interest in the work. Each child should have his own notebook, and keep it carefully all through the course. Remember that the children get new power more rapidly by doing and thinking things out for themselves than they do from the teacher's doing things for them or telling them about what others have done. It takes more time and intelligence on the part of the teacher at first to get the children to do things right themselves, and to think for themselves, than it does to do things for them and think for them; but when the pupils have once learned to think and do for themselves, they progress more and more rapidly as they go, whereas, by the other plan, they get weaker and weaker and need more help as the course advances. A child that has germinated a seed, marked the root and stem according to directions, watched to find out how each grows, and has written a record of his observation, will not only remember the results easily, but will be better able to work out some other new problem for himself; whereas the child that merely reads about how roots and stems grow, or is told about it by the teacher, is likely soon to forget it, and may be very little more able to work out a new problem at the end of his course than he was at the start.

Again, the course in agriculture should be correlated with the other work of the school. The lessons in agriculture should help the spelling and reading work by requiring correct spelling in the notebooks and papers, and by having the children read interesting bulletins and nature poems and literary masterpieces dealing with farm life and the country. They should help the geography study by arousing interest in such questions as the relation of climate to production, the arithmetic study by sums in farm mathematics, and all science classes by pointing out the practical applications of the principles of these sciences in the daily work of life. In fact, agriculture properly taught enriches and enlivens every other study in the school by helping to connect these with the daily life and daily needs of the pupils. It gives the pupils not only facts and principles that will help them to get greater returns from their labor on the farm, but it starts new lines of interest and opens up to the young minds broad fields for future study in the varied sciences, some of whose principles they have learned to appreciate in their study of agriculture.

It is hardly possible for any average child to work all of the problems, perform all of the experiments, make all of the observations, and write all of the reports suggested in this book. Circumstances and local conditions are so varied that no single set would suit all equally well. The large number given makes it easy to find a set suited to practically any condition. A part of the experiments may be performed by the teacher as demonstrations before the class, part may be omitted when first going through the book and given on the review, thus giving freshness to the work. At times it will be well to assign one part of the problems and observations to one set of pupils, and one to another, and let each report to the whole class. Sometimes it would be best for only one pupil to try out a problem in his home plat and let all observe the results. Other observations and reports are better made by the class as a whole in company with the teacher, especially those demanding visits or rambles over the neighborhood. It is especially desirable to have the children work on the problems at home. In this home study, free discussion with parents should not only be allowed but encouraged. When strong differences of opinion are developed, do not try to settle the matter by argument, or by assumption of infallibility, but arrange a definite experiment and test the matter in the experimental plat, and have the boy do the same in his home plat. At times parents will not agree with your teachings. This is no reason for worry. Don't set yourself up to know it all, and don't treat your patrons' views with disrespect. State your reasons for your views and then decline to argue, but agree to put the matter to an experimental test. Making such an experiment will be a most valuable lesson for both pupil and parent.

Always encourage the planting of the home garden and home plat, or the carrying out of a home project in animal husbandry by each child. The work in agriculture should be a uniting link between the home and the school. The child should carry into the home through his textbook, his home problems, and his home projects a large part of the valuable things that he learns in the school, and the school thus become the center of interest to which both children and parents look for information and guidance.

CONTENTS

CONTENTS

LIST OF ILLUSTRATIONS

CHAPTER I

INTRODUCTORY

1. **Agriculture the Most Important of All Industries.** What is the most important of all industries? The industry that supplies all the others with the necessities of life, and without which the others could not exist is the most important one, is it not? If that is true, then agriculture is the most important of all industries. If the farmer produced no crops everybody would starve, except the few who could live on wild plants and dress in animal skins. If the farmer grew no crops the railroads would have very little to haul, the factories would have practically no material out of which to manufacture their goods, the merchants would have practically nothing to sell, and nobody would have money with which to buy, the lawyers would have few clients who could pay them, and the doctors, preachers, and teachers would all have to starve with the rest or live like the savage. The farmer is the foundation of our civilization, for, when he fails to support them, all the other occupations fall to the ground.

2. **How Agriculture Has Developed.** Agriculture is almost as old as the human race. The early savage learned first to eat the plants and fruits as they grew wild, and to catch and eat the wild animals. Then he learned to tame and use, that is, to *domesticate* (dō-mĕs′tĭ-kāt), the wild animals. These herds of domesticated animals were driven from place to place as the food or water supply was used up in one place. Later, man learned to gather the seed of wild plants, and to sow and cultivate patches of those plants that gave him food

or clothing. These were planted on one spot till it would no longer yield a good crop, and then the man or tribe moved over to another spot and let the old one grow wild again. This is the way the savage farmed. Next, man learned that, by plowing the weeds and grass under and letting the land rest a year he could again produce a good crop on it. This is called the *bare fallow.* Then man learned that by putting manure on the soil he could continue to get good crops on the same land. Next, it was found that the planting of a certain kind of crop one year would make the land give a bigger yield of some other crop the following year. This is *rotation* (rō-tā′shŭn) of crops. You will learn more about this when you study the lesson on crop rotation. Finally, man found just what the different plants were made of, and learned that the few chemicals that the growing plant takes out of the soil could be gathered from other places on the earth and put into the ground ready for the young plant to use in growing. These chemicals that are put into the ground for the plant to feed on are called *fertilizers.* Man has also learned that rain and floods wash out of the soil the plant food materials that are there and even wash away the soil itself. The plant food materials and the fertile top soil washed away in a year often are worth more than the value of all the crops raised that year

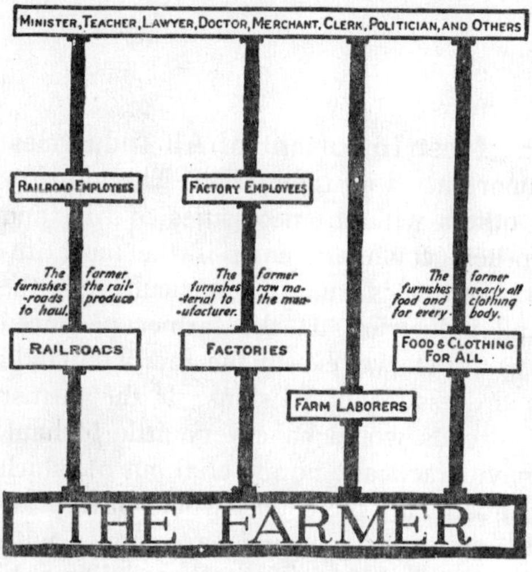

FIG. 2. The farmer sustains all.

on the farm. But the wise farmer is learning how to prevent a large part of this terrible loss of food material and of the soil itself by terracing, by better crop selection for planting

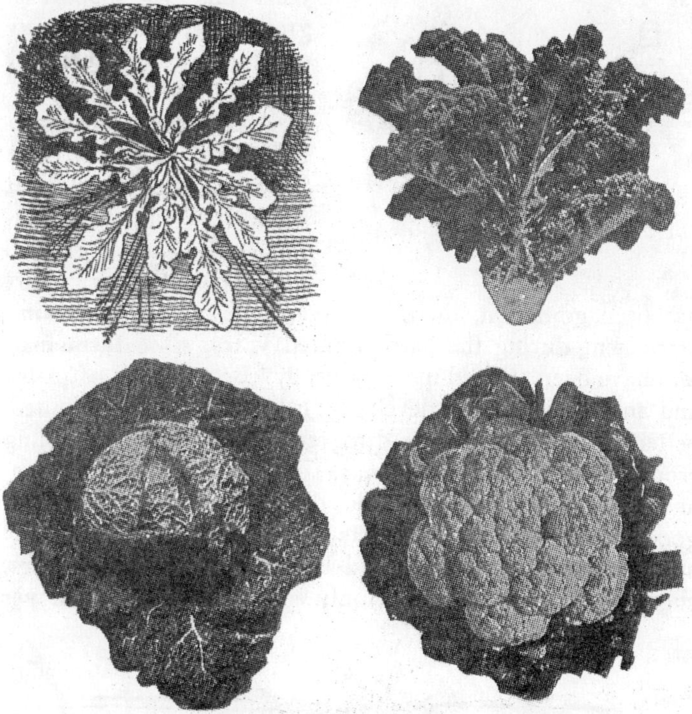

FIG. 3. Above is the original plant from which by variation, selection, and cultivation have been produced the kale, the cabbage, and the cauliflower which are illustrated above.
Adapted from "Bailey's Encyclopedia of Horticulture."

on slopes, by covering steep hillsides with orchards or wood lots or with grasses and clovers and using them as pastures. You will soon learn how this is done.

3. **Agriculture Enables the Earth to Support More People.** The earth could support only a few people when all depended upon wild plants and animals for food. It could support more when cultivated as the savage cultivated it, or

with the bare fallow; after better methods of cultivation, the use of manure, of fertilizers, and of crop rotation had been discovered, it became possible to use all the land year after year, and to support a great many times as many people as could formerly be supported.

FIG. 4. The wooden plow of our early ancestors.

Courtesy of the U. S. Department of Agriculture.

4. **Greatest Improvements in Agriculture Have Come in Recent Years.** While for thousands of years a gradual improvement in agriculture has been going on, there has been an especially rapid improvement during the past hundred years, since farm matters have been studied more carefully. Better kinds of plants and animals have been developed, better farm tools invented, better methods of tillage discovered, new ways of killing crop pests learned, better ways of harvesting, preserving, and marketing the crops, and dozens of other things. A good example of the way plants have been improved by study and careful breeding is found in the sugar beet. The best sugar beets in 1812 yielded only eight pounds of sugar per

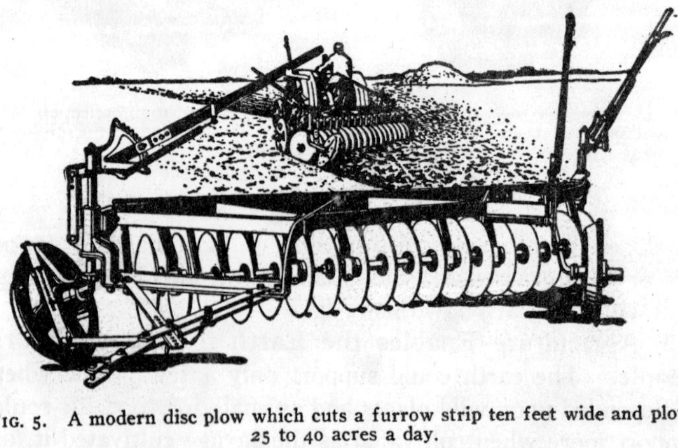

FIG. 5. A modern disc plow which cuts a furrow strip ten feet wide and plows 25 to 40 acres a day.

FIG. 6. 4-H Club Boys and their blooded calves.

hundred pounds of beets. So much better kinds have been bred that the average for the United States in 1928 was sixteen and three-fourths pounds, and an especially fine kind gave twenty-two pounds per hundred.

By applying in similar manner to animals the principles of breeding, we have produced the big Shorthorn and Hereford cattle, and the fine Jersey and Holstein cows in place of the little wild scrub stock, and have the fine Poland-China, Duroc-Jersey, and Berkshire hogs that grow as large in six months as the old wild hogs used to grow in six years.

The American Indian used to cultivate his crop with a clam shell tied to a pole. The early settlers used a wooden plow. Then the iron plow was invented, then the sulky plow, and finally the great tractor plow that opens many furrows at once and at the same time grubs out the brush. Our grandfather thought his new iron plow, with which, by hard labor, he could cultivate an acre a day, was a wonder. Now, his grandson can ride and more easily cultivate fifteen acres a day with his double-row sulky cultivator. By methods used in 1830, it took sixty-four hours of a man's labor to produce an acre of wheat; in 1930 it took just two hours. Our improved methods and machinery enable one man to cultivate on the farm about what it took five to cultivate in 1850. In colonial days, just a hundred and fifty years ago, it took about 95 per cent of all the population, working often twelve to sixteen hours a day at hard labor, to produce on our farms and ranches enough for a bare living for all the colonists. Even a large part of the women and children had to join in this hard farm work. Now, with improved methods, tools, implements, plants, and animals, less than 25 per cent of the population of the United States produces on our farms and ranches with shorter hours of labor more than the whole population can consume, in spite of the fact that the population now is about a hundred times greater than it was in colonial days. These improvements in agriculture have not only provided us with more food, clothing, and shelter, but have elevated the whole of our civilization. They have released from farm labor more than half of our population, who have been, and are, free to build roads, bridges, railroads, automobiles, airplanes, comfortable homes, great factories, stores, harbors, schools, churches, libraries, colleges, and to work in

FIG. 7. The high priced cuts are above the stick. The county agricultural agent shows the 4-H Club boy the good qualities of his baby beef.
Courtesy of the U. S. Department of Agriculture.

producing all the other means of making our lives happier and finer.

5. **Corn Made in Spite of Drought.** By using newly discovered methods of conserving the moisture in the soil, farmers have raised twenty-five bushels of corn per acre with only one-half inch of rainfall between planting and harvest, and with one and a half inches of rain, forty bushels. Fifty years ago not a grain of corn could have been produced with that little rain.

6. **What You Will Learn in This Course.** As we go on from chapter to chapter, you will learn how these things

are done. You will also learn how to use the little *bacteria* (băk-tē′rĭ-ȧ), that are so small you can not see them with the naked eye, to gather plant food out of the air for your plants, how to make your fields richer and richer each year, how to gather and market your crop to better advantage, and many other interesting and helpful things.

7. **4–H Clubs and Future Farmers of America.** Doctor Knapp, the famous government expert in agriculture who many years ago started the farm demonstration work and the boys' and girls' corn, hog, tomato, canning, and other agricultural clubs, said that we can yet make the average yield from our land eight times what it is. His club boys and girls, the members of present 4–H Clubs, and the "Future Farmers of America" are rapidly bringing this to pass.

One of the "Corn Club" boys, Jerry Moore, has grown 228¾ bushels of corn on one acre, and another boy, Walker Lee Dunson, beat this with 232.7 bushels on an acre. Paul Robinson, in East Texas, grew 154 bushels of corn on one acre, and six bales of cotton on five acres. Another East Texas boy, Homer Thigpen, produced 2652 pounds of seed cotton, and an El Paso County boy, Bill Flint, produced 3700 pounds of pedigreed seed cotton, each on one acre. I. J. Thornton, another West Texas boy, gathered 19,392 pounds of grain sorghum from three acres, and cleared $233 an acre. In Arkansas, four club members, Irene Redburn, Blanche Beall, William Cassady, and Elstner Beall, produced 12.6 bales of cotton on four acres. Donald Mitchell in Louisiana raised a litter of Hampshire pigs to weigh 2181 pounds in one hundred and fifty days, and another club boy in Minnesota raised a litter to weigh 3700 pounds in one hundred and eighty days. Isaac Corns, a Texas boy, took over his mother's flock of 330 Leghorn hens and by proper management cleared $1100 in his first year. Mary Lou Harbour, a Texas girl, in four years made net profits of $1,642.34 from her club work in gardening, poultry raising, dairying, and clothing construction. James Childers, an Oklahoma club boy, has completed eleven demonstrations in club work and made a total profit on these of $6560. The average yields of corn, potatoes, and several other crops made by 4–H Club members in a number of states have been more than double the average yield per acre in those states. Robert Burdine produced 455.9 bushels of potatoes on one acre. Carney Davis, a 4–H Club boy, in 1930, when eleven years old, started calf club work and has kept it up ever since. He has in six years made a profit of $1,655.31 in addition to the prizes he has won. One of these prizes was a year's scholarship at his state college. He now has eight pure-bred Guernseys valued at $700.

In addition to producing more per acre and breeding better varieties of farm crops and farm animals, both 4-H Club members and the Future Farmers of America are carrying on valuable projects for building up the soil, balancing and improving farm practice, and making home life on the farm healthier, happier, and more interesting. You will learn more about the 4–H Clubs and the Future Farmers of America further along in your text.

8. **What You Can Do.** What has been done by a few can be done by other intelligent boys, with corn, and with various other crops. But to do this you must learn the secrets of nature, how plants grow, how they feed themselves from soil and air, how they are bred and improved, how to protect them from their enemies, how to use labor-saving tools and devices, how to improve your soil from year to year, how to keep your stock strong and healthy, how to keep your farm home attractive and keep yourselves healthy, happy, and eager for work, how to gather and market your crops, and how to keep your farm accounts so that you will know just which crops pay best for your labor and time. To learn everything about all of these is the work of a lifetime, but in this first course you will learn the most important facts and general principles, and will learn how to study these farm problems for yourselves in the future.

QUESTIONS, PROBLEMS, AND EXERCISES

1. Name all the ways in which the people would have to get food and clothing if all the farms failed entirely for a year.

2. Give an account of some one in your neighborhood who has used the bare fallow. Can you find out how much was produced on the land the year before the fallow, and how much the year following?

3. Can you give an actual case of manuring land, and tell how much manure per acre was used, and what increase it produced in the crop? (Be sure that the manure caused the increase. How could you make sure that the manure caused the increase?)

4. If a farmer had more land than he could use, what would he gain by raising fifty bushels of corn on one acre rather than twenty-five bushels per acre on two acres?

5. Give an actual case of some one who practises rotation, and tell what crops make up the rotation. Can you get any figures showing how much the rotation helped the land?

6. As we learned to farm so much better that 25 per cent of our working population could produce more on the farms than the 95 per cent produced one hundred and fifty years ago, what did the 70 per cent who were no longer needed on the farm do? Make a list of the new things that have been produced for our comfort and civilization by that 70 per cent who were set free by improved farming.

CHAPTER II

PLANT GROWTH

9. **What You Are to Learn.** You have now seen some of the wonderful advances that have been made in agriculture, and learned that still greater improvements can yet be made. This past progress was made possible by learning first how plants grew, how they got their food, and how they multiplied, how they were improved by breeding, how they were affected by favorable and unfavorable conditions, and so on. Let us now learn these laws of plant growth ourselves. Then we shall understand the reasons for doing the things we now do in raising our crops, and shall be able to think out still better methods in the future. In this chapter you will learn how the plant starts from the little seed, how it gets strength to burst its coat, how it gets food materials from the earth and from the air, how it forms its different parts, and how each of these parts helps the plant in its growing, how the roots take the crude food materials from the soil, how this food material is carried through the stem up to the leaves, how it is made into plant food in the leaves and other parts of the plant, so that it can be used by the plant in building more plant substance, and then how the prepared food is distributed to all parts of the plant. You will also learn how the plant produces a new seed or a new plant, and how we can by breeding and selection make this new plant different from and better than the old one.

10. **The Seed.** As most farm plants are raised from seed, let us begin with the seed. Many so-called seeds are not simply seeds, but are fruits, or parts of fruits, containing

one or more seeds. Some are larger than eggs, such as the seeds of the cocoanut or the alligator pear; others are smaller than the mustard seed; some are encased in hard shells and hulls, like the hickory nut; others are inclosed in soft pulpy substance like the orange seed; some have soft down on them and float in the air, like the thistle seed; while others have

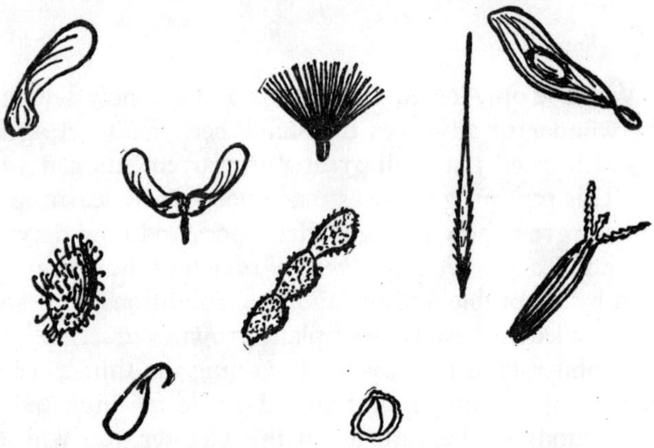

FIG. 8. Different types of so-called seeds, which are in reality true seeds with parts of the fruit adhering.

sharp spines like the cocklebur or the beggar-louse. But all seeds serve the same purpose of producing the new plant, and all true seeds are made up of three distinct parts. There is, first, an outside coat or seedcase, usually thin and tough, which protects the parts inside. Second, there is the little *embryo* (ĕm′brĭ-ō). The embryo is the new plant itself, often showing plainly beginnings of a stem, leaf or leaves (called *plū-mūle*), and root (called *răd-ĭ-cle*). Third, there is a small mass of highly concentrated food, on which the embryo must feed till it can grow roots and leaves strong enough to gather its own food materials directly from the earth and air, and manufacture them into plant food. This *reserve food,* as it is called, is made up chiefly of starch, sugar, protein, and oil, in varying proportions, and is the main part of the nuts and grains that we eat. In the pecan, it is the rich oily meat of the nut; in wheat and corn it is the white starchy part of the grain that makes our flour and meal. We also eat the little embryo. This embryo is easy to see in corn, where we call it the germ.

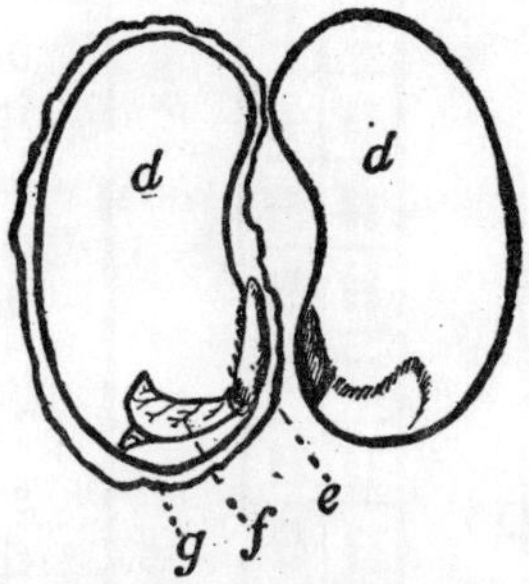

FIG. 9. An opened bean seed that has been soaked till the seedcase has softened and the seed begun to germinate. *d* is the cotyledon; *e* is the sprouting radicle; *f* is the plumule; *g* is the loosened seedcase.

11. The Seed Is Alive. The dry, lifeless-looking seed is not a lifeless thing, like a chip of wood; but its living germ will never grow till the right conditions for its growth are supplied to it. In this state in which it is alive, but not active, it is called *dormant* (dôr′-mant). When the right conditions for growth are supplied the seed, it becomes active and begins to grow. The first growth is called *germination* (jĕr-mĭ-nā′shŭn).

12. What Starts the Seed to Growing. Take a few seeds of the bean or other common plant and place them in a seed germinator. In from six to twelve hours you will notice that the beans have absorbed some water and increased in size. In a short time the seedcase bursts and a white sprout appears, which is the beginning of the new *plantlet,* or little plant. The seed has now begun to germinate. Let us see under what conditions a seed will germinate, and under what conditions it will not germinate; then we shall know what conditions to provide in the soil when we plant seeds.

FIG. 10. This shows a seed germinator made of two plates and two layers of canton flannel or of blotting-paper. The cloth or paper is moistened thoroughly and the seeds laid between the folds. Then this is placed in one dish, a little water is added, and the other dish or a pane of glass used to cover and prevent evaporation. On the right is shown a modification of this, made by inverting one deep dish in another containing water and then placing the seeds between folds of a cloth which is placed on the bottom of the inverted dish in such manner as to allow the edges of the cloth to hang down in the water.

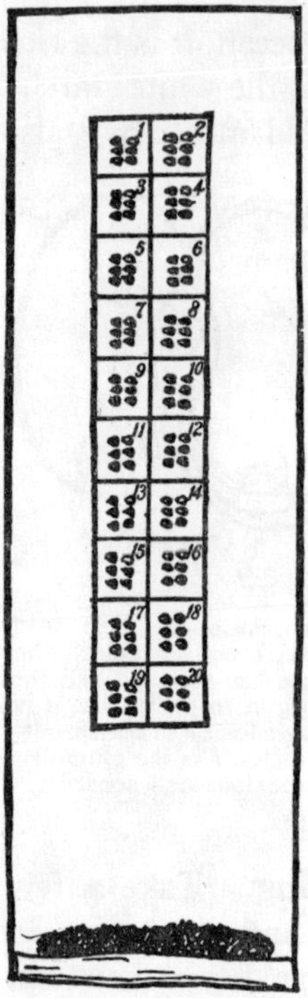

FIG. 11. The rag-doll germinator, consisting of a piece of canton flannel with a number of little squares, usually two or three inches square, marked on it and each numbered. This is first moistened thoroughly, the seeds are laid on the squares, a record is made of what is on each square, and then the whole cloth rolled up, beginning with the darkly shaded part at the bottom. The shading represents a damp sponge or rag, placed there to make the roll larger and easier to roll. When rolled up, the end of the "doll" is placed in a vessel of water, so that the cloth will soak up a constant supply of moisture for the seeds.

We can make three experiments and find out for ourselves what makes a seed germinate.

13. **First Experiment.** Plaoe six beans in a can of moist soil and keep it moist for a few days. Place six in a can of perfectly dry soil and keep it dry. When the beans in the moist soil have germinated, dig up those in the dry soil and see what they have done. If you find that the beans in the dry soil have not germinated, and those in the moist have, what does this show about water being necessary for germination? If seeds must have moisture in order to germinate, then what must the farmer watch for and get his land and seeds ready for so that he can make use of it when it comes? In the chapter on the soil, you will learn how to cultivate your land so as to hold the moisture in it for weeks, and by winter and early spring ploughing make it possible for your seeds to germinate even if no rain should come at the planting season.

14. **Second Experiment.** Take two cans two-thirds full of fresh water and boil one can for ten minutes, and in this way drive out all the little particles of air that are naturally in the water. Then pour mineral oil on top of this boiled water till the water is covered with oil about an eighth of an inch thick. This film of oil will prevent any air getting down into the boiled water again. Now place some seeds of rice or watercress in the can of boiled

and oil-covered water that has no air in it, and the same kind of seeds in the other can of water that has not had the air boiled out of it. Watch these seeds for several days and see which ones begin to germinate. If you find that the seeds in

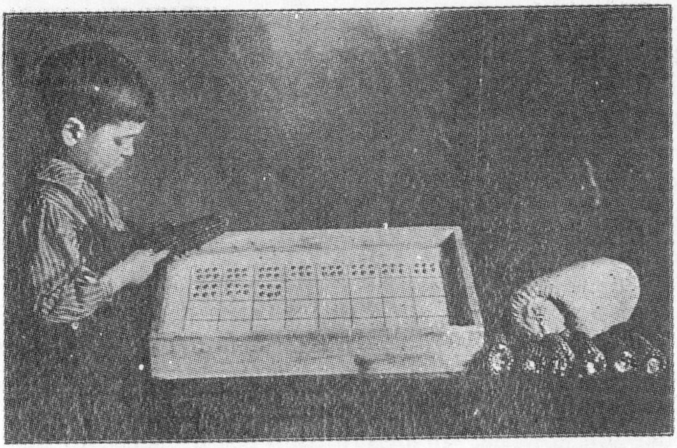

FIG. 12. The sand-box germinator, which consists of a box about four inches deep with holes in the bottom for the escape of water. This is filled half full of sand or sawdust and a cloth is laid over this and marked into squares. The seeds are placed on these squares and covered first with a cloth the size of the box and then with a cloth which is made large enough to extend above the top of the box all around. Then the box is filled with sand or sawdust, thoroughly watered and kept moist. When it is time to examine the seeds, the covering is lifted off by raising the top cloth without disturbing the seeds. If it is not desired to examine the seeds, but merely to see how well they germinate, the cloths are not necessary. In this case the box is laid off in squares by means of string tacked across the top. In cool weather the box may be made eight inches deep and filled with four inches of horse dung and then the sand placed on top of this. The manure serves to keep the box warm.

Courtesy of the Department of Agriculture, University of Minnesota.

the water that had the air boiled out of it have not started to germinate, while those in the water with the air still mixed in it have begun to germinate, what does that show to be necessary besides water to make seeds germinate? Beans, corn, and most other farm plants require more air for germination than is present in fresh water. This being true, what would happen to these seeds if immediately after they were planted such a long hard rain fell that all the free air was all packed out of the soil, and the soil so filled with water that no more air could get down to the seeds? Ask at home if that has ever happened on your farm.

15. **Third Experiment.** Take two cans of moist soil and plant six beans in each. Put one of these where it will keep warm day and night, with a temperature of 60 to 80 degrees, and place the other in the refrigerator at home, or out-of-doors where it will be kept cold, with a temperature near freezing. After three or four days examine the seeds in both cans and see if both are sprouting alike. If you should find that the beans in the warm soil had begun to sprout, and those in the cold soil had not, what would this show to be necessary for the germination besides moisture and air? If seeds are planted in spring when the ground is still cold, what will happen? If, later in spring, after the ground is warm and seeds are planted, there should come a long, cold, rainy spell, what would happen? If the cold weather lasted very long after the seeds were planted, what would probably happen before they got a chance to germinate? Has this ever happened on your farm?

16. **Differences in Seeds.** There are great differences in seeds. Some require a very warm seed bed to germinate, such as cotton, while others, such as oats or *rescue-grass,* can germinate with much less heat. Some seeds require also less moisture and less air than others do. If the cold is not too great, many seeds will germinate in cool weather, but do it more slowly than when warm. Beet seeds that germinated in three days at a temperature of sixty-five degrees took twenty-two days to germinate at a temperature of forty-one degrees.

17. **How Roots Grow.** When your bean seed has the necessary amount of heat and air and water to germinate, you will see that it first swells, and then a tiny white sprout bursts out of the hull between the two divisions of the bean. This little white sprout is called the *radicle* (răd'ĭ-kl). The word radicle is made from the Latin word *radicula,* which means a little root. As soon as it gets out of the seed-coat,

the radicle turns down and makes its way into the moist soil to form the root of the plant. This root must gather food material from the soil for the young plant, or the plant will starve as soon as it has used all the reserve food in the seed. If there is no moisture in the soil, the root can get no food materials from that source, for plants can take in only liquid foods. If the ground is packed so hard that the tender radicle cannot force its way among the soil particles and come in contact with the little films of water covering these soil particles, then the plant can get no food materials and must die, or can get so little that it cannot grow. What does this show us that we must do to all seed beds before the seeds are sown, if we wish the young plants to grow? If the soil is not too hard, this radicle grows on down into it and makes the main root, or *tap-root,* as it is called, while upon its sides grow the branching roots called *lateral* (lăt'ĕr-ăl) roots. Lateral is also from an old Latin word, *latus,* which means the side. From the sides of both the main root and the lateral roots grow out thousands of little fine *root hairs.*

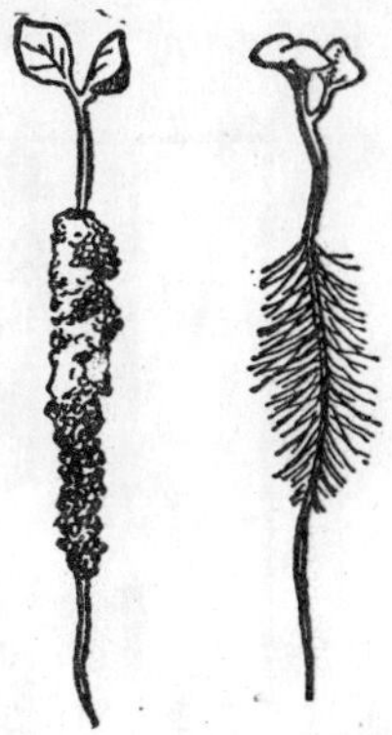

FIG. 13. The root hairs on a young radish-plant. The drawing on the left shows the soil still adhering to the hairs.

18. **What Roots Do.** Neither the tap-root nor lateral roots are able to take in any food material directly from the soil, but serve to hold the plants steady and give a large surface from which root hairs may grow out. These tiny root hairs grow out from the new growth of both the tap-root and laterals. Some of these hairs are too small to be seen with the naked eye, there being in some cases nearly forty thousand growing out from a square inch of root surface. These are so tender that they are usually torn off when we pull a plant out of the ground, but you can see them easily if you will germinate corn or oats in a germinating-dish. Figure 13 shows the mass of root hairs on the roots of a

young radish-plant. It is through these soft root hairs that the plant takes in all the food material it gets from the soil. These hairs reach out between the particles of soil and absorb through their skinlike covering the water and various kinds of food materials that are dissolved in the water that is in the soil. This raw, unprepared liquid food material then

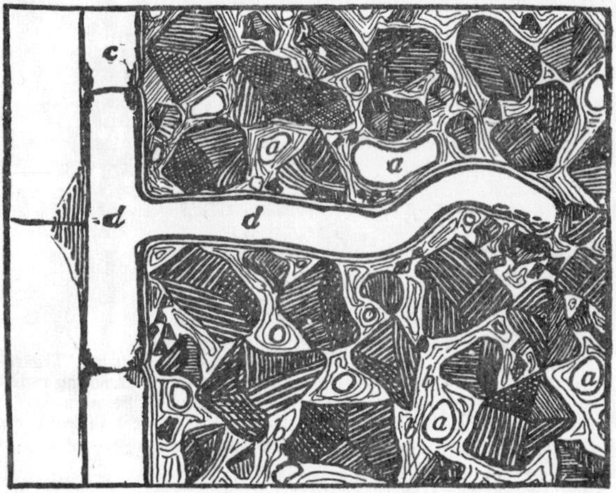

FIG. 14. The root hair (*d*) penetrating the soil, as seen under the microscope. Note the black soil particles with the films of water (*b*) surrounding them. Also note the air spaces (*a*) among the soil particles. Note that the root hair is a continuation of a single cell in the outer membrane of the root.

passes through the root hairs into the roots and on through the stem up to the leaves to be prepared, so that the plant can make it into its own substance.

19. **How Root Hairs Take Food Materials.** The process by which the thin membrane of the root hair, which has no mouths or holes in it, lets the liquid food material come into the root is very remarkable. You can see this same process going on if you will take a lamp chimney and tie a piece of well-washed and softened bladder over the end, or, better still, take a tube shaped like the one in Figure 16 and tie the bladder tightly with a waxed thread over the large

end of this, so that it will hold liquid when poured into it. Then pour into the chimney, or tube, either molasses or strong brine till it is half full or more. Then fasten the chimney in a jar or large bottle of fresh water, so that the water in the jar is just level with the molasses or salt solution in the tube. Leave this for an hour and see if the liquid inside the tube has not risen higher than the water level in the bottle. If it does this, does that show that some of the water from the bottle has passed through the bladder into the tube and increased the amount of the fluid there? This passing of a liquid through a membrane is called *osmosis* (ŏs-mō′sĭs). It is by this process of osmosis that the tiny thread-like root hairs take in all the food materials that the plant gets from the ground. You see that the plant can take in only such food material as is in liquid form dissolved in water in the ground. When all the water is dried out of the ground, then the plant is not merely unable to get any water, but is unable to get what else?

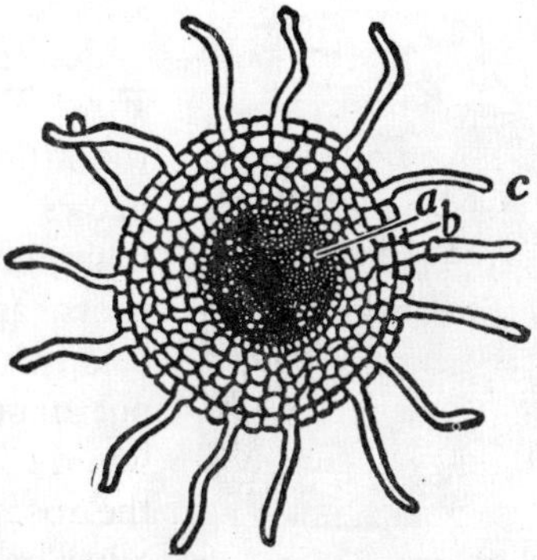

FIG. 15. The cross section of a small root as it appears under the microscope. Note the root hairs (*c*), the epidermal cells (*b*), and the fibro-vascular bundles (*a*) through which the liquid plant-food materials pass up to the stem and on to the leaves.

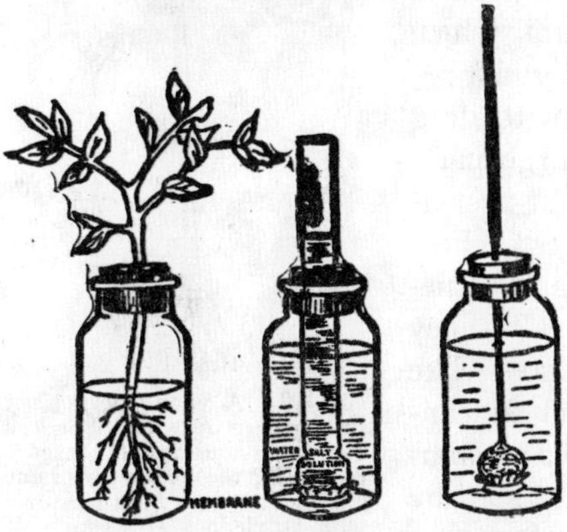

FIG. 16. An easy method of studying the working of osmosis. The dilute liquid in the outer vessels passes into the stronger solution in the inner vessels in the same way that the soil water passes into the root hairs of the plant.

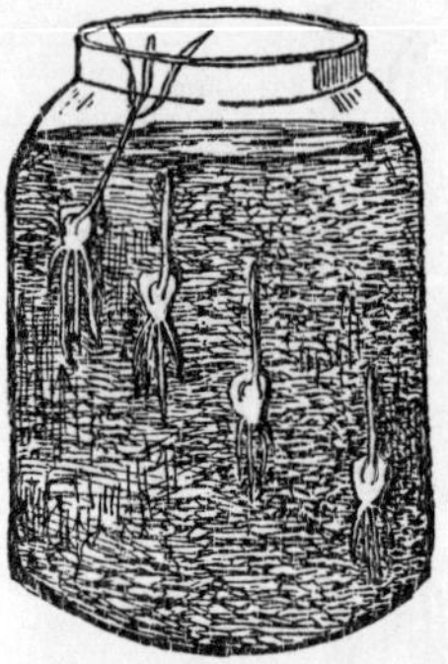

FIG. 17. Shows the method of planting seeds at different depths in such way that their growth may be watched.

20. **How the Plant Gets Its First Food.** We have seen now how the little radicle bursts out of the seed, grows down into the earth, forms the main tap-root, and throws out lateral roots, and how the tiny root hairs develop on the new parts of all roots, grow out among the soil particles, and soak up liquid food material and send it back into the roots. Now let us see next what the other parts of the seed are doing while all this is going on. Let us see how the plant gets ready a stem and branches and leaves to receive the crude food materials gathered by the roots, manufactures these into true plant foods, and distributes this food to all parts of the plant.

21. **Reserve Food of Plants.** If you will watch a bean seed growing in the soil, you will see that soon after the point of the little radicle begins growing downward, the other end of the germ, which is called the *plumule* (plū′mūl), begins to grow upward and to drag along with it the two large parts of the bean, which look at first as if they are thick fat leaves. These two thick oval pieces are sometimes called seed leaves. They do not, however, behave like green leaves, but are full of concentrated food, which was manufactured and placed in the seed by the parent plant. This food reserve supplies the germinating plant with nourishment until it can develop the root hairs, stem,

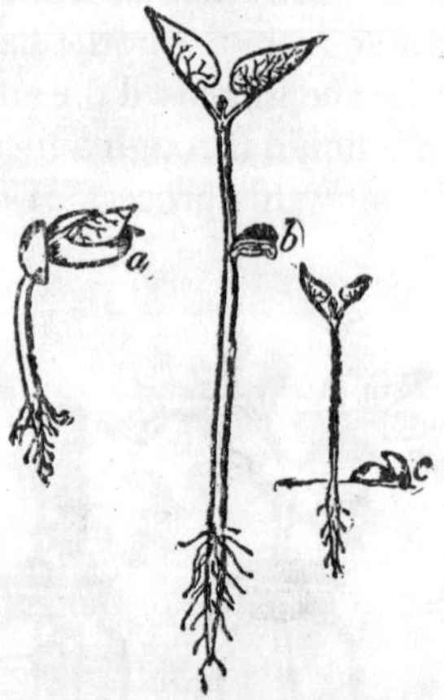

FIG. 18. A young bean plant just coming up with cotyledons (*a*) thick and full of reserve food. In the center the same plant a few days later is shown with cotyledons (*b*) empty of food and wrinkled. On the right is a young plant that has been stunted in its growth by the removal of the cotyledons too early, thus depriving it of its reserve food.

and leaves to gather food materials and manufacture its own food. These thick leaf-like pads of food are called *cotyledons* (kŏt-ĭ-lē′dŭns). All seeds have this reserve food in them, but all do not have the two cotyledons, nor do all plants draw them through the soil when sprouting, as do the beans. The grains and many others have only one package of reserve food. A few plants have more than two cotyledons. Sometimes this reserve food, instead of being inclosed within a part of the embryo, is attached to the embryo, or may even merely surround it. As the plumule of the bean grows upward into a stem and the tiny leaves unfold and grow, you will notice that the cotyledons get thinner and thinner, and after several days shrivel up, and after a week or so more fall off. Can you see why this happens?

FIG. 19. A woody stem with the terminal bud and several lateral buds.

22. **How the Plant Develops Stem, Branches, and Leaves.** The way the stem of this young plant, or *seedling,* as it is called, grows is interesting. If you will take a bean that is just out of the ground, or one that has just begun to grow in the germinator, and will put marks in waterproof ink every eighth of an inch along the stem, and then look at these same marks twenty-four to forty-eight hours afterward, you will see that some parts of this stem are growing much faster than others. The most rapidly growing part is at the tip and is called the *growing point.* The stem elongates most rapidly just back of the growing point. This growing point grows rapidly upward, forming the stem, and soon leaves and branches push out on the side. These side branches are called lateral branches. The bud on the end is called the *terminal* (tẽr′mĭ-nal) bud. This word terminal is made from another Latin word, *terminus,* which means the end. If you will open up carefully and examine the growing tips of main stems and laterals of some woody plant, you will see that on the outside there are close-fitting scale-like modified leaves,

called *bud scales,* which protect the tender growing point and the tiny undeveloped leaves within. As these leaves begin to grow and the growing point pushes out, the protecting scales are burst, and the new leaf or leaves unfold and grow, while the growing point pushes on. With *herbaceous* (hẽr-bā′shŭs) plants like the bean, the growth is similar, except that there are no protecting bud scales.

23. **Peculiar Ways of Growing.** Each variety of plant has its own peculiar way of growing; some stems stand erect, as the oak; some lean on other things, as does the grapevine; and some lie flat on the ground, as do the melon vines. Some send out leaves and branches in pairs, opposite each other, at regular intervals along the stem; some have their leaves come out singly, first on one side and then on the other side of the stem and branches; some come out on all sides at the same level; and some in yet other ways. Why some plants always act one way, and others another, we do not as yet fully understand. That question is not important for us; but why any plant grows at all, and just how it manages to do so, is most important to know if we expect ever to grow crops. In order to understand how the stem can elongate and throw out these leaves, we must look inside the plant and see first how it is constructed and just how it gets its food and makes out of it the new leaves, stem, branches, and other things. By studying this, one step at a time, it will soon all be plain.

24. **How the Plant Gets Crude Food Material.** You know now that the bean has leaves, branches, stem, roots, and root hairs. Practically all plants dealt with on the farm have these same parts. Each part has its own work to do for the good of the whole plant, including itself. We are now ready to see how each of these parts of the plant helps the plant to get food and grow. You will remember that the roots are covered with root hairs, which by osmosis take in the food materials dissolved in the water in the ground. In addition, these root hairs, by means of the small amount of acid that they give out, dissolve some of the substance of the

soil itself, which is then taken up by the water present and absorbed by the root hair. The whole plant is made up of a mass of small cells similar to the root-hair cells, in having a membrane surrounding them, but differing from each other in size, shape, and other ways. The liquid food material, once inside the root-hair cell, passes on into the adjoining cells by osmosis through the cell membrane, just as it passed from the soil to the root hair. From these root cells it is passed through tiny tubes extending in sections up through the stem and branches out into the ribs and veins of the leaves, and finally is spread out into all the tissues of the leaf. The leaf is a wonderful kitchen and chemical laboratory combined for the plant. Here in the leaf the plant does some things that no man has yet been intelligent enough to learn how to do. Here the crude food material that is drawn up from the soil through the root hairs, roots, and stem is mixed with other material taken from the air, and new foods are made that are sent back all over the plant to build up its tissues and to store reserve food. A little later we shall look at the structure of the stem and see by what means it carries the sap back and forth, up and down the plant at the same time.

25. **What Crude Food Material Is Made Of.** Let us now see what the crude food material is made of, what it is changed into in the leaves, and how this is done. The raw food material that comes up from the roots is in large part water, while that taken in by the leaves is a gas which is free in the air. The plant manufactures its food mainly out of water and this free gas which is called *carbon dioxide* (kär′bŏn dī-ŏks′īd). The plant, however, cannot live on these alone. Dissolved in the water taken in by the plant are a number of different food materials. If this food material were not dissolved in a liquid form, it could not pass through the roots and stem up to the leaves to be made into foods and then pass back over the entire plant to feed all the parts as it does. There are just a few of these substances in the soil that plants make their foods from, and all plants and animals use practically the same ones, but in different combina-

tions and proportions. You will learn the names of these in your lessons on the soil. When the crude food material reaches the leaves, a part of the water is used by the plant in making the plant food, which is later made into new plant substance. The larger part passes out of the plant in the form of vapor, mainly through openings in the leaf. This giving off of water vapor by the plant is called *transpiration* (trăn-spĭ-rā′shŭn).

FIG. 20. An easy method of catching some of the water transpired by the plant. The card-board is slitted and fitted closely around the plant, and then the glass is turned over the plant so that very little of the water vapor given off by the plant can escape. Soon the air is so saturated with this transpired water vapor that it is deposited in drops on the inside of the glass.

26. **How the Plant Gives Off Water.** Transpiration goes on all the time, but most rapidly when the air is dry and the sun is shining on the leaf. You can catch some of the water transpired by a plant if you will follow the directions given under Figure 20. In Figure 21 you can see the structure of a leaf as it appears under the microscope, and can see the openings through which the greater part of the water passes out of the leaf. The amount of water passed off from a plant is surprising. In this way a full-grown apple-tree will give off about two hundred and fifty pounds of water in a day, or over thirty-five thousand pounds during one growing season. From three hundred to five hundred pounds of water pass through and out of the ordinary plant for every pound of dry matter left in it. Of the material left in the growing plant, very little is solid matter. Often as much as nine-tenths is still water. This is why green plants weigh so much more than dried ones do. Corn, for example, at roasting-ear season is over 80 per cent water.

27. **How the Plant Makes Starch and Sugar.** While the excess water is being transpired out of the leaf, the leaf

is taking in from the air the carbon-dioxide gas, one part of which supplies the material that will make up fully half of the solid matter of the plant. This material is *carbon,* which we can see left as charcoal when wood is burned without sufficient air. All this solid carbon then comes out of the

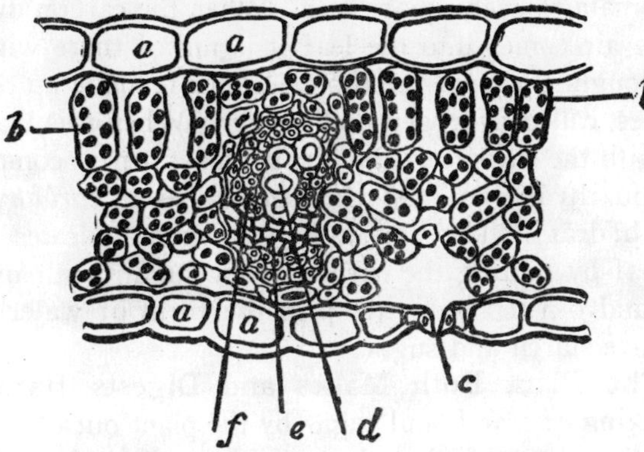

FIG. 21. A cross section of a leaf as it would look under the microscope. Note (1) the outside layer of epidermal cells (*a*) on top and bottom; (2) the stoma (*c*) through which the water vapor mainly passes off and the air passes in; (3) the green chlorophyl bodies (*b*); (4) the water-tubes (*d*) through which the crude food materials come up; and (5) the phloem (*f*) through which the digested food passes back over the plant.

air in the form of a gas and not from the soil at all.[1] The leaf has special little openings through which to take in the carbon dioxide and give off water. These are mainly on the under side of the leaf, and are so small that the naked eye

[1]It may puzzle you to see how an invisible gas can contain a solid thing like carbon, but, if you will remember what you know about water, you will see that the same substance can exist in the form of either a gas, a liquid, or a solid. Cool water below thirty-two degrees and it becomes a solid called ice. Heat it above 212 degrees and it becomes a gas called steam. If you could run a strong current of electricity through a vessel of water, you would see that this liquid could be broken up into two gases, one called *oxygen* (ŏks'ĭ-jĕn) and the other called *hydrogen* (hī'drō-jĕn). In like manner the chemist can put two gases, hydrogen and oxygen, together in the proper manner and form a liquid—water. You see, then, that the same substance may exist as a liquid, a solid, or a gas; and, out of a liquid compound a gas may be taken, or out of a gaseous compound a solid may be taken. Such changes and such making and breaking up of compounds are going on constantly in nature.

cannot see them. Figure 21 shows you how these look under a microscope. They are called *stomates* (stō'māts), or *stomata* (stō'mä-tä). One of them is called a *stoma* (stō'mä), which is an old Greek word meaning mouth. The under side of an apple leaf has more than one hundred thousand of these stomata to each square inch. When the carbon dioxide from the air comes into the leaf it is united there with the water brought up from the roots. Water is made up of two substances, called hydrogen and oxygen. When this water is united with the carbon of the air, it makes a new compound of carbon, hydrogen, and oxygen, called a *carbohydrate* (kär-bō-hī'drāt). The two important carbohydrates made in the leaf by uniting the carbon from the air with the hydrogen and oxygen brought up in the form of water from the soil are starch and sugar.

28. **The Plant Both Makes and Digests Its Food.** This making of starch and sugar by the plant out of carbon, hydrogen, and oxygen is quite a different thing from digestion of foods by men and other animals. It is not digestion at all, but manufacture of food. This manufacturing of foods no animal can do. Man and all animals are dependent upon the plants for the manufacturing of all their foods. We eat these prepared foods when we eat plants or eat animals that live on plants. Then we have to digest these prepared foods. The plant can manufacture its own foods from the crude food materials in the soil and air, and then later it, too, has to digest this food before it can turn it into new plant substance, in a way very similar to that by which man digests these same foods.

29. **How the Plant Turns Starch into Sugar.** In most plants the main part of these carbohydrates in the leaf first appears in the form of little starch grains, but as these cannot be dissolved in water they cannot be taken up in the sap and passed back from the leaf down the stem into the roots and other parts of the plant to feed it. The leaf then has to digest this starch, or turn it into a form in which it can be dissolved in the sap of the plant and passed around, just as

we have to digest food in our stomach and intestines before it can pass into our blood and be carried in our blood over our body to feed every part. The starch is therefore now turned into sugar, which is dissolved by the watery sap in the leaf. In this form it is passed back along through the inner bark down the stem and branches to every part of the plant.

30. **How the Plant Uses Carbohydrates.** This sugar food is used in many ways by the plant. Some is later changed back into starch, some is modified by the addition or subtraction of certain things and made into oil, and into another very important set of compounds which we shall study later. A part of these products is used by the plant in growing, in building new buds, leaves, stems and roots; a part is deposited in seed as reserve food to start a new set of plants growing; a part is deposited as reserve food within the plant itself for future use. The peach trees, for instance, must deposit each year enough reserve food to support them next spring, while putting out their blossoms and getting their leaves started and developed enough to commence again the manufacture of food out of the raw food materials sent up in the sap from the roots. You can often taste the plant sugar in the sap that is passing down the inner bark of a tree. It is this sweet sap of the maple-tree that is caught and boiled down to make maple-syrup and maple-sugar. The tuber of the potato is largely a mass of reserve starch, put there by the plant for future use. This shows very briefly and incompletely the way the plant takes its raw liquid food

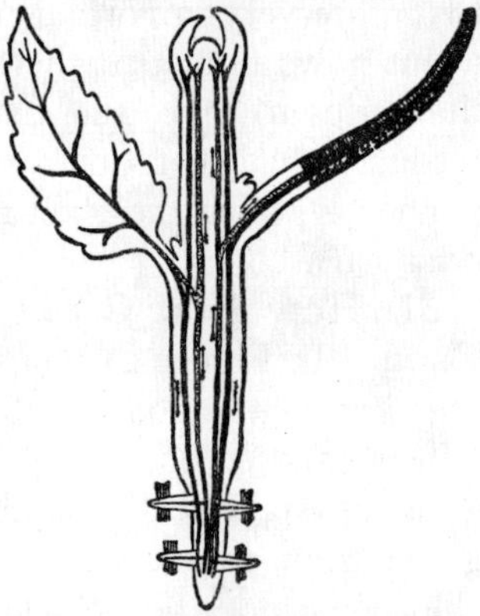

FIG. 22. A very much simplified diagram to illustrate roughly how the crude food materials are carried up the layer of outer new wood, and the digested plant food is brought back down the phloem just outside the cambium layer. Adapted from Stevens's "Introduction to Botany."

material from the soil, passes it up the stem to the leaves, and there adds material taken from the air and makes up new compounds which are dissolved again in the sap and passed back down the stem and around to every part of the plant. To tell all the details of the work done by the plant in getting its food would take a long time and confuse your mind now. All of this you can learn later. There are just two more very important points that you should know more about now: first, that there is a special substance in the leaf that helps to make the starch and sugar for the plant; and second, what that other very important set of food compounds is which the plant makes. These two things we shall now learn about.

31. **How Leaf Green Helps the Plant Manufacture Food.** In Figure 21 you will see that the leaf has within its cells some little bodies called *chlorophyl* (klō'rṓ-fĭl) bodies. When exposed to sunlight these bodies develop in them a green substance called chlorophyl. It is this which gives the green color to the leaves. If the plant gets no light, this chlorophyl does not develop. That is why plants grown out of the light are pale. It is the action of this chlorophyl and the sunlight which splits the water and the carbon dioxide to pieces in the leaf, and makes possible the formation of the new compound, the carbohydrate, out of the carbon, hydrogen, and oxygen. Without chlorophyl and sunlight, then, the ordinary plant could manufacture no carbohydrate. As we shall see later that all other plant foods are made by changing or adding to these carbohydrates, it is plain that the whole growth of the plant is dependent upon this action of the chlorophyl and sunlight in the leaves. There are certain plants that have other ways of getting carbohydrates, and all plants can make a little in green parts outside their leaves, but the ordinary farm plant gets practically all its carbohydrates from the action of sunlight and chlorophyl in the leaves.

32. **How the Plant Builds New Living Substance.** While the chlorophyl bodies are helping part of the carbon taken from the air by the leaf to join with the water and

make carbohydrates for the plant, a part of these carbohydrates is being combined with a substance called *nitrogen* (nī'trṓ-jĕn) and some of the other substances that we saw are also in the sap which comes up from the roots. The new compound which is made by this process is called *protein* (prō'tḗ-ĭn). There are many kinds of proteins, made in different parts of different plants. This protein is carried along with the carbohydrates in the sap to all parts of the plant, and helps to nourish the plant *protoplasm* (prō'tō-plăzm), which is found in every living part of the plant. Protoplasm is the most wonderful substance in the world, for it is the basis of all plant and animal life. When the plant is growing, the protoplasm is in an active state; when the plant is dormant, the protoplasm is still there, but it is not active. When the parent plant forms a seed, a small portion of the protoplasm of the old plant goes into the seed. When the seed germinates and grows, this protoplasm grows; and whenever a cell of the plant divides and thus forms a new cell, the protoplasm in the old cell divides and part goes into the new cell. If it did not do this, the new cell could not live. This protoplasm likewise cannot live and continue to grow in the cells unless the plant has protein. Thus, you see, the plant cannot live without nitrogen and the other substances brought up from the soil. If you will strip the green bark from a tree or bush and run your hand over the inside of the bark, you will feel the thin, slippery film which is a mixture of sap and protoplasm that is spilled from the broken cells there.

33. **The Layers of the Stem.** We have seen that raw liquid food material is passed up the stem to the leaves, that it is then mixed with the raw food material taken from the air, and then this newly prepared food is passed back down the stem and out to the various parts of the plant. Let us now see how the stem is constructed so as to make possible this process. Figure 23 shows you one-fourth of the stem of a tree cut crosswise. You will notice that first on the very outside of the stem is a hard outer bark (*a*), and at the center

are rings of hardwood (f). The hard outer bark is dead, and the central hardwood has practically no share in carrying the sap. The tough outer bark serves to protect the delicate inner bark, and the dense hard center, or heart, of the tree serves mainly to strengthen the stem so that it can hold the great masses of leaves and young branches up to catch sunlight and air. At one time this heart wood was composed of active cells through which the sap passed, but as the plant became older these cells became clogged with various substances which rendered them inactive. Often you can see a tree in which this central part has become diseased and rotted away, leaving only the bark and the new outer part of the growing wood standing. Such trees may continue to grow, because the layers that pass the food supply up and down the tree are still active. Looking again at Figure 23, you will see that next to the rough outer bark, there is an inner fibrous bark (b), also dead, but inside this is a thin layer of soft bark, too thin to show well in this cut. Just at the boundary line between the bark and the wood is a very thin compact layer (c), called the *cambium* (kăm′bĭ-ŭm). Just inside the cambium is the layer of soft sappy new wood (d). When you peel the bark off a young sapling, the split occurs at the cambium. The soft new wood is left on the sapling, the watery cambium is split, and the soft inner bark is left inside the hard bark. The soft inner bark, the cambium, and the ring of soft new wood are the three important layers con-

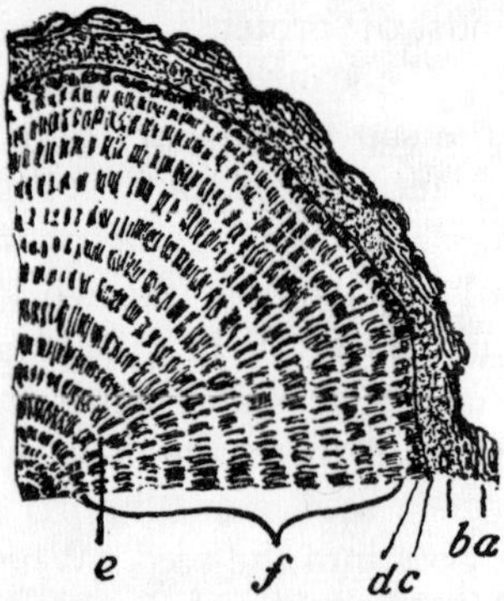

FIG. 23. A quarter of the stem of a tree. Some of the details are exaggerated for the sake of clearness. a represents the rough dead outer bark, b the dead fibrous inner bark, and c the cambium layer. The phloem is not represented, but is just outside the cambium and on the inner surface of b; d represents the layer of new wood, and f the numerous annular rings of old wood. The medullary rays going from the center to the bark are not plain, though traces of them are seen, especially at e.

cerned in growth and are the ones we shall study. The raw sap coming up from the roots to the leaves usually and mainly passes up tubes in the new wood, and the prepared food passes down tubes in the soft new bark. This being true, what would be the result if you cut through the bark of a tree down to the cambium all the way round, but did not cut into the new wood? What different result would follow if you cut on through the new wood all around?

34. **How the Crude Food Material Passes Up the Stem and How the Prepared Food Passes Down.** You remember that the liquid food material passes from the soil into the root-hair cells by osmosis, and then passes on in by osmosis through other cells till it reaches some tiny tubes which conduct the liquid up through the layer of new wood from the root to the leaf. The exact make-up of these tubes is too complicated to go into here. We need only to know that they are made of a special type of elongated cells which by overlapping and fitting end to end afford an easy line of passage through the new wood for the up-going sap. In a similar way, there are special elongated cells in the soft inner bark which make up a series of tubes through which the prepared food, or *elaborated* (ē-lăb′ō-rā-tĕd) sap as it is called, comes back from the leaves and is distributed over the plant. These tubes are not evenly distributed through the new wood and bark, but occur in groups or bundles that are called *fibro-vascular* (fī-brō-văs′kū-lär) bundles. These vascular bundles show plainest in a young stem. Figure 24 shows how these look in the year-old stem of the *Dutchman's Pipe* cut crosswise. The drawing is not complete, and is exaggerated in some respects to make the matter clearer. You see the young bark (*e*) with the tough layer of fibres in it (*b*), and, inside this, the thin cambium layer (*c*). The seven fibro-vascular bundles are seen lying, each one with part of its tubes just outside the cambium and part just inside (*a* to *w*). The tubular part of the fibro-vascular bundle that is outside the cambium in the soft inner bark, and through which the prepared food is conducted, is called the *phloem* (flō′ĕm),

and that inside the cambium in the layer of new wood and through which the crude food material passes up is called the *xylem* (zī'lĕm).

35. **Movements of the Sap.** Unfortunately, the circulation of the sap is not so simple as this account makes it, nor are all the vascular bundles situated exactly as these are shown, but this gives a rough, general idea of the usual, main flow of sap. It differs in different types of plants. In addition to the flow up the xylem and down the phloem, there is some passage constantly going on from cell to cell in all directions by osmosis, and at times some prepared food gets into the xylem, and likewise some raw food material gets into the phloem. For instance, in early spring prepared reserve food rushes up in the xylem to start new leaves, and the sap also regularly passes in and out from center to outside of the stem through passages in the *medullary* (mĕd'ŭl-lā-rў) rays that are shown in Figure 23. Again, food has at times to pass up the phloem as well as down. All this you will learn about later. Now we need to keep in mind especially that the main general flow of raw food materials is up to the leaves through the tubes of the fibro-vascular bundles in the new wood, and the main general flow of prepared food is from the leaves back down the phloem in the inner soft bark.

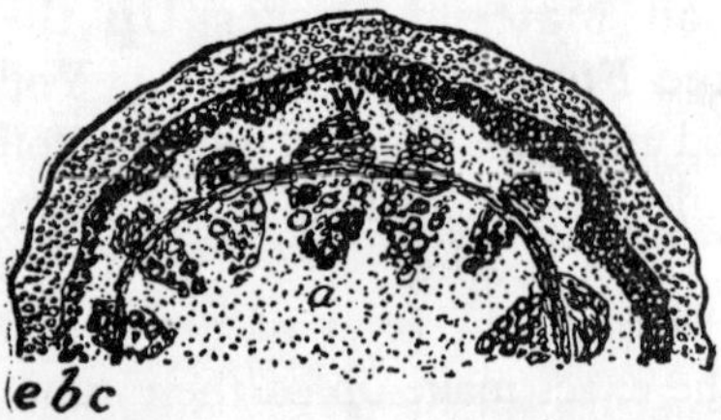

FIG. 24. The cross section of a young stem. *e* represents the outer layer of bark; *b* the layer of tough fibres in the bark; *c* the cambium. Seven fibro-vascular bundles are shown lying across the cambium, with the phloem at *a* and xylem at *w*.—*After Bergen and Caldwell.*

36. **What Forces the Sap Up the Stem?** The next question is, What forces the sap up the stem along these tiny tubes in the fibro-vascular bundles? Where these tubes have joints in them at the junction of two cells, the fluid passes by the same process of osmosis by which it passed into the root hair. In addition to this, three other forces are working.

First, the outer cells of the root become filled with fluid by osmosis. Their membranes are stretched and they press in on the cells inside them and tend by this *root pressure,* as it is called, to force the fluid out of the roots and up the tubes. Second, the water is being evaporated out of the leaf constantly at the top of the tubes, and this produces a sort of suction which helps to draw the water up the tubes. Third, these tubes are exceedingly fine, so fine that they are called *capillary* (kăp´ĭl-lā-rў) *tubes.* This is from the Latin word *capillus,* which means a hair. In all such fine tubes there is a force called *capillary attraction,* which tends to force liquid to climb up higher and higher on the sides of these tubes. Figure 25 shows you how you may see this force working in a glass tube. It is this same capillary attraction which causes the oil to rise in the fine tubes produced in a lampwick by the closely twisted cotton fibres.

FIG. 25. The working of capillary attraction in tubes of different diameters.

You will see later that this force is very important for you to keep in mind, for capillary attraction not only helps to circulate the sap in the plant, but helps also to circulate the water in the soil. By understanding this law, you can learn to keep the water in the soil and save your crops from drought.

We have obtained a general idea of how the plant gets its food and passes it round to make growth possible. Now let us see just how the well-fed plant goes about increasing its size and adding new stem, branches, and leaves.

37. **How the Stem Increases Its Size.** You have learned that the plant is made up of a mass of little cells of various shapes and sizes. All plants and all animals are made up of such cells. Some of these cells are so small that many thousand laid end to end would not extend an inch. Figure 26 shows you how some single cells look under a powerful microscope. The growth of a plant (or animal) is brought

about by the cells of the plant dividing and each thus forming two new cells and then each cell growing large, dividing again, and so on. Not all cells in a plant are doing this, but the cells in the cambium layer surrounding the woody part of the stem and the cells in the growing point

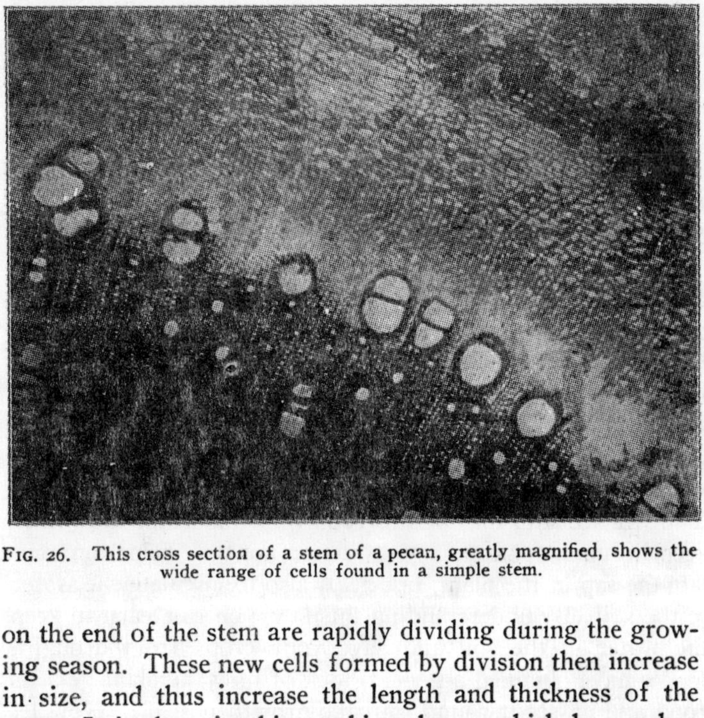

FIG. 26. This cross section of a stem of a pecan, greatly magnified, shows the wide range of cells found in a simple stem.

on the end of the stem are rapidly dividing during the growing season. These new cells formed by division then increase in size, and thus increase the length and thickness of the stem. It is then, in this cambium layer, which has such a rich food supply, that the cells divide and produce the new cells which increase the thickness of the tree. On the inside new wood cells are formed next to the wood cells already there, and on the outside new bark cells next to the bark already there. The main lengthening of the stem comes from the addition of new cells by division at the growing point on the end, and by the elongation of cells just back of the growing point. Many people have the idea that a tree, as it grows, pushes upward from its roots and that the limbs consequently get higher from the ground from year to year. Careful observations show that this is not true, but that a limb re-

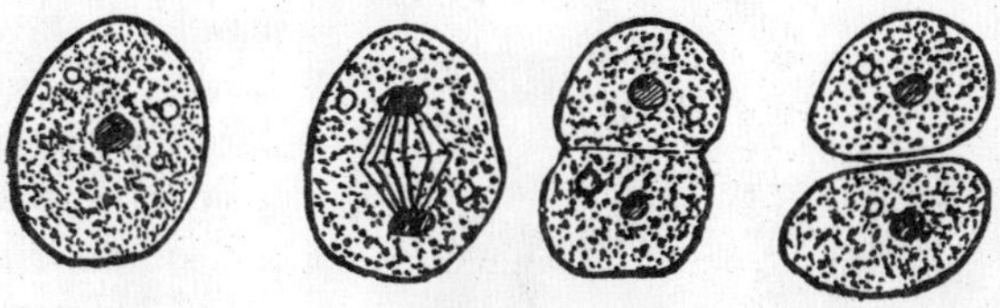

FIG. 27. This shows in a very condensed form how the single cell divides. This division of cells followed by an increase in size of the young cell is the process by which all plants and animals grow.

mains at the same height from year to year no matter how tall the tree grows to be.

38. **Outer Bark and Inner Cells.** As the new cells are added to bark and wood, and the stem thus enlarged, the old outer bark is burst by the pressure. The new bark will later be burst by the addition of still newer cells, and thus the rough dead outer bark with its deep cracks, which we see on old trees, is gradually formed. A layer of large new cells is developed during the rapid-growing seasons, and one of small cells during the slow-growing season. This makes part of the wood more dense than the other part, and in this way are formed the rings which we see when a tree is cut crosswise. These rings each represent a season of active growth, and, as this usually happens only once a year, they are called *annual* (ăn′nū-ăl) rings, from the Latin word *an-*

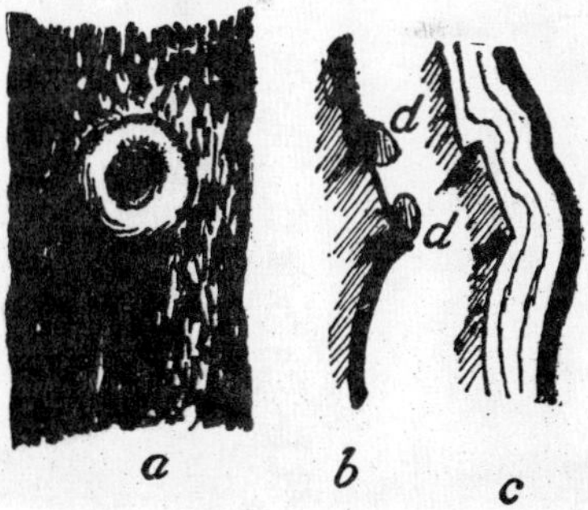

FIG. 28. This shows the healing tissue that the plant is throwing out to cover the wound made by sawing off a small limb. At *b* is represented a longitudinal section, showing the masses of healing tissue (*d*) thrown out by the cambium. At *c* the same plant is shown with the bark perfectly healed and the wound covered completely by new wood and bark.

nus, which means a year. By counting these rings it is usually possible to tell how old the tree was when cut down. In plants that have no cambium layer, of which we speak later, there would, of course, be found no annual rings.

FIG. 29. A pecan tree on which, after being cut back for top working, numerous dormant and adventitious buds sprouted out. Forty-four of these sprouts have been budded to two kinds of improved varieties of pecans as a variety test.

39. **How the Plant Heals Wounds.** This same cambium layer which forms the new cells for regular growth has the power of doing two other things that are of the greatest importance in the life of the plant. First, when a plant is wounded the cambium layer produces a healing tissue which soon fills up and heals over the wound with new growth unless the wound is very large. Even then the plant may for years steadily fill in the wounded places by a constant building in of new tissue. This is a very important matter that you will learn more about when you study pruning, budding,

and grafting. All budding and grafting are made possible by this power of the cambium layer to make healing tissue and fill up cuts and unite two separate surfaces.

40. **How the Plant Saves Its Life.** The other thing that the cambium layer does, which often saves the life of a plant, is to form new buds. At times a tree or other plant gets broken, or cut off back to a single stem, below which there are no branches or buds. In this case, and at times in less serious cases, new buds are formed in the cambium layer. These buds force themselves out through the bark and thus give the plant a new set of lateral branches on which are again developed the leaves, which the plant must have to gather and digest its food. These new buds formed in an emergency in the cambium layer are called *adventitious* (ăd-vĕn-tĭsh'ŭs) buds. The clusters of sprouts that grow out on a sawed-off tree trunk may be partly from adventitious buds and partly from ordinary buds that were before dormant. Recently scientists have been able to cause a new bud to be formed and a new branch to grow at a selected point on the stem of a plant by the local application of certain chemicals. This new process of artificial stimulation of bud formation by chemicals is at present only in the experimental stage.

41. **Plants That Have No Cambium Layer.** In a great many of our farm plants there is no definite area of growth or cambium layer, nor does the sap circulate through certain layers, as with the kinds of plants we have studied. The raw food material passes up through vascular bundles that are irregularly distributed throughout the stem, and the food passes back down through different tubes in the same bundles. If you will break a cornstalk and pull it apart, you will see these vascular bundles appear as long, tough fibres, pulling out of the pith. In this class of plants are all the grasses, including corn and small grain, which are merely grasses grown primarily for their seed. It is an interesting fact that all these plants without the definite cambium layer have only one cotyledon. Plants with one cotyledon, such as corn and cane, that do not have the special cambium layer,

cannot heal cuts in themselves in the same way that plants do which have the cambium layer, nor can they develop new adventitious buds and save themselves if they are broken or cut off below the terminal bud. For the same reason such plants cannot be grafted or budded.

QUESTIONS, PROBLEMS, AND EXERCISES

7. Name some seeds having down on them; some that have wings on them. What purpose do down and wings serve the seeds?

8. Name some seeds having hooks or spines on them. What purpose do hooks or spines serve the seeds?

9. Germinate beans and corn in a germinator, cut them open and point out the seedcase, the reserve food, and the embryo in each. What is the difference in the way the reserve food is stored in these two?

10. If the tiny soft root hairs take in all the food materials for the plant from the soil, what happens to the plant when it is torn out of the soil and replanted in another place? In what ways does the plant show the effects of this? When the transplanted plant gets plump again and begins to grow, what do you know must have taken place down on the roots?

11. If the root hairs which take in the water and food materials grow mainly at the tips of the roots and on the new roots, where should you put water or manure for a large tree—near the trunk of the tree or farther out? Why so?

12. Which is the better in transplanting, to pull the plant and roots out of the ground, or to take up the roots together with the soil surrounding them and put both the soil and roots into the new place? Why is it best to do as you say?

13. Take a pan of water and put it on the fire and watch it closely. What makes the little bubbles that you see forming in the water and coming to the top before the water gets hot enough to boil? Where was this air before the water was heated?

14. Take a glass or jar of water and gently put into it some lumps of soil. As the water goes into the pores of the lumps of soil and fills them, what happens that proves the soil has air in it?

15. Let us see in which direction the liquid always passes in

osmosis. Could it pass out of the root hair into the soil instead of from the soil into the root hair? Take one ounce of saltpetre and dissolve it in a pint of water. This solution is stronger than the sap in a potato. Call this solution one. Take one-eighth of an ounce of saltpetre and dissolve it in a gallon of water. Call this solution two. This solution is weaker than the sap in a potato. Now cut some slices of Irish potato, about one-eighth inch thick, and put some of these in solution one and some in solution two. Osmosis can go on through the cell walls of a potato just as it does through the membrane of a root hair. Look at these potato chips after a while and see if both lots are still plump, or if one is plump and full of water and the other is limp, and some of the water has passed out of it into the solution. Did the juice pass to the weaker or the stronger solution?

16. Fill a jar with good, moist, loose soil and, as you fill it, plant against the side of the jar, so that they can be seen from the outside, five grains of corn, one at one inch from the top, one at two, one at three, one at four, and one at five inches. Keep the soil moist and watch the grains from day to day, making notes of the process of germination as you see the grains through the glass. Through how many inches of soil can the corn-plant grow when depending upon the reserve food alone?

17. Plant a dozen each of radish, bean, and other seeds in moist soil; cover with soil and press down the soil on the seeds. Plant the same number of seeds at the same time in the same soil alongside these seeds, only cover these loosely with soil. Note which seeds come up first, those packed or those not packed. Why do these come up sooner than the others? (Remember what the seed must get before it can germinate.)

18. Remove a half-grown oat plant and a wheat plant from the earth with a shovel, taking up with each a large amount of the soil still in position. Then wash this soil away by pouring over it gently a stream of water. When the roots and root hairs are clean, examine them carefully. Note if there are any differences in the roots of the wheat and the oat.

19. Plant six bean seeds. When they come up clip both cotyledons off two plants, one cotyledon off two, and leave two untouched. Note the results.

20. Plant three bean seeds in good soil in each of three small cans or pots. As soon as they germinate, put one in a dark closet or box, one in the room near a window, and one out in the open

sunlight. From day to day make a note of the results. Can you tell what makes each plant behave as it does?

21. Explain why it is that when an old field grows up thickly with trees the weeds and grasses that were before in the field die out.

22. Place a glass jar over a plant that is growing in a can or pot, and after twenty-four hours note the drops of water deposited on the inside of the jar. Where does this water come from?

23. Plants often wilt in sunshine and look plump again the next morning. What effect has this wilting on transpiration? Why?

24. If there are forty-eight full-grown apple trees on an acre of land, how many gallons of water will be transpired by them in a season? It takes eight pounds to make a gallon.

25. Weigh six good-sized cornstalks. Hang in the air till thoroughly dry and weigh again. What per cent was water?

26. Where do trees get the food with which to put out new leaves?

27. How do potatoes sprout and grow on the floor without being planted?

28. Why are these sprouts pale yellowish instead of green when the potato sprouts in the dark cellar or closet?

29. Name some farm and garden plants that store starch, some that store sugar, some that store oil.

30. Animals, including man, take oxygen out of the air when breathing, and put into the air carbon dioxide. The plants take carbon dioxide out of the air and put into it oxygen. What would finally happen to man and all animals if the plants did not do this?

31. Girdle a young sapling, cutting completely around the tree down to the cambium layer, but not into the new wood. Watch this tree carefully till the end of the growing season and note what results. Watch it also the following year. Explain the results.

32. Cut a young sapling in the same way, only cut down through the layer of new wood all around. Note the results and explain.

33. Cut off a short section of the stem of some young plant, such as the bean, with leaves growing. Stick the lower end of this stem in red ink and notice how the fluid passes up into the leaves.

CHAPTER III

HOW PLANTS ARE REPRODUCED

42. **Three Ways of Reproducing a New Plant.** We have seen how the plant takes food material, manufactures its food, and grows as a single plant. Let us now see how it produces a new plant. This is done in three ways: by seed, by spores, or by division. Some plants reproduce themselves in more ways than one. We shall not consider spores at this time, but shall study reproduction by seed and by division.

43. **The Parts of a Flower.** In addition to having buds which open and develop into leaves, most plants have other buds which open and develop into flowers. It is in these flowers that seeds are formed in a most interesting way. Figure 30 shows a peach flower split in two. You will notice on the outside the small half leaf and half scale-like *sepals* (sē′pa̱l). These were greenish in color, covered the flower, and protected it before it opened. All these sepals taken together are called the *calyx* (kā′lĭks). Next inside the calyx are the large *petals* (pĕt′a̱l), which make the pretty white or pink showy part of the flower. All of these petals taken together are called the *corolla* (kŏ-rŏl′la). While these two parts of the flower, especially the corolla, are the ones usually noticed, they are not the important parts. Many kinds of flowers fail to have one of them and some fail to have both. The calyx serves merely as a protection, and the corolla serves partly as a protection and partly to attract insects, which are needed to help some flowers make seed, as we shall see later. Inside of the corolla you will notice the *stamens*

(stā′mĕnz) and the *pistil* (pĭs′tĭl). It is through these that the plant produces a seed.

44. **How the Seed Is Made.** If you should cut open one of these pistils you would find in the base of it a little thing that may become a seed. This is called an *ovule* (ō′vūl). This ovule, however, can never become a perfect seed till it is fertilized by a powder, called *pollen* (pŏl′lĕn), thrown out by the stamens. The plant makes this ovule, and at the same time makes also this pollen powder, but it makes one inside the pistil and the other inside the stamens. They must get together before the seed will develop. As soon as the ovule is ready to be fertilized by the pollen, the pistil exudes a sticky substance on its upper end, and the stamens split open and spill out their pollen grains, which fall on the sticky end of the pistil. There these small grains of pollen germinate somewhat as a seed would, and send down through the pistil a tiny thread-like tube, which is too small to be seen with the naked eye. This tube sent down from the pollen grain

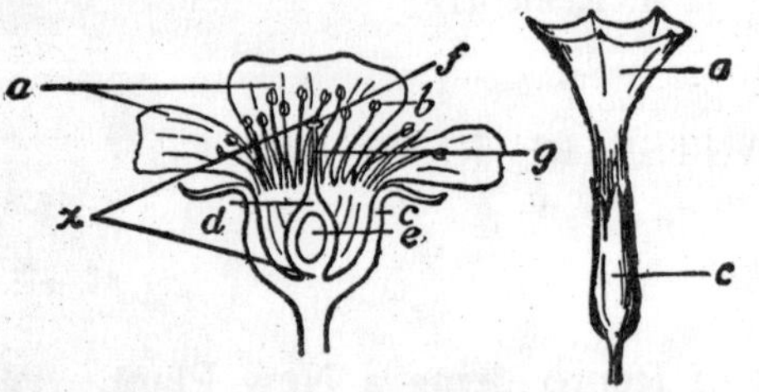

FIG. 30. A peach-blossom cut in two, and beside it a morning-glory. Note the pistil (*x*), made up of the stigma (*f*), the style (*g*), and the ovary (*d*), with its ovule (*e*) within. Note also the anthers (*b*), the petals of the corolla (*a*), and the calyx (*c*).

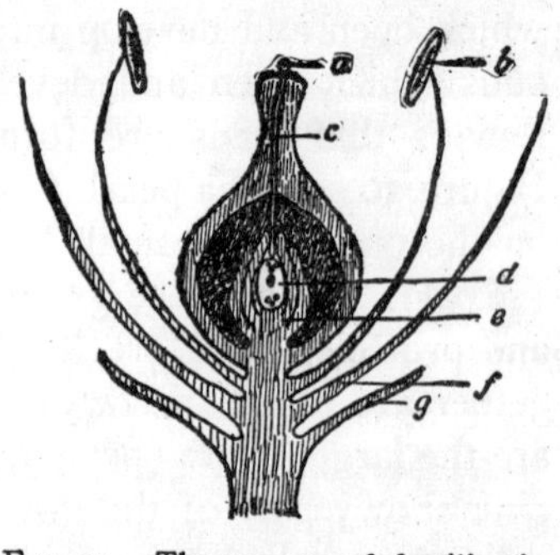

FIG. 31. The process of fertilization of the ovule by the pollen. Note how the pollen-tube extends down into the ovary and comes into contact with the ovule. When more than one seed is to be developed, a tube must be sent down from a pollen grain for each ovule. *a* represents the pollen grain; *b* represents the anther of the stamen from which the pollen falls; *c* represents the pollen-tube; *d* represents the ovule with several cells in it; *e* represents the ovary; *f* represents the corolla and *g* represents the calyx.

will, in a short time, reach the ovule, pierce its wall, and allow the contents of pollen tube and ovule to mix and thus to fertilize the ovule, so that it can complete the making of the tiny embryo and the storing of the reserve food for the embryo in its seedcase. The falling of the pollen on the stigma is called *pollination* (pŏl-lĭ-nā′shŭn). If, then, the pollen tube goes down and mixes its contents with the ovule, it is a complete fertilization. The lower end of the pistil in which the seed is formed is called the *ovary* (ō′vȧ-rȳ). This

FIG. 32. The male flowers (*y*) and female flowers (*x*) of the pecan.

word comes from the Latin *ovum,* which means an egg. These ovules, as you see, serve a purpose in plants similar to that served by eggs in animals. The top end of the pistil is called the *stigma* (stĭg′mȧ), and the slender supporting stem is called a *style.* Some plants have no style. In these, the stigma is directly on top of the ovary. The stamen likewise has distinct parts. On the top is the little box that holds the pollen, called the *anther* (ăn′thẽr). Below this is the slender supporting stem called the *filament* (fĭl′ȧ-ment). These stamens and pistils in different plants are of all sorts of shapes and sizes, and in varying numbers, just as the calyxes and corollas of different plants vary in many different ways. All that is necessary to produce a seed is that there be a stamen that develops pollen and a pistil that starts the development of one or more ovules, and then that somehow the pollen grain gets on the top of the pistil and sends its little thread-like tube down and fertilizes the ovule. As soon as the pollen has by some means reached the stigma, and the pollen tube has gone deep enough to reach the ovule, the seed in the ovary develops rapidly. If, however, no pollen falls on the stigma, or the tube sent down by the pollen is unable to reach the ovule, no seed is produced. Usually in these cases the ovary dries up, the flower soon dies, and no fruit is produced. This is a frequent cause for failure in the crop of plums, pecans, and other fruit and nut crops. Some fruits, however, will develop in spite of a failure in fertilization and consequent lack of seed, as, for example, the banana, or seedless grape, or seedless orange.

45. **Male or Female Flowers.** Usually, both stamens and pistils are made on the same flower, but some plants make their pistils in one set of flowers and their stamens in another set, as the pecans and walnuts do. Some other kinds of plants, such as the willows, the date-palms, and some of the wild grapes, make flowers with only pistils on one plant and flowers with only stamens on another plant. In both these cases there may be uncertainty about the pollen getting from the *staminate* (stăm'ĭ-nȧt), or male, flowers over to the pistils of the *pistillate* (pĭs'tĭ-lȧ̆t), or female, flowers. In such cases the pollen, which is usually a very fine powder, is blown by the wind from the male to the female flowers, or the insects, which go in and out of the flowers to gather nectar, carry the pollen from the anthers of the male flowers and brush it upon the pistil of the female flowers. In our field corn, for example, the pollen is usually blown from the tassels, which are masses of male flowers, or stamens, upon the silk of the corn, which is a mass of female flowers, or styles and stigmas, leading down to the corn ovules below. Sometimes, even when plants have both stamens and pistils on the same flower, these are so situated that it is difficult for the pollen to get from the anther to the stigma. When the anther stands above the stigma the pollen falls easily on

it, but when the anther is lower than the stigma, either the flower must hang downward, or the wind must blow the pollen, or some insect, as the honey bee, must carry it from one to the other on its body. The flowers with white and yellow or other light-colored corollas attract night-working insects especially. The odors of some flowers and the little drops of sweet liquid called nectar in the blossoms of others also attract insects. It was impossible for the Smyrna figs to bear their delicious fruit when cultivated in America until the little insects, also from Smyrna, that feed in their blossoms and carry to them the pollen from male fig blossoms were brought over here.

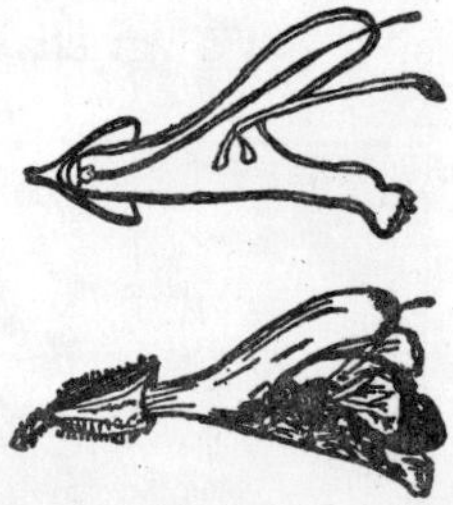

FIG. 33. The manner in which insects help some flowers to get fertilized by brushing pollen on stigmas that would not otherwise be easily reached by the pollen.

In order then to produce seed, the plant must have food enough to produce the flowers, the germs, and the little stores of reserve food in the seed. We shall see later that certain foods are used in larger proportion by the plant in making seed, while certain other foods are used in larger proportions in making stems and leaves. When the plant has the right food and makes both the ovule in the ovary and the pollen in the anther, the ovule still may not get fertilized, and hence no seed ever develop. In your questions and problems you will have a chance to think out how some of these failures may come about.

46. **Crossing and Improving Plants.** If the perfect seed is a result of the union of the substance from the pollen and the substance in the ovule, we should naturally expect that, if we put the pollen of a round, flat squash on the stigma of a long-necked squash, the seed coming from this union would have in it a mixture of the characters of the two. This is just what happens. These mixtures of two varieties of the same plant resulting from the pollen of one variety falling on the stigma of the other, and fertilizing the ovule, are

called crosses, or *hybrids* (hī′brĭds). Carrying the pollen of one plant to the pistil of another is called *cross-pollination.*

You can see what splendid opportunity this gives us to improve our varieties of farm plants. One can cross a big-bolled, heavy-yielding cotton plant that is not early enough, on an early one that does not have so good a boll, and possibly get a variety that has a mixture of the earliness of one

FIG. 34. Dwarf Yellow Milo at the left; Black-hull Kafir at the right, with the F-1 hybrid in the center.
Courtesy Texas Experiment Station.

parent and the big bolls and big yield of the other. Or, he can cross a variety of corn that has a fine root system that enables it to gather plenty of food material and resist drought, but which makes only fair-sized ears of corn, on another variety that does not have such good root system and drought-resisting qualities, but has large ears. By doing this he may get a hybrid that both resists drought and has big ears. It should be remembered, however, that in such hybrids, undesirable characters may combine. For example, in crossing large poor-quality watermelons with small high-quality ones, there is a possibility that the resulting hybrid will be small and poor in quality instead of large and high in quality.

Plant breeders develop many worthless hybrids, but they are always on the alert to discover one that has a combination of desirable qualities. In most cases it is necessary further to improve the strain by using such superior plants as a basis for new crosses. It was by following such a procedure that Mr. J. W. Stubenrauch developed the Carman peach from

FIG. 35. In the center is Ajax, a new variety of grain sorghum produced at the Chillicothe, Texas, station by a cross between kafir and feterita.

the Elberta and Family Favorite varieties. Luther Burbank followed the same procedure in developing the many varieties of fruits, flowers, and vegetables that bear his name. The late Mr. T. V. Munson of Denison, Texas, crossed different kinds of wild grapes upon fine cultivated varieties and from them developed a dozen, or more, promising grape varieties adapted to the South.

An objection to hybrids is that seeds from them may not come true afterward; that is, the seed of a fine hybrid melon may produce a melon that is not like the parent, but like some one of the grandparents, just as a child may not be like either his father or mother, but resemble one of the great-grandparents. This failure to come true to the parent is called *variation.* When the variation is due to the cropping out of some quality that belonged to one of the ancestors back of the immediate parents, it is called *reversion* (rė-vĕr′shŭn). This can some-

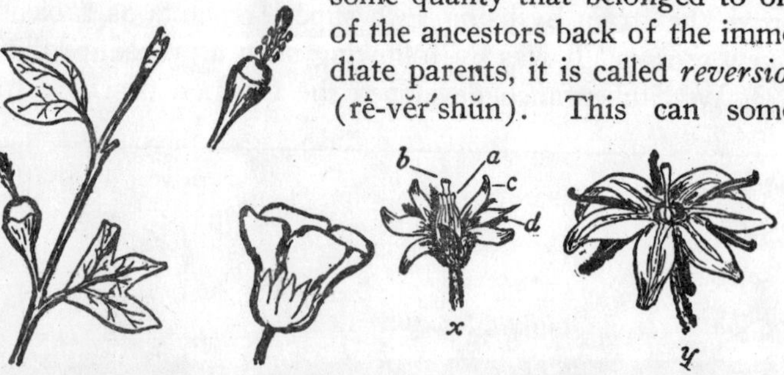

FIG. 36. On the left is a cotton bloom with corolla and stamens cut away ready for cross fertilization and the flower ready to be covered with a bag. On the right is a tomato bloom. *x* shows the plant before the unripe stamens have been cut away. *y* shows the flower ready to bag with stamens removed.

times be bred out of a hybrid and often it can be made to come true, but the way of doing it will have to be learned later.

47. **How to Cross Plants.** Let us now learn how to cross two plants. Select two plants belonging to the same species, the flowers of which ripen at the same time. Just before the corolla of the flower has opened, and before the anthers have opened to let any of the pollen spill, either open up or cut away carefully the corolla, so that you can get at the stamens. Carefully cut away with small scissors or very sharp knife the anthers, making sure that no pollen is left in the blossom. Cover this flower immediately with a small, thin paper bag, and tie it so that no pollen can come to it, either from the wind or an insect. Then examine this flower

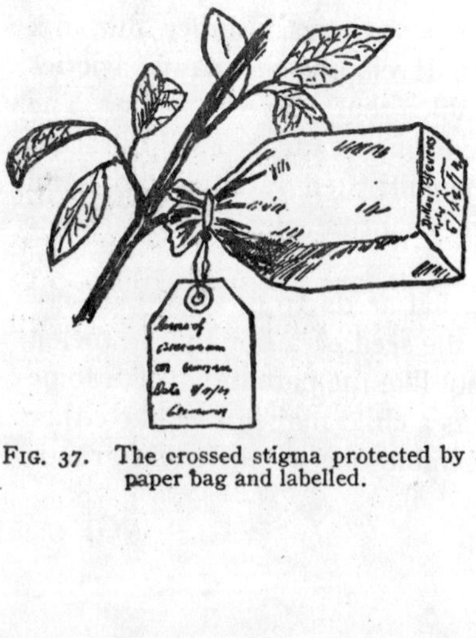

FIG. 37. The crossed stigma protected by a paper bag and labelled.

daily until you find it has ripened enough for the pistil to exude the sticky matter on its stigma. Now take the pollen from a flower of the variety that you wish to cross on it, and gently dust this pollen on the stigma, and immediately cover it again with the bag. Then label this flower and make a record of what you have done. Leave it covered for several days until the ovule or ovules are fertilized and the fruit or seeds begin to form. Since crosses often fail, it is advisable

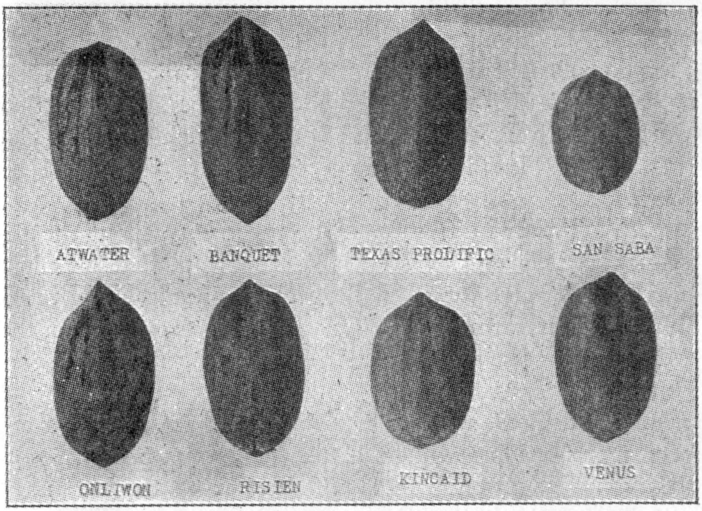

FIG. 38. All of the varieties of pecan shown above came from seed of the original San Saba tree and crosses made from it by Mr. E. E. Risien.

to make several, and to make them on different days, so as to make more sure that you get one or more to live. The larger the number of crossed seeds you prepare, the greater the probability is that one of the plants coming from these seeds will contain a desirable combination of the various qualities that were in the two parent plants.[1] These seeds

[1]The plants resulting from crosses do not in the first generation show mixtures of single characters, as, for example, of color; but they show new combinations of characteristics of one parent with other characteristics of the other parent. For example, the first generation may show the shape of one parent combined with the color or flavor of the other. The production of blends of two differing single characteristics, such as a mulatto skin from the mixture of white and black, is rare and not well understood.

should be planted well separated from any patch of related plants and carefully watched. If a specially favorable cross is secured, the flowers of this should be protected from the pollen of neighboring plants to increase the chance of its seed reproducing the same fine plant unmixed. By continually selecting and protecting the plants that come true each year, and planting only their seeds, you may soon have seeds that will breed true practically all the time. When you take an advanced course in plant breeding, you will learn a shorter and surer way to make your hybrid come true to seed, but it is too complicated for you to learn now. In making crosses, the best results have usually come from crossing plants that do not differ very widely, and both of which represent desirable types. In this way many good varieties have been produced.

48. **Variation in Plants.** One peculiarity about the reproduction of plants by seed is so important that we must study it carefully, for it is the greatest means we have of improving our farm plants. This fact is that even when the seeds of a plant have been fertilized by pollen from the same plant, the seeds will not all produce plants exactly like the parent plant. Some will produce plants that are just the same as the parent, some will be better, and some not so good. The next generation will be different from the parent in many ways. We have already learned that this failure of the offspring to reproduce the parent exactly is called variation, and have seen that the variation may be due to the cropping out of a characteristic of one of the ancestors of the parent plant. It has been discovered recently that under certain circumstances the exposure of some kinds of seeds and eggs to the action of X-rays, or the rays given off by radium, will cause the plants and animals coming from these seeds and eggs to vary more than will those coming from similar seeds and eggs that are not so treated. Some of these artificially produced variations are less desirable than either parent, a few may be better. As yet the scientist does not know just what stimulation to give in order to produce exactly the kind

of good variation that he desires. Therefore, we have to watch for the favorable variations made by nature or by artificial stimulation and propagate these to improve our plants and animals. In a row of cotton or any other plants, even when the whole row is planted from the seed of one plant, you will notice various types of plants. These differ-

FIG. 39. The above picture shows the result of five years of selection in corn for high and low position of the first ear on the stalk. The white tape marks the height of the ears in the first rows of the high-ear and low-ear blocks.

Courtesy of Professor C. M. Woodward, the College of Agriculture, the University of Illinois. From "Genetics in Relation to Agriculture," by Babcock and Clauson.

ences are due partly to variation. If now you take the seeds from the best stalk in the row, while they too will vary somewhat, and may be partly cross-pollenized from some poor stalk near by, they will tend to reproduce this specially fine stalk and even a few better ones. By constantly repeating this selection, and always taking the best specimens for your seed, you will constantly get a better and better variety.

49. **Results of Selection.** It is by this process of selection of favorable variations and breeding and multiplying them that practically all of our finest varieties of farm and garden plants have been produced. As each different soil and climate will best suit a somewhat different variety, there remains still a great work to be done by each intelligent farmer in watching for favorable variations in the crops on his land,

and then protecting these, saving the seeds separately, planting them in a separate place, and, by repeated selection, breeding up the variety best suited to his particular locality. This is a particularly interesting and valuable work for our boys and girls to do. Two excellent new varieties of sweet corn, Honey June and Surecropper Sugar, were recently

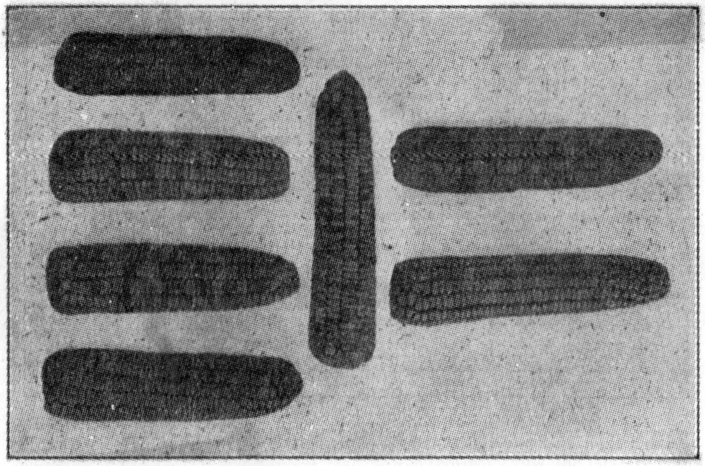

FIG. 40. Wrinkled sweet corn at the top left is crossed with smooth field corn at bottom left, which is not sweet, but is very hardy. Grains of both types occur on the same ears that year. In following generations there will be pure field corn, pure sugar corn, and hybrids, some with desirable combinations of qualities, some with undesirable ones. By selection and continuous reproduction of desirable types due to crossing, many fine new varieties of sweet corn, such as Texas Honey June, have been developed.

Courtesy of P. C. Mangelsdorf.

produced at the Texas Agricultural Experiment Station by crossing the starchy Mexican June and Surecropper field corns with Country Gentleman sugar corn. Of course, all of the resulting crosses did not have the sweetness desired. These were discarded and only the hybrids kept that had both the sweetness of the Country Gentleman and the hardiness and productivity of the Mexican June and Surecropper. All the kinds of roses we have are but variations of the simple wild rose that have been selected and carefully bred. All of our varieties of cotton are the results of variations being carefully selected and properly bred. In Wisconsin they have selected and bred a variety of oats that increased the State's yield fifteen million bushels a year. By careful selection it would be possible in a few years to secure varieties of cotton suited to different sections that would add many bales a year to the crop of the cotton states, without increasing the amount of land cultivated a single acre. The same is true to a greater or less extent of every state and with every farm crop.

FIG. 41. See how the blue grass reproduces itself by stolons.

50. **Plants May Be Reproduced By Parts.** We have seen how plants may be reproduced by seed. Other parts of the plant may also give rise to new plants. When a part other than seed is used in developing a new plant, it is called vegetative reproduction. There are certain advantages in reproducing plants from vegetative parts. Some plants, such as potato and banana, produce few seeds and it is necessary, if they are to be reproduced, that other means than seed be employed. Others, such as the peach and pecan, seldom come true from seed. Vegetative reproduction is practiced generally in order to preserve good varieties. Many of our farm, orchard, and vegetable crops are of this kind and it is important that we learn more about the reproduction of them.

There are seven principal ways by which vegetative parts of plants can be used to produce new plants:

(1) *Natural Layers.* Many plants develop horizontal stems called stolons, which develop roots and new shoots at certain points along the stem. These generally develop at points called nodes, though roots frequently develop at other points, along the internode. One reason that Bermuda

grass spreads so rapidly is because it produces natural layers. After such layers have taken root and have made additional growth above ground, they are often indistinguishable from the mother plant.

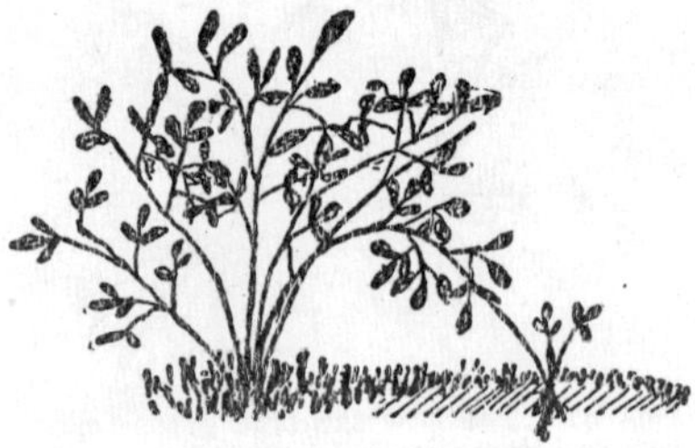

FIG. 42. Note how the tip of the raspberry takes root and grows when layered.

The strawberry and raspberry produce natural layers. They produce them so abundantly that unless thinned from time to time the layers become crowded and discourage growth and make cultivation difficult.

(2) *Artificial Layers.* The grape and many other plants do not ordinarily produce natural layers. Yet if vines or stems are bent to the ground and covered with moist soil, they will take root and grow. The rooted part may then be separated from the mother vine and when planted out in a new location will grow. Layers that develop roots at one point only are *simple layers.* With the grape, however, it is possible to make *compound layers.* The best plan is to dig a long trench about two inches deep, and after the vine is laid in this, the trench is covered over with soil, leaving only the tip of the vine out. The vine will develop roots and send up stems at each joint and each of these may be separated and planted elsewhere the next year.

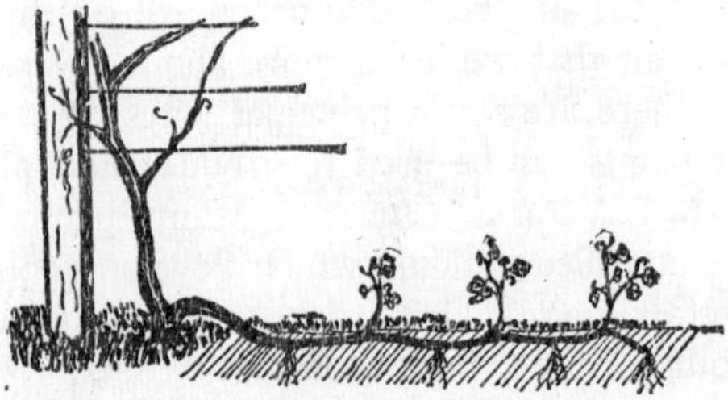

FIG. 43. The method of propagating grapes by layering.

(3) *Modified Stems.* Some plants produce underground stems, which when separated from the par-

ent plant will serve to reproduce it. The *rhizome* of the canna and Johnson grass, and the *tuber* of the Irish potato and the Jerusalem artichoke are examples of such modified stems. The "eyes" on the potato are really modified buds, which under favorable conditions will develop a new plant. The tuber and rhizome consist largely of reserve food which supports the new plant.

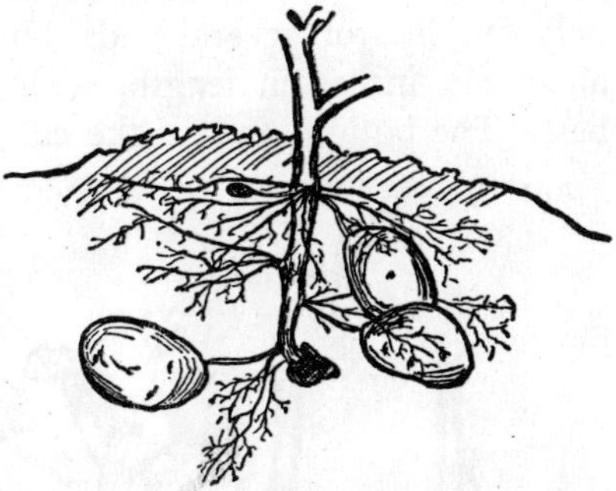
FIG. 44. The Irish potato and the enlarged underground stems or tubers.

(4) *Fleshy Roots.* The sweet potato is an example of a fleshy root. It has reserve food material like the Irish potato, but no "eyes" or buds. It will be recalled that roots do not have true buds, and hence sweet potatoes being modified roots would not be expected to have buds on them. They, however, do develop special buds which grow into young vines, called *slips.*

FIG. 45. A rooted begonia-leaf cutting.

(5) *Cuttings.* A cutting is a piece of stem, leaf, or root, which will form a new plant if cut off from the parent and planted under favorable conditions. It develops roots from the base, and the buds already present grow. Usually such cuttings are planted out in sand. The sweet potato vine will root readily in water, and roses are commonly planted out in ordinary soil. Plants that do not have a cambium layer seldom grow well from cuttings. Of the plants that have a cambium layer, some will grow readily and others will not. The grape roots easily, while the peach and pecan rarely, if ever, grow from cuttings.

In some cases, new wood is used for cuttings; in others, old wood. Cuttings are usually made six or eight inches long. A cutting may be longer or shorter than this, and it may have only one bud or several buds, but usually cuttings that are about six inches in length, with two or three buds, grow best. The bottom end of the cutting should be cut off just

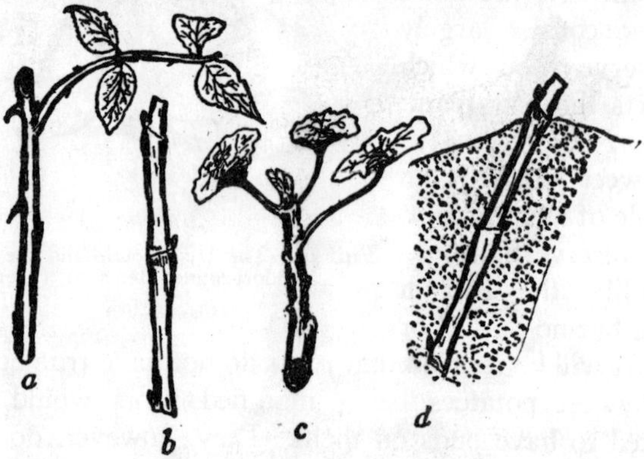

FIG. 46. Cutting of rose (*a*), grape (*b*), and fig (*c*). At *d* the proper position of the cutting in the soil is shown.

below a bud, and the top end from one-half to one inch above a bud. Figure 46 shows how the cutting should be placed in the soil. As soon as the growing season has begun, these cuttings will throw out roots at the lower buried joints, or buds, and the exposed upper bud will start a shoot. Cuttings usually grow better in soil that has very little organic matter in it, as the little bacteria and *fungi* (fŭn′jī) living on the organic matter often attack the exposed cut surface and cause decay. For this reason cuttings are often rooted in coarse sand. The soil should be well drained. The air should be moist and of uniform temperature also, for best results. With cuttings that are hard to root, bottom heat is frequently applied with good results. After cuttings are started they should be carefully cultivated and kept free from weeds and grass, as their roots are near the surface.

(6) *Grafting.* Instead of cutting off a piece of the plant and planting it in the soil to make it grow, we can insert it in the body of another plant and let it grow there. If the inserted piece is a part of a stem, it is called *grafting.* Nearly all grafting work is done when the plant is dormant. The plant upon which the piece is grafted is called the *stock,* and the part that is transferred is called the *scion* (sī′ŭn). Plants must be closely related, else it is not possible for one to live when grafted on the other. Usually, they should belong to the same species, but sometimes even different genera may be grafted, as, for example, the pear may be grafted upon the quince. When the transplanted graft lives and unites with the stock all the limbs, leaves, and fruit developing from the graft remain true to the variety from which the graft was taken, in spite of the fact that the raw food material is furnished by the root of the plant in which the graft was planted. This makes it possible to put grafts from fine varieties on trees or vines that bear naturally poor fruit, and thus force them to bear good fruits instead of poor. Most of our orchard trees have long been treated this way, and now the nut trees are being treated in the same manner. Grafting is easy to learn.

There are many forms of grafting, but the three most important are *tongue grafting* (or whip grafting, as it is also called), *cleft grafting,* and *bark grafting.* Tongue grafting is used mostly on young seedling stocks less than an inch in diameter. For plants larger than that, and especially in top-working old trees, some form of cleft grafting or bark grafting is generally used.

(*a*) For the *whip graft,* the stock should be about the size of a pencil. The scion should be as near the same size as possible, and should have two or more buds on it. Cut the stock off with a slant, so as to give a cut surface about three times as great as it would have been if cut square across. Then set the knife blade about one third the distance down from the top of the cut surface and make a vertical incision about one-half inch long. (See Figure 47.) Cut the scion

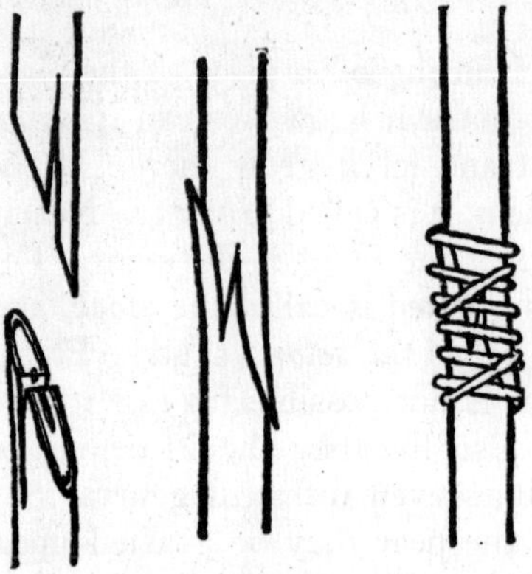

FIG. 47. The method of making the whip, or tongue, graft.

in similar manner, join the two together as shown in Figure 47, and wrap with a string to hold the two in place. Knives should always be thoroughly cleaned before cutting into a plant and, as far as possible, neither the hands nor anything else should be allowed to touch the cut edges. As it is the cambium layer that throws out healing tissue and unites the stock and scion, thus allowing sap to flow from one to the other, it is necessary to use care in fitting the graft so that the cambium of the scion is placed in contact with the cambium of the stock. When this is done, only a small amount of healing tissue is necessary to unite these, after which the circulation of sap from one to the other is possible.

(*b*) The *cleft graft* may be used on small nursery plants also, but it is usually employed on the large plants in putting

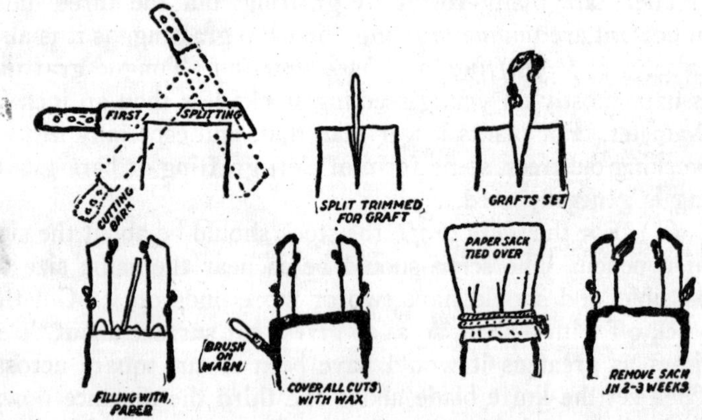

FIG. 48. The process of cleft grafting.

tops of fine varieties of fruits or nuts on common trees. In cleft grafting, the stock, for best results, should not be over three inches in diameter, while the scion should be the same size as in tongue grafting. In top-working an old tree cut back the central limbs, with a square cut, to stubs eight or more inches long. Smooth the end, then drive down a grafting knife one or two inches as shown in Figure 48. Withdraw the knife and keep the incision open with a sharpened stick. The scion should now be trimmed to a wedge shape as shown, with the inside thinner than the outside to make a perfect fit, and one bud left on the outside near the top. The split edges of the limb should now be cut away in such shape that when the stick is withdrawn the scion will fit tightly in the cut as shown in Figure 48. The scion should now be quickly but gently forced down into the cut, the cambium layer of the scion being carefully placed directly against the cambium of the stock. The stick is then withdrawn. As soon as the scion is in place all cut surfaces should be covered with warm grafting wax, or melted paraffin, and a string tied around the stump, so as to help hold the grafts in place. A good grafting wax is made by using four parts of rosin, two parts beeswax, and one of tallow, by weight. These are cut in small pieces, melted together over the fire in a vessel, and poured into water to cool. It is then made into balls, and heated if necessary later as wanted. Heating may not be necessary on warm days.

(*c*) *Bark Grafting.* In the bark graft, the scions should be dormant, but the stock must be in growth, so that the bark will slip. The small seedling or the limb of the large tree that is to be used as a stock is first sawed off smoothly and the very rough outer bark, if present, is trimmed off on the side where the graft is to be placed for three or four inches back from the end, so that the bark can be loosened and bent as necessary with more ease and with less danger of splitting it. Next, a dormant piece of grafting wood is trimmed, as shown in *A, B, c, d, e, f* of Figure 49. Then the bark at the end of the stock is slit and the graft pressed down in between the bark of the stock and the new wood, thus bringing the cambium of the scion into contact with the cambium of the stock, as shown in G, Figure 49. The scion is inserted gently, so as not to tear the point of the scion or the bark of the stock severely. Next, the graft is tied in place, as shown in I, Figure

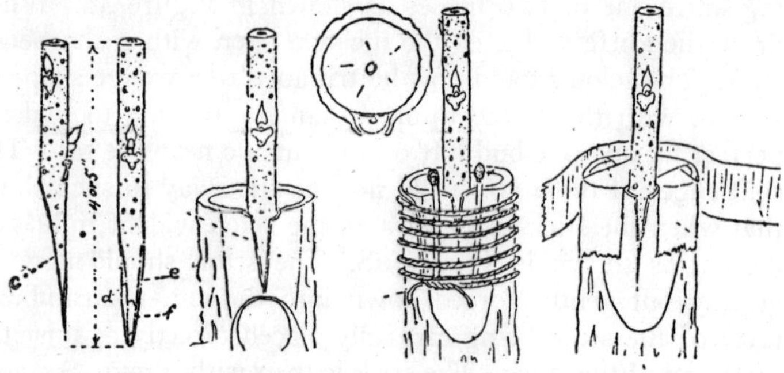

FIG. 49. Method of making a bark graft. *A* is a side view showing *c,* the sloping cut of the scion; *B* is a view of the back of the scion that will go next to the bark of the stock, showing the edges of the bark cut away slightly at *d, e,* and *f* to expose the cambium layer; *G* and *H* show side and end views of the stock with the scion inserted; *I* shows the method of tying the graft in, with two match stems, placed one on each side of the scion, so placed outside the bark of the stock that the string presses the bark closer to the scion and to the stock; *J* shows the method of covering with waxed cloth strips, instead of using the string and melted paraffin or wax.

Courtesy of D. V. Schuhardt.

49, and covered with melted paraffin or grafting wax, or with strips of waxed cloth, as illustrated in J, Figure 49.

(7) *Budding.* If the part cut off and inserted in the other plant is only a bud with some bark surrounding it, the operation is called budding. Budding accomplishes the same results as grafting. It is difficult to tell whether a tree has been grafted or budded after a few years. Budding is usually best done during the plant's active growing season. As in grafting, it is necessary that the two plants be closely related, and that the cambium layer of the bud be brought into connection with the cambium layer of the stock. The three most important forms of budding are the *shield bud* (or T bud), the *patch bud,* and the *chip bud.* Nearly all fruit and orna-

mental trees are propagated by the shield bud, while the chip and patch buds are usually best with nut trees.

(*a*) *The shield bud* is used mostly on young nursery stock about the size of a pencil, though it is sometimes used also in top-working old trees. In budding, usually a branch about the size of the stock and containing several leaf buds is cut from the tree you wish to propagate, and the leaves are at once cut off so as to leave about half an inch of the stem of

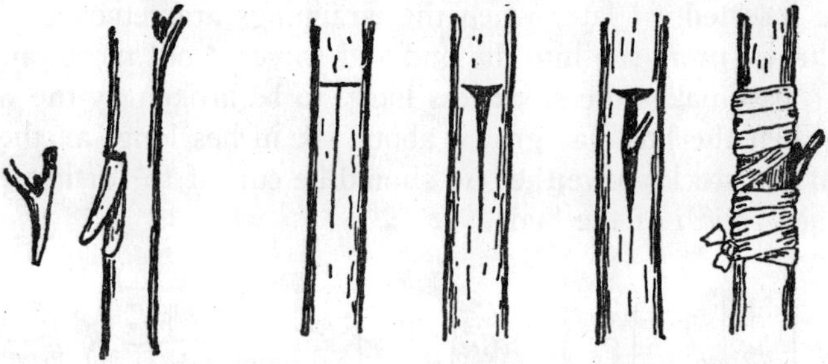

FIG. 50. The steps in the proper method of shield budding.

the leaf, or *petiole* (pĕt'ĭ-ōl), as it is named. This branch is called the bud stick and must be kept wrapped in damp cloth or moss. When ready to begin work, first pick out a smooth place on your stock and make a slit through the bark for about one inch in length up and down the stock. Then at the top of this incision make a crosscut, about one-quarter inch long, giving your incision the shape of a T. In making these incisions be careful to cut through the bark and cambium, but not into the young growing wood. Then cut a bud from your bud stick by placing the blade of a sharp knife about one quarter of an inch below the bud and cutting upward to a point about the same distance above the bud, but leaving the cut strip still adhering to the bud stick at its upper end. Then withdraw the knife and cut through the bark entirely, as shown in Figure 50. Then, by catching hold of the petiole of the leaf, lift the bark entirely free from the wood, as shown in Figure 50. Then open the cut on the stock by lifting up the bark in both directions from the crosscut, and slip the bud from the scion under the bark of the stock, as shown in Figure 50. The bud should then be wrapped as shown, with either raffia, twine, waxed cloth, or similar material, so as to hold the two cambium layers close together and to keep out water, air, and dust. In ten days or two weeks the bud should have united with the stock, and the wrapping should be removed to allow circulation of sap and growth. Part of the stock above the bud and other buds close to the inserted bud should be removed when the bud is inserted, or later when the wrappings are removed. This throws more sap into the bud and forces it out more rapidly. It also makes the stock less likely to be broken by the wind. When the bud has grown about six inches long, all the top of the stock above the bud should be cut off to further force the growth of the bud.

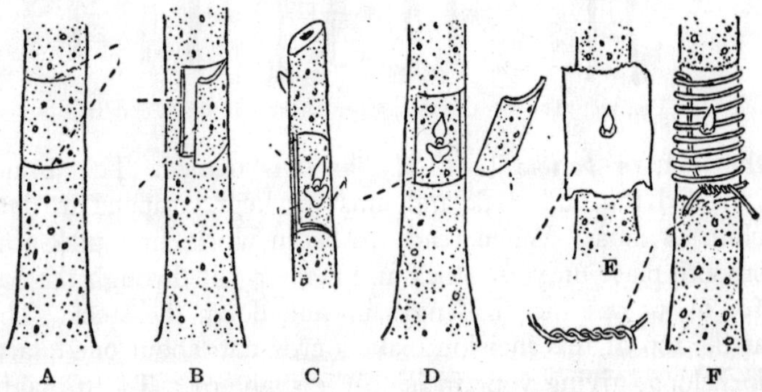

FIG. 51. Method of patch budding. *A* shows the stock with the two parallel crosscuts made in the bark of the scion by a two-bladed patch-budding knife; *B* shows the patch slit vertically on one side and the patch of bark partially lifted out; *C* shows the scion with the bud patch similarly cut and partly peeled from the scion; *D* shows the bark patch peeled out of the stock and the bud patch inserted in its place; *E* shows the inserted bud patch covered with a paraffined or waxed cloth ready for tying; *F* shows the bud tied in and complete.

Courtesy of D. V. Schuhardt.

(*b*) *Patch Budding.* Patch budding is the most common method used in propagating the pecan. It is a modification of the ring-budding method and may be used on nursery stock, stump or root sprouts, or on the sprouts that come out on the limbs of large trees that have been topped. The sprouts on trees that have been topped during the dormant season may be patch budded late in the following summer or during the spring and summer of the next succeeding year. Patch budding may be done with an ordinary pocketknife, or more easily with a Jones patch-budding tool, but it can be done most rapidly and successfully with a double-bladed patch-budding knife that can be bought ready-made or easily made at home out of a piece of wood and two knife blades or old safety-razor blades. In patch budding, two parallel cuts are made on the stock, as shown in Figure 51. Then a vertical slit is made between these cuts and the piece of bark is raised, as shown in *B*. Leaving this bark still in place for a moment, similar cuts are made on the bud stick, as shown in *C*. Then the patch of bark with the bud in it is lifted from the bud stick, the loosened bark is lifted back from the stock, and the bud patch is slipped in place upon the stock. The surplus flap of loose bark of the stock is now cut off, so that the bud patch fits neatly into the place from which the bark was taken on the stock, as is shown in *D*. Next, a patch of waxed or paraffined cloth with a small hole in the center of it is slipped over the bud and wrapped securely, as is shown in *E* and *F*. This protects the bud to a great extent from wind, water, and insects.

(*c*) *Chip Budding*. Chip budding differs from other methods of budding in that it can be done successfully in late winter and early spring before trees have begun to grow. It may be done also for three weeks or a month after growth has begun. The stock should be from one-half to one inch in diameter, and the scion should correspond roughly in size to the stock into which the buds are to be placed. Buds should be placed near the base of the stock. Select a smooth place on stock and cut downward and slightly inward for a distance of one and one-quarter inches. Let a second cut, made obliquely downward, intersect the lower part of the first cut. Thus the base of the matrix is roughly V-shaped. The bud is cut from the scion by making precisely the same cuts. It

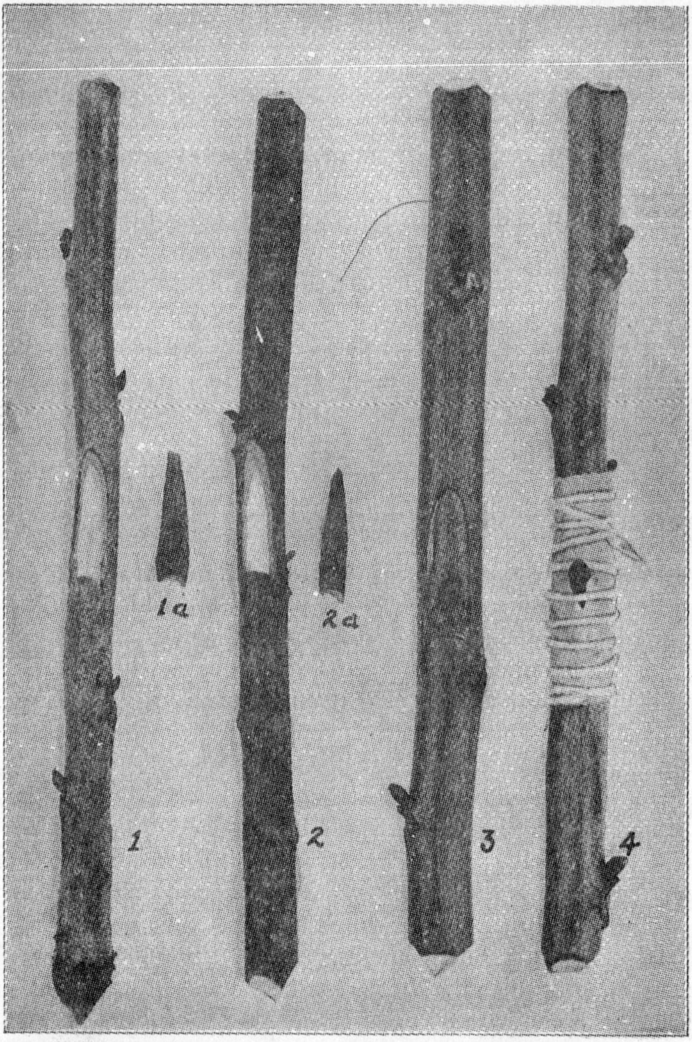

FIG. 52. Chip budding. 1. The bud stick with the chip bud (*1a*) cut out. 2. The stock with a chip (*2a*) cut out to make a bed for the bud. 3. The stock with the chip bud from the bud stick fitted into the bed cut for it. 4. The stock with the bud covered with waxed or paraffined cloth and tied in place ready to grow.

is placed into the matrix in a way that insures contact of the cambium layers on at least one side. Tying and waxing complete the job, except for forcing the bud. This is accomplished either at time of budding or somewhat later by cutting

FIG. 53. This shows three years of growth from twelve buds placed in the top of a seedling pecan-tree that had been cut back properly for the purpose of being top-budded.

stock off four inches above the inserted bud. The chip bud is used at times on pecans, peaches, plums, apples, and pears.

51. **How to Succeed in Budding and Grafting.** In all kinds of budding it is especially important that the knives be kept clean, that the cut surfaces and inner bark be not

touched by the hands or other things, that the work be done quickly in order to expose the cut surface as little as possible to the air, and that the cambium layers be carefully brought into contact. If these directions are followed, if fresh budding wood and vigorous stock are used, and if the buds are watched afterward, the wrapping not removed until the buds have started growing, and all sprouts that would rob the bud of its nourishment are kept cut off, you are sure to have success in budding and grafting.

QUESTIONS, PROBLEMS, AND EXERCISES

34. Examine a peach, pear, or plum tree before it has budded out in spring, and see if there are more than one kind of bud on it. Draw the branch, showing the buds, and describe each kind. Note later into what each kind develops.

35. Examine the branches of a budding pecan tree carefully, and find both the male and female flowers. Can you tell now why the pecan crop fails if there is a long rainy spell during blooming season? Can you also see why pecan seeds do not usually come true, but are mixed?

36. Find a male grapevine and a female grapevine. Tell how they differ in appearance.

37. Bring in a flower in which the pollen would fall easily on the stigma. Bring in another flower in which it would be difficult for the pollen to get on the stigma, and find out how the pollen is carried in this last case.

38. Select in the field, make notes on, and bring to school especially desirable variations found in one of the following plants: cotton, corn, oats, wheat, cane, milo, kafir, peas. Plan with your teacher a scheme for breeding and developing this desirable variation. (See the lessons on cotton and corn.)

39. Make and root cuttings of each of the following: rose, fig, grape. Root, by layering, a blackberry and a grape.

40. In September and October plant apple, peach, and plum seed, to have stock on which later to bud and graft. Plant pecans in January and February.

41. While the tree is still thoroughly dormant, cut some nice smooth twigs a foot or two long and about as large as your finger from the pecan tree in your neighborhood that bears the best

pecans and bears regularly. Be sure to select twigs that have on them some plump dormant buds. Keep these bud sticks buried in the earth in moist sand in a cool shady place, until needed. When the buds upon the pecan trees begin to swell, saw off, about seven feet from the ground, the tops of two vigorous young pecan sprouts or seedlings that are not more than three inches in diameter, or saw off the ends of the limbs of a larger tree, sawing these out where the limbs will be about three inches in diameter or less. Now, take out your dormant budwood and place a bark graft in each sawed-off end, in accordance with the directions in this chapter. Watch those that grow, pruning and bracing as necessary to prevent their being blown out. On any sprouts or limbs in which the graft dies, allow the new sprouts to grow; and, either late in the summer or early the next spring, bud desirable ones of these sprouts by the patch-bud method.

CHAPTER IV

THE SOIL

52. The Study of the Soil. Without being planted in soil at all, plants have been raised to maturity and crops of fruit or seed gathered from them. This is done by placing

FIG. 54. Giant tomato plant produced in fertilized water.
Courtesy Professor W. F. Gericke, Agricultural Experiment Station, University of California.

the seeds, tubers, or cuttings in a vessel of water or in loose non-nutritive material, such as moss, placed on top of the water in which are dissolved the plant food materials necessary for life, growth, and full development. Professor W. F. Gericke of the University of California has already raised many surprisingly large crops in his experimental tanks. One tank with exactly a hundredth of an acre of water surface produced 25.6 bushels of potatoes; others produced remarkable crops of tomatoes and other vegetables. When planted in water, or in some absorbent non-nutritive material placed on top of the water, the plants while growing must be held upright by strings or other devices. Provision must be made in some cases for some substance, such as moss or excelsior, to be supported on top of the water. In this substance tubers and similar parts of the plant may grow. This brings out two important facts about the soil. First, the soil serves to hold plants up. Second, the soil has to provide the plant with a supply of liquid food, which is the only kind of food the plant can take in from the soil. While in the future this kind of farming or horticulture may be practicable, under certain conditions for certain plants, it is at present in the experimental stage.

The practical method of producing farm and garden crops is to plant the seeds or young plants or cuttings in the soil, under such conditions that the plants can take directly from the soil such food materials as they have to take in through their roots. Let us see now what it is that the plants take out of the soil, how they take it, of what the soil is composed, how soil is produced, and what we can do to keep the soil supplied with the food materials needed by the plants.

53. **The Earth's Surface Once Had No Soil.** For unknown thousands of years there was no soil at all upon the earth. The surface of the earth was everywhere either rock or water. All soil had its beginning in the breaking and pulverizing of the rock. The main forces that have been and still are working upon the rock and pulverizing it and producing soil are the sun, water, air, plants, and animals.

54. **How the Sun Helps to Make Soil.** The rock crust is heated by the sun by day and cools off again by night. Some parts of the rock expand more than others. The

expanding at different rates of the various substances in the rock causes these substances to pull loose from one another. Likewise, those parts of even the same substance which are more exposed to the sun heat more quickly and expand more rapidly than do the parts less exposed. Because of this the heated parts pull away from the other parts that are not so heated. This constant expansion and contraction produced by the varying heat of the sun has always been cracking and pulling apart the exposed surface of the rock crust just as you have seen it crack a cement sidewalk.

55. **How Water Helps to Make Soil.** The water, first falling as rain, has passed for millions of years over these rocks, and has worn them by rubbing, and dissolved them, or otherwise changed them by chemical action. When the water has carbon dioxide in it, as you will soon see that it frequently has, it dissolves the rock much faster. The water further breaks up the rock by getting into the cracks and freezing there and bursting the rock. After the rocks are broken, the water grinds them finer by rolling them against each other, and frequently carries them great distances. You have often seen the mass of well-worn stones of all sizes that are deposited along the beds of our rivers and creeks. However, the greatest amount of material carried by the water is the lighter and more finely powdered soil which is suspended in the water as mud and deposited in the valleys and river-bottom lands. In addition to the wearing and grinding and carrying of rock and soil by the stream of water, there are, in high cold mountains, rivers of ice and snow that flow along, though extremely slowly, dragging and grinding rock as they go. These rivers of ice are called *glaciers* (glā'shẽrs). At a very remote date, when the climate of the world was very different from what it is now, there were much larger caps of ice at each pole than there are now, and these huge caps of ice spread out and flowed in the form of many glaciers toward the equator. These immense glaciers broke off great masses of projecting rock and ground them to pieces as they dragged them along. In this way they

helped to make a considerable amount of soil in the northern part of our country. As they melted long before the Southwest was reached, we have no glacial soil in the Southwest.

56. **How the Air Helps to Make Soil.** The air wears the surface of the rocks, by blowing piece against piece, just as the water does, only more slowly. It is also constantly changing some of the rock by a chemical combination of a substance in the air with some substance in the rock, just as the oxygen of the air unites with all exposed iron surfaces and forms rust. When a rock is being worn and in other ways changed by water and air, it is said to be *weathering.* Rock that has been changed and broken up by water and air, or by other means, until it is all in fine pieces, is said to be *disintegrated* (dĭs-ĭn′tḗ-grā-tĕd).

57. **How Plants and Animals Help to Make Soil.** The plants that first began to grow on the earth were of a low order, such as could live in the powdered and ground-up rock. Upon the death of these plants, this vegetable matter was added to the soil. This added not merely that much matter, but the decaying vegetable matter held water and air in the soil better than they had been held before, and also gave rise to an acid which helped to dissolve the rock faster. The acid given out by the roots of the plants also helped to dissolve the rock, and the roots growing into cracks soon expanded and split the rocks farther apart. The bodies of the animals that ate the plants likewise went into the soil after their death. These decaying animal bodies had a similar effect to that produced by the plants. In addition to this, the worms and other animals that live or burrow in the soil open it and move enormous quantities of soil from one place or one depth to another.

58. **Of What the Soil Is Composed.** The soil, then, is this finely divided surface of the earth in which plants may grow. Its composition is nothing like so simple as most people suppose. It is by no means mere dead matter. Besides the various sized particles of the ground-up and disintegrated rock, and the decaying or decayed bodies of plants and ani-

mals, the soil contains innumerable millions of microscopically small living plants and animals that feed upon the dead and living organic matter in the soil. In the innumerable pores in the solid matter are vast quantities of water, air, carbon dioxide, and other gases.

59. **How Soils Are Named.** Soils are named and classified in various ways. The most common way is according to the size of the particles composing the soil. Soil composed of the finest particles is called *clay* soil. The particles of a purely clay soil are so fine that they cannot be distinguished separately with the eye, nor can they be felt separately when the soil is mashed between the fingers. Some clay soils have particles less than $\frac{1}{25000}$ of an inch in diameter. Clay has a sticky feel when wet and becomes cloddy or granular when dry. Next above the clay in fineness comes the *silt*. Silt has a smooth talc-like feel when moist and crumbles easily when dry. Then come, in order of fineness, very fine sand, fine sand, medium sand, coarse sand, and gravel. These need no explanation. Names are given to soils also in accordance with the amount of vegetable and animal matter in them, and the condition of decay of this matter. If there is a great mass of almost pure vegetable matter, not very much decomposed, it is called *peat*. This soil is found in swamps where rich vegetation has fallen year after year into the water and has been so covered that it has not thoroughly decayed. This decaying mass is often dug up, dried, and used as fuel. Such soil is not used for farm crops. When the vegetable matter is very plentiful but is more decayed, it is called *muck*. This makes a rich, black, loose soil that holds much water and, if properly drained, supports finely a few special crops. A soil that has some clay, and enough sand to make it loose, is called a *loam*. Practically all soils have more or less vegetable matter mixed with the other particles. Many soils are mixtures and have compound names, such as sandy loam, gravelly loam, clay loam. These are so easily understood that they need no explanation. A sticky clay soil that is hard to plow is called a *heavy* soil, and a loose sandy soil that is easy

to work is called a *light* soil. As a matter of fact, a yard of clay soil is lighter than a yard of sand; that is, it weighs less.

Fig. 55. A soil profile, or picture of a cross section of Houston Black Clay soil. You will see that there are nearly four feet of black clay, which shows deep cracks due to drought. Below this is a yellow marl, or clayey marl. Often, especially on hillsides, the surface or top soil is only twelve or eighteen inches thick.

A soil that warms up quickly is called a *warm* soil. A sandy soil that drains well and is open for the free circulation of air is usually warm, while a sticky, tight, clay soil is usually a *cold* soil. The upper and more porous layer of the soil, which has the organic matter in it and has been greatly modified by the action of the air and water, is usually spoken of as the soil, while the compact, hard, usually lighter colored layer below, which has little or no organic matter in it, and has been less affected by air and water, is called the *subsoil.*

60. **The Soil Profile.** A soil is not just a mass of "dirt" or rock particles. It consists of a series of well-defined layers of soil material and organic matter. Different soils have different kinds of layers. That part of the soil that is turned over in plowing is called the *surface soil.* It is usually dark on account of its organic matter. The earth next below the surface soil, that is, the subsoil, is usually a clay or other heavy-textured material. Below the subsoil is the *substratum.* This is usually a layer of unaltered geologic material and is light in color because it contains no organic matter, except a few deeply penetrating tap roots of larger plants. The surface soil is called the *Horizon A,* the subsoil the *Horizon B,* and the substratum *Horizon C.* The *A* and *B* horizons are called the *Solum,* because they are affected by the climate. Material is forced by rain and other effects of the weather from the *A* horizon into the *B* horizon. Below is a diagram of a soil profile:

Horizon A Surface Soil Leached out	Solum
Horizon B Subsoil Weathered in	
Horizon C Substratum	

61. **Color in Soils.** A red soil indicates a well drained soil that is well aerated and is therefore well oxidized. It is the oxidized iron in the soil which gives it its red color. It may occur in the surface soil or the subsoil. A yellow soil

indicates good drainage, although not as much aeration as a red soil. A black soil indicates comparatively high content of organic matter. When the organic matter is burned out of a black soil it usually becomes red, due to the iron that is in it. A gray or white soil is usually low both in organic matter and in plant nutrients.

62. **What Purpose Each Part of the Soil Serves.** We have seen now that the soil is made up of rock particles and decayed or decaying vegetable and animal matter. It is filled with water and air, and contains billions of microscopic living things. Let us see now what part each plays in the production of our farm crops and how we can use each to the best advantage.

63. **The Rock Particles in the Soil.** The rock particles make up the bulk of most soils and give 65 to 95 per cent of the weight. These particles, when disintegrated and dissolved in the soil water, furnish the plant the original mineral food materials. They also act as a reservoir for holding the water, air, decaying vegetable matter, and other things. As it is the thin film of water surrounding each little particle of soil that does not flow away at once, and is left for the use of the growing crop, you can see that the finer the soil the more of this valuable soil water it can retain, because there is a greater amount of surface for holding the water in a cubic foot of small particles than in a cubic foot of large particles. The sum of the surface areas of all the particles in any soil measures its water-holding capacity. It has been calculated that the sum of the surface of all the particles in a cubic foot of soil is 37,700 square feet, when each particle has a diameter of only $\frac{1}{1000}$ of an inch. This is about the actual diameter of the particles in a coarse river-bottom silt. Clay is much finer than this. Just as the finely divided soil exposes more surface to hold water, so it exposes more surface to be acted on by the water, air, and other things in the soil. In this way more food material is constantly being disintegrated for the plants that are growing in the soil.

64. **Water in the Soil.** The most important element in the soil is the water, because it is itself a most important food material, and it is the means of dissolving all the other food materials so that they can be taken in by the root hairs of the plants. After a heavy rain, every pore in the soil is filled with water, but very soon all the free water in the larger pores drains away, leaving a thin film surrounding and adhering to each tiny soil particle. This mass of films of water left is called *capillary* water, because it is held and moved from place to place in the soil by capillary attraction, about which you have already learned. It is this capillary water, filled with dissolved food materials, that is the mainstay of our farm crops. This water moves very slowly in every direction, passing from particles that are wet to particles that are drier. In this way, as the surface particles of the soil are dried by the sun and wind, the water passes up from below to these dry particles by capillary attraction, just as oil passes up a wick as fast as the blaze burns it off at the end. In the same manner, as fast as the root hairs soak up the soil water surrounding them, the soil water moves by capillary attraction from the neighboring moist particles to these dried particles and thus gives a continuous supply to the plant as long as there is any capillary moisture near by, and as long as the pores of the soil are not too large for the water to pass by capillary attraction.

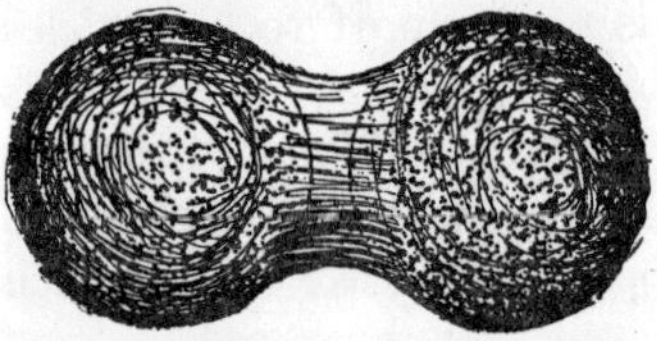

FIG. 56. This shows how the films of capillary water pass from particle to particle of soil, passing always toward the dry particle.

65. **Air in the Soil.** Most good soil is about half air space. After a rain, or where not properly drained, the water fills this space, pressing out the air. Plants, as you have learned, must have air to live. A few plants can get enough air out of the soil water to live, but most farm plants demand more air than is contained in water, and will as surely drown in water-soaked soil as a man will in a pond,

though they drown more slowly. The air in the soil serves several purposes. It contains free nitrogen which is in part changed into a soluble form in the soil, so that the plant can absorb it. You will soon learn that there are in the soil some especially helpful little microscopic *organisms*[1] (ôr′gan-ĭzmz) that are able to take the free nitrogen of the air which the plant cannot use, and work it into a soluble nitrogen compound which the plant can use. Unless there is a plentiful supply of air in the soil these little organisms are not active. Not only is this true but, when the soil is full of water, another group of organisms in the soil that tear down soluble nitrogen compounds gets especially active and destroys the soluble nitrogen food materials that are already in the soil. The yellowing of plants when the water stands long on the soil is thought by some to be due to the lack of nitrogen.

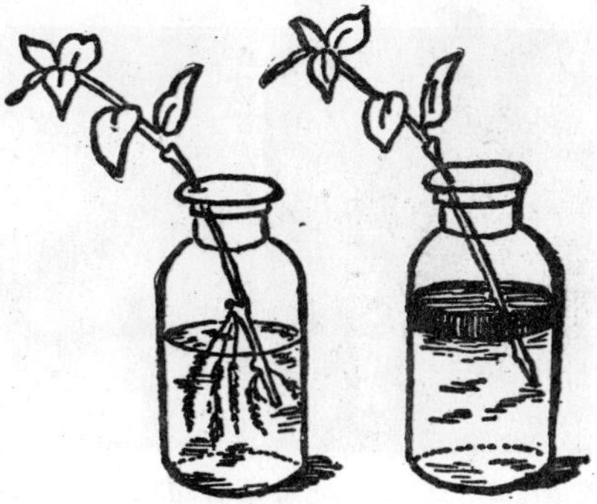

FIG. 57. The water in the bottle on the left is fresh, that in the one on the right has had the air boiled out of it and other air is prevented from entering by the film of oil on top. The cuttings were put into the bottles at the same time. Note the effect of air in the water upon growth.

66. **Organic Matter in the Soil.** The decaying organic matter in the soil is called *humus* (hū′mŭs). It is usually the humus which gives the dark color to the soil. While it is possible to grow a plant in pure sand, if all the food materials are added in chemical form, it can be said that for practical field purposes humus is necessary for all successful crop production. Humus serves many good purposes. As dead bodies contain practically the same substances that they do when living, they give back to the soil a good part of what

[1] *Organism* is a general term which may refer to either a plant or an animal, and *organic matter* is a general term which refers to the matter of the bodies of either plants or animals.

the plant took from it when growing. In addition to this, humus serves four other good purposes. First, it increases the water-holding capacity of the soil; second, when a soil is too tight, it helps to loosen it up and get air into it, and when it is too loose it helps to fill the large pores and bind the soil together; third, it furnishes food for and encourages growth of helpful bacteria that change the insoluble nitrogen into soluble nitrogen compounds; fourth, while it decomposes, it sets free carbon dioxide, which, when mixed with the water in the soil, helps it to dissolve more food materials for the plant. You see then that the value of humus is far greater than the mere value of the food material contained in the bodies of the dead organisms that compose it.

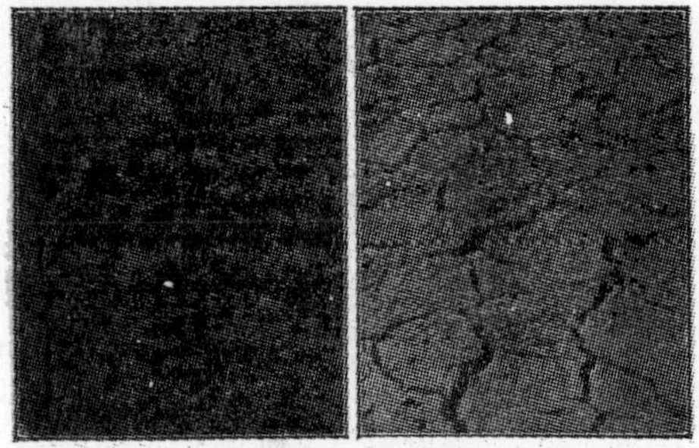

FIG. 58. If humus had been added to the soil on the right, it would have held its water as did the soil on the left.

67. **Living Organisms in the Soil.** While worms help to make the soil porous and to decompose some of the vegetable matter, the greatest work done by living organisms in the soil is that done by very small plants—moulds, yeasts, and bacteria. The little bacteria are so small that they can be seen only with a strong microscope. It takes about 150,000 of the smallest of them to stretch an inch, and it takes about 25,000 on the average to measure that much. They are one-celled plants, and can multiply every few minutes by each dividing into two, just as you saw that the cells in the cambium layer of the tree do. The number of these little plants in the soil is astonishing. A soil poor in bacteria would have over 20,000,000 per ounce, while a rich soil might have many billion in an ounce. Some bacteria are very harmful, destroying the useful nitrogen compounds in the soil, but the

vast majority of them are of the greatest use. They cause the decomposition of the humus in the soil. Some tear down especially the carbohydrates, some the fats, and some the proteins. The insoluble proteins are broken down and part of the nitrogen is changed to ammonia which is in turn changed to a soluble nitrate which the plants can use. If it were not for the action of these bacteria, all plant and animal life would soon cease. The plant, as you have seen, takes the crude food materials (water, carbon dioxide, nitrogen compounds, etc.) and makes them into sugar, starch, fat, and protein foods, which the animals, including man, must have to live on. These foods the animals eat and return at once in large part to the soil as manure. Later on, all of the remainder not returned as manure is returned to the soil in the dead bodies of the animals. In this way the soil gets back everything that was taken from it. But the roots of the plant cannot take in the fats and proteins and other compounds in the manure or in the bodies of the dead animals or even of the dead plants until these are changed. If something did not step in to break up and change these insoluble organic compounds into simple soluble crude food materials again, the soil would soon become a mere mass of corpses and all plants would starve for want of food materials on which to live. Here is where the little bacteria come in. They tear down the dead organic matter and help to prepare the crude food materials for the use of the growing plants again, and thus complete the circle, so that the round of nature can go on and on forever. In addition to tearing down the organic compounds, the action of the bacteria has a valuable indirect result. During the process of decomposition of the humus, acid gases are produced which help with the decomposition of the rock particles. Some of these bacteria also take free insoluble nitrogen out of the air and make from it soluble nitrogen compounds.

68. **How to Improve the Soil.** We have now seen that the soil is composed of finely divided rock particles, of organic matter in various stages of decay, of living organ-

isms, and a varying quantity of water and air, which fill the pores and take up about half of the space of a good soil. Let us now see how the soil can be treated so as to make it most favorable to the growth of the plants rooted in it. If we will keep in mind what we have learned about the way plants feed, and the composition of the soil, we can soon reason out what is necessary to do in order to favor the growth of plants.

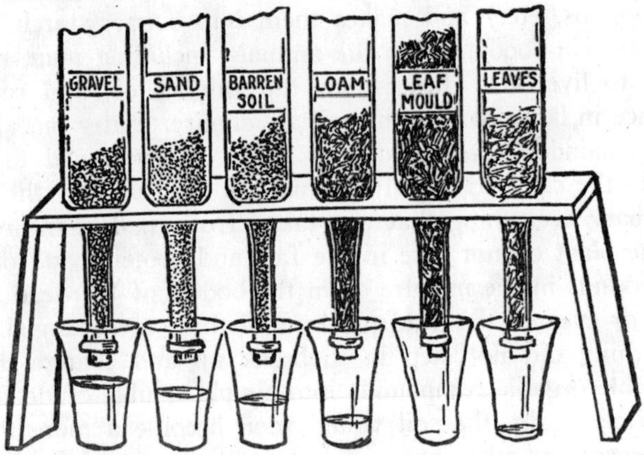

FIG. 59. An inexpensive equipment for testing the water-holding capacity of soils. *Courtesy of the U. S. Department of Agriculture.*

69. **How to Make the Soil Hold More Water.** We have seen that plants can take food materials from the soil only in liquid form, and hence that, without a supply of water, the plant can get no food material at all from the soil, no matter how much is there. Many of our arid Western lands are rich in food materials, but crops starve to death in them from want of water. Then, the first essential of good farming is to keep plenty of moisture in the soil. We have seen that after a rain the soil has in it not only capillary water, but free water that fills the larger open spaces between the soil particles. The valuable water for the crop, as we have seen, is the capillary water left surrounding the tiny soil particles after the free water is drained away. In

order to increase the amount of this water left in the soil, the first thing to do is to break the soil into as fine particles as possible and thus give more surface for the films of water to stick to. Some fine clay soils can hold as much as forty pounds of capillary water in a hundred pounds of soil, while some very coarse soils hold as little as five pounds per hundred. Breaking the land also makes more large pores, and hence, when a rain falls, less of it runs immediately off. Of this water that is caught in the large pores, a part runs off into springs and streams, a part may go down and be left as a reservoir of free water, or may diffuse itself as capillary water further in the soil. However, land that has been broken and opened up tends to pack together again. One of the best things to prevent this and to help keep such land porous and capable of holding water is a plentiful supply of humus in the soil. Some land has the opposite trouble. It is coarse and open, so that the water drains out too rapidly, there being only the small amount of surface of the large soil particles for the films of moisture to stick to. In such land humus helps to fill the pores, delay the water, and furnish surface to which the films of capillary water can adhere. In order, then, to increase the water-holding capacity of soils, we should break our land deep and thoroughly, and put into it plenty of organic matter such as manure and turned under vegetation.

70. **Capillary Water Moves Toward the Dry Particles.** As soon as the water is in the soil, it begins to come out. The free water is carried down by the force of gravity, and the capillary water begins to move slowly toward the surface of the soil. As rapidly as the sun and wind evaporate the water that is on the soil particles at the surface and these become dry, the capillary water on neighboring particles moves up from the wetter particles below to these dry particles. This water is then evaporated and still more water comes up from below by capillary attraction and is in turn evaporated. This continues as long as there is any capillary water in the soil, for capillary water moves constantly,

though slowly, toward the dry particles. While the fact that capillary water moves always toward the dry particles causes the water to be lost from the soil by evaporation, it is the salvation of the plants, for, as we have seen, in the same way, as fast as the root hairs take up the water from the soil particles next to them, the soil water from other particles near by moves by capillary attraction over to these dried particles and thus keeps the root hairs supplied with water. In this way a twenty-five-bushel-per-acre crop of wheat uses on the average about five thousand pounds of water per day, or a million pounds in a season, for each acre. While the plant takes an immense quantity of water from the soil in growing, the loss of water from evaporation of the capillary moisture from the surface of the soil may be much greater. It has been estimated that on a hot, dry, windy day as much as 40,000 pounds of water may be lost by evaporation from the surface of one acre of ground. That is as much water as is used by the plants in producing between 400 and 500 pounds of green corn or wheat.

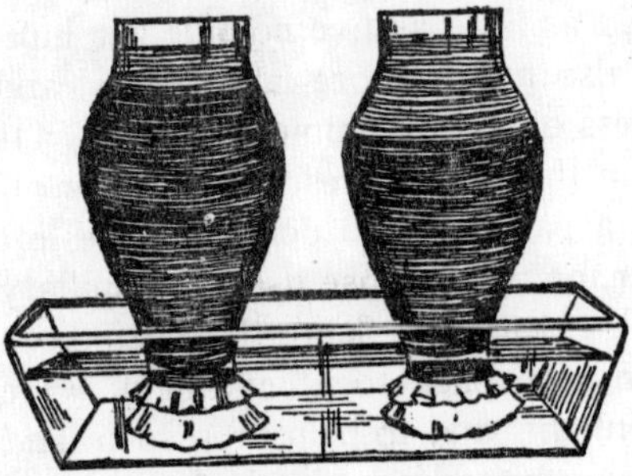

FIG. 60. An inexpensive equipment for testing the capillary rise of water in soils. The chimneys should have fluted tops in order to admit water freely to the soil.

Courtesy of the U. S. Department of Agriculture.

71. **Controlling Losses of Soil Moisture.** Crops probably suffer more during the growing season from lack of moisture than from any other single cause. One of the large problems of crop production is to prepare the land, plant, and cultivate in such a way that moisture can be held in the soil through periods of dry weather with the least risk at the same time of having the crop drown in rainy periods.

The moisture-holding capacity of the soil may be increased by deep plowing, by properly selected methods of planting and cultivation, and by increasing the amount of soil humus. Planting the crop in a deep furrow, or "listing," is a common practice in the Southwest. This practice is not suited to regions in which heavy rains are likely to occur, because on sloping land the crop would be in danger of being washed out by the roots, while on the flat land the water would stand in the furrow and drown the crop. Level planting and tillage is a good practice to follow in holding soil moisture, provided the land has slope enough to drain well. Ridge planting is wasteful of soil moisture; but on poorly drained soils or on level lands in regions of heavy rainfall, this method is preferable.

Mulches consist of any material applied to the surface of the soil to prevent the soil from drying out rapidly or to keep down weeds. Boys often look for fish bait in the moist earth under rubbish piles or boards. Here we have a good example of a mulch that shaded the soil from the sun's heat and checked the loss of moisture. Manure, straw, hay, leaves, or similar material may be used as a mulch. As long as the ground is kept covered, evaporation and weed growth are greatly reduced. Some remarkable crop yields have been obtained by the use of mulches. A specially prepared paper mulch was developed in Hawaii in growing pineapples. This form of artificial mulch has recently been used for vegetables. For vegetables grown in drills, strips of the paper are spread on each side of the row and held in place by raking soil along the edges. In using paper mulch for transplanted vegetables, a single strip of paper is spread along the row and the plants set through holes made in the paper.

A natural mulch formed from pulverized soil has long been advocated as an effective means of reducing the loss of soil moisture. Recent investigations, however, seem to indicate that a layer of loose soil has no effect in checking the rate of evaporation. Authorities are now of the opinion that the benefits formerly credited to the soil mulch resulted from the destruction of weeds and the filling of cracks in the soil rather than from holding moisture by the blanket of loose soil. Weeds draw heavily on moisture, while deep cracks allow moisture to escape from the lower layers. The recent change of opinion on the soil mulch really does not suggest any change in farming methods, because the chief aim of cultivation is, and always has been, to keep the crop free from weeds and to keep the soil from packing around the plant roots and cutting off air.

72. **Dry Farming.** This term is applied to the farm practices in conserving soil moisture followed in sections where the rainfall for one year is not enough to grow a crop. Each cultivated field is planted every other year. The land is broken before the short rainy season so as to catch and soak up all the moisture possible. Weeds are kept down by cultivation as needed. Where the soil is of loose texture, great care must be taken to avoid blowing. High winds prevail in these sections, and much of the soil is light and fluffy; therefore, the problem of blowing is often serious. By ridging or listing the land crosswise to the direction of the prevailing winds and leaving the surface as rough as possible, the tendency of the soil to blow may be checked. During the time the land lies plowed without a crop, it is said to be "lying fallow." When planting time comes, the crop is usually spaced wider than in sections where rainfall is abundant. Such methods make it possible to grow good crops every other year on land that would likely fail to make a crop every year. In Northwest Texas farmers have found that terraces without any fall are very efficient in holding the rainfall on the land long enough for the water to soak into the soil. Otherwise, much of the water ran off as drainage, besides causing serious washing.

73. **Supplying Water to Crops Artificially.** In addition to the above methods of keeping a supply of water in the soil, it is often possible to add by artificial means a great deal to the natural supply of water furnished by rainfall. Over a third of the land of the United States is too dry to

produce a crop without some artificial means of providing water. The furnishing of water artificially to the crop is called *irrigation* (ĭr-rĭ-gā′shŭn). A very large part, though by no means all, of this waste land may be made to yield fine crops by irrigation. Lands that were before worth only a dollar or two an acre have, since irrigation has been provided,

FIG. 61. A concrete irrigation canal in the Rio Grande Valley.

brought two or three hundred dollars per acre. Irrigation has been practised for thousands of years. The laborers of Egypt used to carry the water from the River Nile in vessels and pour it on the plants. Later, wheels were so placed that the current of the stream would turn the wheel and by machinery lift to a higher level a part of the water, which would then be led by pipes and ditches to the field. Some of the Indians practised irrigation in our country before the white men came, but the great progress in irrigation has come in very recent years. Fifty years ago there were less than 100,000 acres irrigated in the United States. Now there are over 20,000,000 acres under irrigation and the rate of increase is rapid.

74. **Not All Sections Can Be Irrigated.** In a large part of our arid land irrigation is not possible, because there is not a sufficient supply of underground water to be pumped from wells, nor is there enough rainfall to supply surface water for irrigation even if all of it were saved. In other sections there is plenty of water, but it contains substances which would accumulate in the soil if it were used for irrigation and would soon poison the land so that no crops would grow. The water from many of the flowing wells contains so much of salt or soda or of certain sulphur compounds that it cannot be used for irrigation. Before using water for irrigation one should always have it carefully analyzed to see if it has harmful substances in it that would accumulate in the soil and ruin it in a few years. Occasionally, even when the water itself is harmless, it cannot be used for irrigation because of the nature of the land. This is true at times of soils that have underneath them a layer of *alkali* (ăl′kȧ-lī), or other substance injurious to plants. The water when flooded over the field goes down to this poison layer, dissolves some of it, and brings some of this poison up to the surface by capillary attraction. The poison, being in this way brought up where the plants will absorb it, destroys the crops. When unintelligently used, irrigation is as great a danger as it is a blessing when properly used.

75. **Methods of Irrigation.** The methods of irrigation are many. There is space here only to give a very general idea of a few methods. At times a small stream or part of a river is led by a canal from its regular channel and carried along until there has been fall enough in the land for the bottom of the canal to be about level with the top of the ground. The water is held in the canal by banks built partly above the level of the ground. From this large canal smaller canals branch off and distribute the water to different fields. Then, each field has running through it a series of smaller ditches coming from the canal. Into these ditches the water from the small canal is turned whenever the crop needs water. Sometimes these ditches are close enough together

for the water, by soaking through the banks, gradually to wet all the land. More often the ditches are broken at certain places when water is needed and the water allowed to pour over the field. The field must be nearly level, and the ditches laid off with care. Often the water, when let out of the ditch, is led down the rows in the field. At other times the field is simply flooded all over. In many cases the water is pumped by engines from a stream or lake through pipes to a canal or to the field, and then spread over the field by ditches or by other means. In many places wells are bored and the water pumped or allowed to run into a large tank, from which it is led by pipes or small canals and ditches over the field. In some places, especially on truck farms, a network of pipes is raised on poles over the field. These pipes have holes bored in them, so that when the water is turned into them they sprinkle an artificial rain over the crop. In other places the pipes are laid under the ground and the water turned into these so that the supply of water goes directly to the roots of the crop, and less of it is lost by evaporation.

76. **Irrigation in the Southwest.** Land is being placed under irrigation more and more in certain sections of the Southwest. Along the Rio Grande River below El Paso, also above Brownsville and near Laredo, great quantities of formerly almost waste land are now irrigated from the river and produce profitable crops, principally of fruits, vegetables, cotton, and alfalfa. In the Fort Stockton, Del Rio, and Toyah Valley regions, water for irrigation is secured from springs and small streams. In the "Winter Garden" section below San Antonio, the water is secured principally from wells, while the Medina Lake northwest of San Antonio, Lake Wichita near Wichita Falls, and other large lakes in West Texas impound immense quantities of water for irrigation. Around Barstow the water is taken from the Pecos River, while near the coast the Brazos and Colorado supply the water for the irrigation of rice.

In Arkansas the only crop grown to any extent under irrigation is rice. This crop is produced in the southeastern section of the State, where the water for irrigation purposes is secured principally from wells.

Rice is also the only crop grown under irrigation in Louisiana. The water is secured both from streams and from wells. The irrigated lands are found only in the southwestern part of the State. In both Arkansas and Louisiana there is sufficient rainfall for all crops except rice, which must have an abundance of water at certain periods.

FIG. 62. Irrigating a young citrus orchard in the Rio Grande Valley.

There is very little land under irrigation in Oklahoma.

Irrigation is practised quite generally in New Mexico. The water is secured from streams, wells, and the reservoir from the Elephant Butte Dam. Practically all cultivated crops produced in the State, including general farm crops, fruit, and vegetables, are grown under irrigation.

77. The Need of Conserving and Using Wasted Water. The saving of wasted water and applying it to the fields is one of the most important economic matters before the people of America. Enough water goes to waste in floods in our country to add millions to our annual production. Each year more and more of this water should be conserved and used. A good way to learn more about irrigation by practical experience on a small scale is to study the bulletins on this subject and then prepare a garden spot near a tank on your place and irrigate a vegetable garden. Every farm in a dry section that has a tank should have at least an irrigated garden.

78. **How to Keep Air in the Soil.** We have seen that all plants must have air to live, and that in a good soil about one-half of the space is taken up by air. As the soil settles down and is packed by the rains, the pores in it are made smaller and smaller, and the air is slowly squeezed out. The result is that the favorable bacteria in the soil do not flourish, as they too need air; nor is the free nitrogen of the air changed into soluble nitrogen compounds as rapidly. The remedy for this is simple. First of all, before the crop is planted, the soil should be broken deeply and turned again until the particles are well broken apart and plenty of air is mixed with the soil. Then, after each rain, when the patter of the water on the surface has run the top of the soil together and largely closed the pores, this tight crust, which tends to shut off the entrance of air and the circulation of air in the soil, must be broken by cultivation as soon as the land can be worked.

79. **The Injuries Resulting from Water-Soaked Soil.** Whenever water enters the soil, this water takes the space that has been occupied by air and drives out that much air. A completely soaked soil has therefore no space left at all for air, and contains only so much air as is contained in the water. We have seen that only a very few crops can live with so little air. It is therefore necessary to get the surplus water out of the soil in which most crops are growing in order to allow the air to get to the roots. The water-soaked soil also encourages the growth of the injurious bacteria which tear down and destroy the valuable nitrogen compounds already in the soil. Fortunately, in most soils the free water goes down rapidly to a point below that reached by the roots of ordinary farm crops, and rests in the permanent bed of ground water, or it goes down until it strikes a layer that it cannot penetrate, and runs along over this layer

until it finds an outlet in some spring or stream farther down the hill. There is, however, a great deal of land which is so close that water penetrates it so slowly that the average farm crop dies for want of air before the free water escapes after long rains, or before the water which runs into this soil from the soil of neighboring higher ground can find its way out.

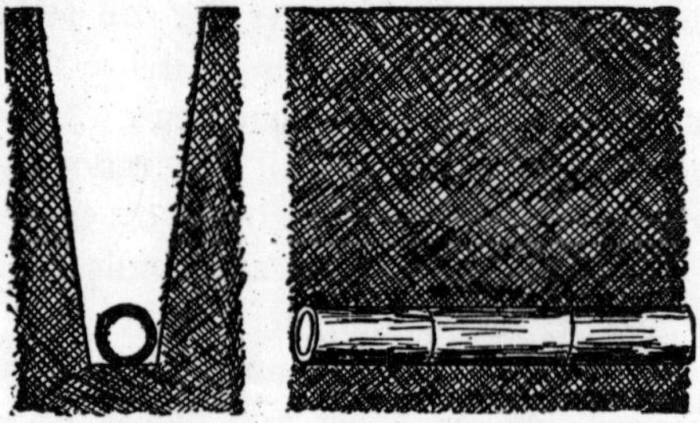

FIG. 63. The method of laying the drains.

Such soil often has below the top soil a still closer subsoil, which makes the passing down of the water impossible. In all these cases of soils that are soaked with water near enough the surface to shut the air from the roots of farm crops, it becomes necessary to drain the soil in order to let in air.

80. **Soil Drains.** The simplest method of draining surplus water from land is to dig ditches in the field, so that the water in the soil will seep into these open ditches, and to so plan the ditches that they lead the water off to a creek or other natural drain near by. The depth of these drainage ditches and their distances apart in the field should vary according to the nature of the land. The usual ditches are from two to three feet deep and located from fifty to one hundred feet apart. The planning of these ditches may be easily learned from books or bulletins. While open ditches will drain the land, they take up a deal of space, interfere with cultivation, and require frequent cleaning out. They should be made with sloping sides and when not very deep should be made so sloping that they can be driven across. In order to avoid the disadvantages of open ditches, underground tile drains are coming to be used more and more. These tiles are usually made of earthenware, in short, open joints, and are laid in trenches at about the same level

that the bottom of an open ditch would be placed. The tile is then covered completely, and the trench filled up even with the surface of the soil so that the entire field may be cultivated. The free water as it settles down goes into these drains, which are so planned as to lead the water gently off down the hill to some natural drain. The method of laying these is easily understood from Figure 63. When the drains are laid, a carefully prepared diagram should be kept showing the exact location of each drain, as occasionally these tiles become choked by roots, and have to be opened and cleaned out. If no chart is made when they are laid, it is difficult later to find a pipe when repairs are needed. In certain districts great drainage canals are dug, and all the surrounding fields are drained into these. The details of these large drainage plans you can learn, too, from books and bulletins.

81. **Effects of Drainage.** Drainage has several good effects. First, it lets air into the soil and thus promotes growth; second, it makes the soil warmer, and because of the warm air being able to circulate deeper in drained soil it warms up quicker and is sooner ready for planting in spring; third, it enables the crop to stand drought better. At first this seems strange, but it is easy to understand. In the poorly drained soil the roots stay near the surface, as they cannot get sufficient air lower down. Later in the season, when drought comes and the water is dried out of the top soil, the plant starves because it has no roots down in the deep, moist soil. When the free water has been properly drained out of the soil, the plant roots go deeper down into the soil, and hence, when the drought dries out the top and the upper roots can get no food materials, these lower roots deeper down in the still moist soil can continue to supply the plant.

82. **How to Keep Bacteria and Plant Food Materials in the Soil.** We have seen how the needed supply of water and air can best be kept in the soil. If we can learn now how to keep a supply of bacteria and of plant food materials in the soil in such form that the plants can use them, the

growing of our crops will be put upon a safer basis. Let us now see how this can be done. The supply of bacteria and the supply of plant food material are so closely connected with each other that they can best be considered together.

83. **How the Soil Is Exhausted.** Before we can intelligently plan to keep a supply of food materials in the soil and prevent its becoming exhausted, we must learn what it is that causes exhaustion of the soil. The first step in remedying an evil is to remove the cause of the trouble, but before we can do this, we must find out what the cause is. Most people think that the taking of the crop from the land is the cause of its exhaustion. The crop does take food materials from the soil, but this is only one of four main causes of loss of fertility. Soil unwisely handled may lose a great deal more from other causes than from the removal of the crop. The four main causes of soil exhaustion are: (1) surface washing, or erosion, (2) leaching, (3) loss to the air, and (4) loss through removal of vegetation. Let us now see how each of these takes place, and how it may be prevented.

SOIL EROSION AND ITS CONTROL

84. **Enormous Losses from Soil Erosion.** As a nation, the United States is just beginning to realize the need of a permanent plan for erosion control. Perhaps no other nation in the world has been so neglectful of her soil resources as has our own. The chief reason for such waste may be attributed to the fact that our pioneer forefathers could always move on to new land as soon as the old land was exhausted. Today, the good land is practically all occupied and much that was good is now ruined. It is estimated that approximately 50 million acres of land in the United States have been destroyed and 50 million acres more have been seriously damaged by erosion. Only about one-fourth of the entire area of this country has escaped erosion altogether. On steep hillsides in many areas the top soil has been removed entirely by erosion. Professor Salisbury says: "It has been estimated that the Mississippi River carries to

Fig. 64. Notice the way in which the rain has washed away great quantities of rich surface soil and started a series of gullies that will ruin the land if not properly attended to.

Courtesy of the Ohio State University Agricultural Extension Service.

Fig. 65. Notice that a silt bed twelve to eighteen inches thick has been washed down from slopes above and deposited upon this field, although the slopes are only two or three per cent slopes: the hat is resting upon the original soil surface.

Courtesy of the U. S. Soil Conservation Service.

the Gulf more than 400,000,000 tons of sediment each year, or more than a million tons a day. It would take nearly 900 daily trains of 50 cars per train, each car holding 25 tons, to

FIG. 66. The disastrous results of erosion on farm land that was not properly protected.

Courtesy of the Ohio State University Agricultural Extension Service.

carry an equal amount of soil to the Gulf." Doctor H. H. Bennett estimates that the cultivated fields and pastures of this country lose in general 22 times more plant food materials yearly through erosion than is removed by the crops grown. Such a loss from erosion translated into money value is well over $2,000,000,000 annually, and no doubt represents a net reduction in crop yields of at least $200,000,000 yearly.

85. **Classification of Erosion.** Erosion may be classi-

fied under three general types, namely: (1) sheet erosion, (2) gully erosion, and (3) wind erosion. Sheet erosion is usually the most destructive of the three. After every rain the muddy water running off the fields is carrying soil and valuable plant food materials. All of the top soil may be removed by sheet washing before the erosion is noticed. A farmer seldom realizes that erosion is taking place until gullies appear in the fields. One man may run his rows up and down the hill because in doing this his land does not have large gullies formed. Yet this man is losing all his valuable top soil and doesn't realize the loss because he sees no gullies in his field. Gullies are very common both in the North and South. For example, an 80-acre farm, on which there were no gullies thirty-five years ago, has been practically ruined by the formation of a large gully with many branches. This gully grew to a depth of 15 to 20 feet and a width of 30 to 60 feet. Wind erosion also escapes the attention of many farmers. However, this type of erosion is most common in the northwestern part of Texas and the Great Plains region of the United States. Wind erosion has affected 322 million acres with approximately 88 million acres seriously damaged or practically ruined for agricultural purposes.

86. **The Control of Sheet Erosion.** The practices most commonly recommended for prevention of sheet erosion are: (1) terracing, (2) strip-cropping, (3) combination of terracing and strip-cropping, (4) contour cultivation, and (5) the use of vegetative cover, including (*a*) pasture grasses, (*b*) meadow grasses, and (*c*) trees. With all of these practices recommended for erosion control, the maintenance of good physical condition of the soil should not be overlooked. A soil in good physical condition will allow more of the water to soak into the ground instead of washing on down the slope. The physical condition of the soil can be greatly improved by plowing at the proper time, turning under crop residues, barnyard manure, and green crops of various kinds.

87. **Terracing.** There are two primary aims in terracing a field. One is to prevent the loss of water which can be used by the growing crops, and the other is to prevent soil losses. In semi-arid sections, the chief reason for terracing is the conservation of moisture; whereas, in regions of heavy rainfall, the primary purpose is to prevent loss of soil. Where the rainfall is less than 30 inches a year, level terraces are most common. In sections where the rainfall is greater than

FIG. 67. A broad terrace as it appears when first thrown up.

30 inches, the gently sloping terraces, which carry the water from the field slowly enough to prevent erosion, are recommended.

Since erosion depends directly upon the amount and speed of the run-off water, protective and preventive measures should diminish the amount of run-off and the rate at which it flows over the ground. If the rain water can be made to sink into the soil or flow slowly over the surface, there will be slight erosion. To this end, the Mangum terrace can be of great value. The Mangum terrace is a ridge of soil built up to a definite grade. It has a broad base and relatively flat sides, with a broad, flat waterway or channel at the upper side. When finished, the terrace looks like a graded road running across the slope of the hill. In terracing an entire field, a series of such ridges should be built at certain intervals from the top to the foot of the hill, each of these terraces usually being 24 feet wide, with the center 14 to 24 inches higher than the edge. Like other types of successful terraces, the Mangum terrace is designed to retard the flow of surface water, thereby allowing more of it to be absorbed and preventing it from gaining enough speed to wash the soil. It is best adapted to more gentle slopes where sheet washing is most prevalent. The degree of slope above

FIG. 68. Terraces and strip-cropping on land with heavy slopes. *Courtesy of the Ohio State University Agricultural Extension Service.*

which land should not be cultivated will depend upon the type of soil, climatic conditions, intensity of rainfall and other factors. Usually in the more humid portions of the Southwest, land with a slope of more than 10 feet per hundred feet (10 per cent slope) should be retired from cultivation.

In determining the interval between terraces, a simple rule may be followed. Let the vertical drop in feet from terrace to terrace equal the fall of the land in feet per hundred feet, plus 2, divided by 2. For example, if the average slope of the field is four per cent, or four feet in one hundred feet, then four plus two divided by two, or three feet, would be the correct vertical drop from terrace to terrace. As the percentage of slope increases, the terraces will be closer together. This is very important, since the speed of the run-off water becomes more rapid with the increase in slope.

In regions of little rainfall, terraces are often run level in order to retain all the water on the field. Where the rainfall comes in torrential showers, or amounts to more than 30 inches annually, the terraces should have sufficient grade to carry the excess water from the field at a rate slow enough to prevent washing of the terrace channel. The fall along

FIG. 69. Strip-cropping between terraces sowed to red top cane. *Courtesy of the U. S. Soil Conservation Service.*

the terrace should not exceed three inches per hundred feet.

88. **How to Plan and Make Terraces.** In planning a system of terraces, the first thing to do is to look over the field and get in mind the general lay of the land in relation to other fields. If the terraces are to be graded so as to carry water off the field, then a suitable outlet for the water at the ends of the terraces must be located.

Next, with the aid of a level and rod, as shown in Figure 70, find the average slope of the land and determine the vertical interval to plan between the terraces. Then locate the first terrace below the top of the hill with about the same vertical interval that you would allow between the other terraces on the hillside. From this starting point, run a line around the hill, keeping it always level, if a level terrace is desired. Stakes should be set about fifty feet apart along the proposed terrace. When the first line is run, then lay off the second line, keeping the same vertical interval if the slope is uniform, starting at the same part of the field as with the first terrace. Repeat this with each successive terrace until the bottom of the hill or field is reached. After the terrace lines are staked out, run a furrow following the line of stakes. If there are sharp curves or angles in this line of stakes, these are somewhat smoothed out and the terrace planned with easy curves. The terrace is then built upon this furrow line with plows, scrapers, regular terracing implements, or whatever is available and necessary. Figure 67 shows a completed broadbase terrace. Often terraces should have a fibrous rooted crop sown upon them the first year in connection with a strip-cropping system, in order to lessen the danger of breaks which are most likely to occur during heavy rains before the terraces are well settled.

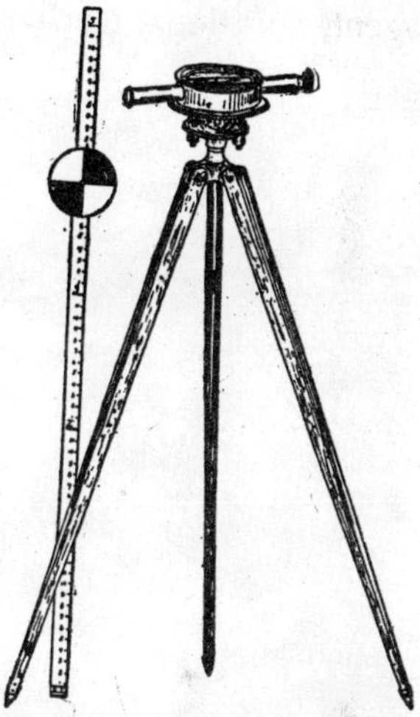

FIG. 70. A farm level and rod suitable for laying off terraces.

If a graded terrace is needed, the same kind of procedure is followed, except instead of being level each terrace is laid out and built with a slight slope so that the surface water will flow off along the upper edge of the terrace. A pasture or well sodded drainage way is needed as an outlet for the water at the end of the terrace. In any case, the outlet should be protected by vegetation or a mechanical structure. These outlets must be both provided and maintained to prevent a gully being formed in the terrace channel, or down the slope where the terrace water is emptied. Figure 71 shows one method for carrying drainage water from a terraced field. The broad drainage ditch is designed to carry the run-off water, and is then sodded solidly with Bermuda grass. For more detailed information, one should consult his county agent, vocational teacher, or extension specialist.

FIG. 71. A sodded terrace channel and outlet that withstood a rain of 14½ inches in 24 hours. Note the width and sodded floor of the outlet.
Courtesy of the U. S. Soil Conservation Service.

89. **Strip-cropping.** Where the slope of the field is slight, erosion may at times be held in check without terraces by strip-cropping. By strip-cropping is meant the planting of bands of fibrous rooted crops sown broadcast on contour lines run around the hill in the same way that terrace lines are run. Then other cultivated crops are planted between the contour strips, with the rows running parallel to strips. The broad-cast strips should be 24 to 48 feet wide and the cultivated strip twice as wide as the broadcast strips. The cultivated

rows running parallel to the broadcast strip tend to hold the water and slow up the flow toward the broadcast strip. The broadcast strip, with its fibrous rooted vegetation, tends to still further slow up the flow of surface water, so that more of it will sink into the ground and more of the soil carried along by the water will be deposited in this strip. Strip-

FIG. 72. Rows run on the contour may prevent fifty per cent or more of losses of both water and soil by holding the water in the rows until it has time to settle into the soil.

Courtesy of the Ohio State University Agricultural Extension Service.

cropping cannot be depended upon where the rainfall is very heavy or where the slope of the field is more than five or six feet per hundred feet. It is less expensive than terracing, but where the rainfall or the fall of the land is too great, it should be combined with terracing.

On slopes where the erosion is likely to be impossible of control or too expensive to control under cultivation, the fields should be retired from cultivation, planted in strong rooted grasses and clovers, and used as pasture or meadow, or should be planted in trees.

90. **Eroded Pastures.** Overgrazed pastures are subject to both sheet and gully erosion. These areas should be resodded to adapted grasses and protected from sheet erosion

by contour ridges, as shown in Figure 73. These ridges are built by throwing up a ridge with a long-winged mouldboard plow. The lines should be run level and turned uphill on each end in order to hold all the rainfall on the pasture. As gullies are approached, turn the ridge uphill since the ridge is too small for holding water that may be concen-

FIG. 73. Contour ridges in a pasture conserving rainfall and checking erosion. *Courtesy of the U. S. Soil Conservation Service.*

trated by a gully. The ridges should be 12 to 16 feet apart to allow for convenient mowing of weeds.

91. **Control of Gullies.** Where gullies are shallow and scattered, they may be controlled successfully by terracing and all evidence of gullying may disappear in a few years. If the gullies are deep and numerous, the gully banks should be plowed in or otherwise sloped and the entire area planted to grass or trees. The control of gullies by the use of grasses, trees, and other vegetation is usually both economical and effective.

92. **Farm Management and Erosion Control.** No system of farming can be profitable over many years if the soil is being lost by erosion. If properly planned, erosion control methods can be practised without loss of farm in-

come. Putting the complete program of erosion control into effect over the entire farm may extend over a number of years, much as a farm should "grow" into a livestock business. Some erosion-resisting crops may be utilized as feed for livestock or for soil building purposes after they have rendered their service of preventing run-off losses. Some retired acreage should be put in grasses and legumes to prevent erosion and improve the fertility, and be kept free from weeds in order to furnish a good pasture. Other areas should be planted to trees to hold the soil, provide cover for game birds and animals, and to provide timber and fuel for the farm. An erosion control program will bring about a diversified system of farming, which experience has taught us is the most satisfactory program for progress in agriculture in the Southwest. In this way we can live in prosperity, and the next generation will not have to suffer because of our failure to conserve the soil on which it will be dependent.

93. **Loss by Leaching.** In addition to the surface water, the free water that fills the pores of the soil after each rain and passes on down dissolves great quantities of soluble food materials present and carries these down below the reach of the roots of ordinary farm crops, or carries them out to the valleys and empties them into the streams. Disintegration is going on all the time in the soil, and soluble food materials are being formed. If no crop takes these up before a heavy rain comes, they are dissolved in the free water and largely carried away. In the winter months when the heavy rains usually fall, many fields have no crops growing on them to utilize the soluble food materials present, so these are leached out and lost. In many cases more is lost in this way each year than is consumed by the crop.

94. **How to Prevent Leaching.** The means of preventing leaching are very similar to those for preventing washing. Deep breaking of the land and filling it with humus so that it will hold more of the water in its pores by capillary attraction is the first step. In addition to this, we should see to it that at practically all seasons of the year some crop is

growing on the land, so that the soluble food materials may be taken out of the soil as soon as they are formed and utilized by the plants, and not left to be leached away by the rains. A mild climate favors the action of bacteria and the rapid disintegration of the soil and the making of soluble food materials during fall and winter and early spring. We should therefore keep our fields covered during these sea-

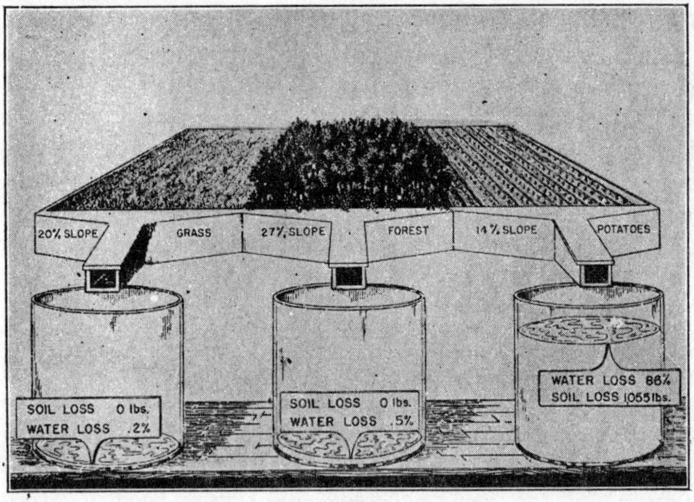

FIG. 74. Notice the amounts of loss of soil and of water with different degrees of slope and with different crops—grass, forest, and potatoes.

sons with grains, other grasses, clovers, and similar cover crops, in order to save our land from leaching.

95. **Loss of Nitrogen to the Air.** In addition to the loss of food materials to the water, at times large quantities are lost to the air. The harmful bacteria which tear down the soluble nitrogen compounds[1] set free a quantity of nitrogen which escapes into the air. These denitrifying bacteria flourish in soil that has an excess of water and a poor supply of air, and in soil that is acid. The means of preventing this loss are obvious. If wet, the soil should be well drained and

[1]These are called *denitrifying* (dê-nī'trĭ-fī-ĭng) bacteria.

opened up so as to hold an ample supply of air. If acid, the soil should have lime added to it to correct this acidity. The method of testing a soil to see if it is acid is simple. Dig down into the soil and press a piece of blue *litmus* (lĭt′mŭs) paper against the moist soil. If the soil is strongly acid, the litmus paper will turn red or pink. Litmus is not sensitive enough to change color in most soils, as they are not strongly acid or strongly alkaline. When a soil is only slightly acid or medium acid and we wish to grow lime-loving crops such as alfalfa or clover, a more accurate test can be made by any commercial tester such as the "Soiltex," or the "Teskit." The amount of lime needed depends upon how acid the soil is. From five hundred pounds to a ton or more per acre are used. After a certain quantity has been applied, and time allowed for it to be diffused through the soil, another test should be made, and the lime added until the soil is either *neutral* (nū′tral) or slightly *alkaline* (ăl′kȧ-līn). Alkaline means the opposite of acid. Such a soil will turn red litmus paper blue. Neutral means neither acid nor alkaline.

96. **Plants Take Material From the Soil in Growing.** Let us now see what the crop takes out of the soil. We plant about ten pounds of seed corn on an acre. If everything is favorable and a hundred bushels of corn are produced on this acre, that will give 5600 pounds of corn and about 6000 pounds of stover. The tiny embryos in that ten pounds of seed corn have therefore taken about 11,590 pounds of material from the soil and air. Plainly we cannot continue to take such enormous quantities of material out of the soil and air year after year and put nothing back without finally exhausting the supply. But before we can plan intelligently to put back, we must know what the substances are which the plant uses.

97. **How to Find Out What the Plant Uses in Growing.** It is not easy to find out of what a plant is made. You or I can tear a pie to pieces and see that it is made of apple and sugar and flour, but if we then try to find what the flour is made of, we have to use a microscope to recognize the tiny

starch cells, the gluten, and other parts. There we have to quit, but a trained chemist can take the starch or gluten, or a drop of the water in the apple, and tear each of these apart by delicate operations and learn what they are made of. He can, as you know, run a current of electricity through the water and split it up into the two gases, hydrogen and oxygen. At last, even the chemist comes to something that he cannot split any further, as, for instance, the hydrogen and oxygen. The tearing up of a compound and finding what it is made of is called *analysis* (ȧ-năl′ĭ-sĭs), and the place in which such work is done is called a *laboratory* (lăb′ō-rȧ-tō-rў). You know that any substance that can be analyzed into two or more simpler things is called a compound, and one that is absolutely simple and cannot be analyzed further is called an *element*. Iron, silver, gold, and carbon are some of the elements with which you are familiar.

98. **Only Ten Important Elements in Plants.** There are only about ninety-two known elements in all the world, everything we know being one of these elements or a combination of them. Strange to say, it has been found that all plants and all animals are made of the same elements. Of these elements there are ten especially important ones.

Carbon Oxygen Hydrogen	Nitrogen *Phosphorus* (fŏs′fō-rŭs) *Potassium* (pō-tăs′sĭ-ŭm)	*Magnesium* (măg-nē′zhĭ-ŭm) *Calcium* (kăl′sĭ-ŭm) Iron Sulphur

For many years botanists and chemists held that the ten elements given above were all that were needed for plant growth. In recent years, it has been found that several others are also necessary, though in very small amounts. Among these are copper manganese, boron, and zinc.

99. **Vitamins.** Besides the elements spoken of above, another group of very powerful materials called vitamins, which are not elements, have been discovered in recent years in plants and animals. These are so complicated and so little known that their study must be put off to an advanced course.

100. **The Three Elements Ordinarily in Most Danger of Being Exhausted.** The carbon, oxygen, and hydrogen make up 95 per cent of the plant. As these are secured from the air and water, we need not consider them further here. The supply of carbon dioxide is practically inexhaustible, as all animals are constantly breathing out a fresh supply into the air. It is estimated that the human race alone gives off more than 50,000,000 tons of this gas per day. The supply of water has already been considered. The calcium, iron, sulphur, and magnesium are used only in small amounts, and are usually in the soil in practically inexhaustible quantities, so that these four also need not concern us. Occasionally calcium is needed. This is easily supplied in the form of lime, which is a calcium compound. The three elements, nitrogen, phosphorus, and potassium, are used in considerable quantities, and all soils are liable to be exhausted of one or more of these if not intelligently handled.

101. **How Plants Exhaust the Soil of Nitrogen, Phosphorus, and Potassium.** Every hundred-bushel crop of corn takes out of the soil 150 pounds of nitrogen, the amount of phosphorus found in 52 pounds of phosphoric acid (a compound of phosphorus), and the amount of potassium found in 85 pounds of potash (a compound of potassium). The cotton crop which produces a 500-pound bale takes out of the soil 100 pounds of nitrogen, the phosphorus found in 40 pounds of phosphoric acid and the potassium found in 65 pounds of potash. Similarly all other plants take these elements in large quantities out of the soil. On the other hand, analysis of soils has shown only a limited quantity of these substances in the soil. Analysis made of 49 soils in different parts of America showed in the top six inches of soil an average of 3000 pounds of nitrogen, 4000 pounds of phosphoric acid, and 16,000 pounds of potash per acre. A bale-to-the-acre crop of cotton takes out 100 pounds of nitrogen. You can see that at this rate such a crop would exhaust the soil of nitrogen absolutely in thirty years, if it could be grown that long, and if no fresh nitrogen were put into the

soil. A hundred-bushels-to-the-acre crop of corn would, under similar conditions, exhaust the nitrogen in twenty years. If the nitrogen were exhausted, no plant could grow, no matter how much of other food materials remained, as plants cannot live without nitrogen.

102. **The Nature of the Soil Tends to Prevent Permanent Exhaustion.** Fortunately, the complete exhaustion of the soil is not as easy as the above would suggest. Two things tend to prevent this: the nature of the soil itself, and the work of the wise farmer. The nitrogen and other elements in the soil are never all in a condition in which they can be used by the crop at one time. The plant can use only so much of the material as is in a soluble form, so that it can be taken in by the root hairs. The material that is in a condition to be used by the plant is called *available food material.* Only a part of the total food material in the soil is at any one time available. If a field that has been exhausted by continued cropping is allowed to rest a few years, it will produce again, because disintegration will have gone on in the soil and some more of the food material will have been changed into available form. Food materials will have been prepared also by the bacteria. The soil thus tends to save itself and renew its own fertility. This is, however, a very slow and expensive process. The farmer can, by intelligent handling, prevent the land ever needing a rest. Indeed he can gather profitable crops each season, and still make his soil richer and richer each year, if he will arrange to supply the soil with the needed nitrogen, phosphorus, and potash.

CHAPTER V

MANURES, FERTILIZERS, AND ROTATION

103. **How the Farmer May Add Plant-Food Materials to His Soil.** Let us now see how the farmer may most economically add to the supply of food materials in the soil. The principal methods of doing this are: 1, turning under stubble and other vegetation; 2, adding manures; 3, adding fertilizers; 4, growing special crops that encourage nitrogen-fixing bacteria. We shall now study each of these methods.

104. **Turning Under Stubble and Other Vegetation Adds Food Material.** Fortunately, the part of most of our field crops which is sold contains only a portion of the food material taken from the soil by the plant. In cotton, for example, only about 1 per cent of the material that made the lint came from the soil, so that if the farmer returns the stalks and seed, the soil will get back nearly all that it lost. The stalks which bore the lint in a 500-pound bale alone contain food materials that would cost about nine dollars if bought as fertilizer to add to the soil. The seeds are, however, much richer in the needed food materials. Fig-

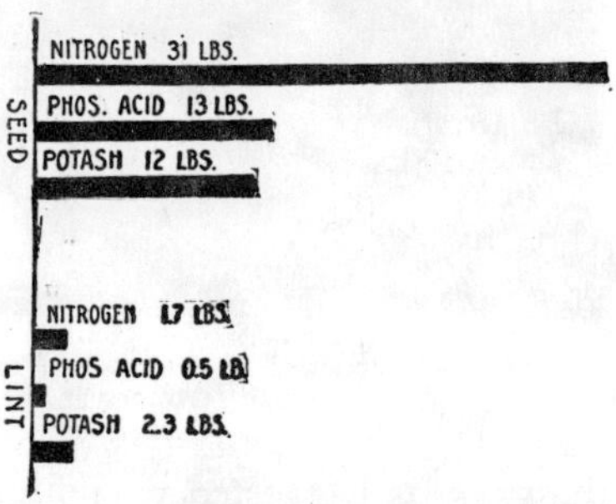

FIG. 75. Showing the large amounts of nitrogen, phosphoric acid and potash used by 1000 pounds of cotton-seed and the very small amounts used by 500 pounds of lint cotton.

ure 75 shows you the large amounts of nitrogen, phosphoric acid, and potash taken away in the seeds. The plant-food materials in the stalks of a hundred-bushel corn crop after

FIG. 76. No fertilizer on the rows of corn and tobacco on the right; a complete fertilizer on the rows on the left.

the grain is harvested would cost, as fertilizer, over sixteen dollars. The plant-food materials in the stubble and straw of a thirty-five-bushel crop of oats are worth over three dollars. The facts are similar in the cases of other crops. This shows how very important it is to turn back under the soil all stubble and stalks before they lose a great part of their value by decay and by giving off nitrogen into the air. In addition to the plant-food materials added directly by the turned-under vegetation, we have already seen that by encouraging the growth of bacteria, and through other effects on the soil, the humus adds perhaps even more to the available supply of food materials indirectly than it does directly. Recall these effects and consider them again. The farmer, then, who burns his stubble and straw is burning money, for when vegetable matter is burned nearly all its fertilizing value is wasted, leaving little except the small amount of potash in the ashes.

105. **Manure: What It Is and What Are Its Values.** A large part of our farm crops is fed to animals. Of the elements in this food which the plants took from the soil dissolved in water, the animal retains in its body only about 15 per cent, giving back in its manure 85 per cent. The manure consists of the solid dung and the liquid urine. The urine contains more than twice as much of the valuable elements per ton as does the dry manure. The value of manure for fertilizing depends upon the animal from which it comes and the food which the animal has eaten. Horse manure is richer than cow or hog manure, but not so rich as sheep or poultry manure. A ton of horse manure contains from seven to twelve pounds of nitrogen, five to eight pounds of phosphoric acid, and nine to twelve pounds of potash, depending largely upon the foodstuffs used. At the price now paid for these fertilizing materials, the amount in a ton of manure would be worth from $2.25 to $3.60. You have already seen that the manure, in addition to the value of the food materials which it contains, is of perhaps greater value to the soil in holding moisture, keeping the pores open, adding useful bacteria, supporting those already there, and in giving off acid gas that helps in the dissolving of the rock particles. In experiments carried on for several years in New York and Ohio, it was found that the crops of hay and oats yielded

$2.58 worth of additional produce for each ton of manure put upon the land, while crops of wheat, clover, and potatoes yielded $2.96 worth for each ton. These figures by no means measure the full value of the manure, because a large part of the fertilizing value of manure remains in the soil many years. This is proved by experiments at Rothamsted, England, where a field continued to give an increased yield from the effect of long use of manure for thirty years after the manure was applied. Two fields, as nearly equal as could be found, were cultivated alike for twenty years. On one, fourteen tons of manure per acre were used annually. On the other no manure was used. For the following thirty years both were cultivated alike again, no manure being applied to either. At the end of this time the effect of the manure was still being shown. The land which had been manured produced on the average for the last ten years 2900 pounds of grain to 1300 pounds produced by the unmanured land.

106. Amounts of Manure from Different Animals. The manure produced each year for each thousand pounds weight of animal is seen in the following table compiled by Lyon and Buckman:

	POUNDS A DAY	TONS A YEAR
Horse	50	9.1
Cow	70	12.7
Steer	40	7.3
Swine	85	15.5
Sheep	34	6.2
Poultry	23	4.2

These figures, it must be remembered, are for 1000 pounds of animal, which is not a very big horse, but which represents two or more of the smaller animals.

107. How the Value of Manure Is Lost. The first waste of manure results from the failure to save the liquid manure. If the urine is not saved, over half of the value of the manure is lost. The next waste occurs when the manure is left out in the weather or is not kept properly covered or sufficiently wet. A large part of the valuable food materials in the manure is in soluble form, so that if the manure is left in the rain these are leached out and carried away in rain-water. Some of the nitrogen is changed to ammonia and passes off to the air in the form of a gas. A large part of the other materials of the manure which are so valuable in loosening the soil and supporting soil bacteria is slowly changed by the oxygen of the air and lost when manure is left exposed. If the manure is allowed to become dry, these changes and this waste go on more rapidly. In tests made at the New Jersey Experiment Station manure exposed to the weather lost over 50 per cent of its value in four months. At the Ohio Station exposed manure when used on a crop was found to have a value of $2.15 per ton, while the value of stable manure was $2.96. When twenty-three cents' worth of superphosphate was added to the stable manure its fertilizing value was $4.80 per ton. At Cornell 4000 pounds of manure were exposed from April 25 to September 25, at which time it weighed only 1730 pounds. The nitrogen in this manure had fallen from 19.60 to 7.70 pounds, the phosphoric acid from 14.80 to 7.70 pounds, the potash from 36 to 8.65 pounds. The value of the plant-food materials had fallen from $6.46 to $2.38, a loss of 63 per cent.

108. **How to Save Manures.** The first thing to do toward saving all the value of manure is to save the liquid manure, either by having a waterproof floor in your stable or by keeping sufficient litter in the stable to absorb all urine. All manures should be kept under cover until hauled to the field, and never allowed to lie exposed to the air and rain. Even under cover the manure needs attention. It should be packed down to press out the air and retard the action of bacteria, and kept wet enough to prevent heating, which drives off nitrogen. Even when properly wet, there will be some giving off of nitrogen, and in order to save this, the manure heap should be covered with loam, sawdust, or straw. Loam is best, as this absorbs thirteen pounds of nitrogen to the ton, whereas sawdust absorbs eight and straw only four. Still better results are obtained from manure if a compost is

made. The United States Department of Agriculture gives the following directions for making a compost heap and applying the compost to the land.

109. How to Make a Compost Heap. "Locate the compost heap in an old shed, or build a shed, with any kind of cheap material for a roof. Spread on the ground a layer of stable manure 8x10 feet, 6 inches deep. Over this spread 100 pounds of superphosphate or ground phosphate rock. The phosphate rock answers as well as the superphosphate and costs about half as much. Continue these alternate layers until the manure is used up, or until the pile has become inconveniently high. To these layers might be added straw, leaves, mould, or other litter, adding 100 pounds ground phosphate rock to each ton of material used. Be sure to wet all thoroughly. When the compost heap is completed, cover it about 4 inches deep with good loam or with forest mould.

FIG. 77. The top picture shows the usual method of saving manure, by which about one-half of its value is lost by leaching and by giving off nitrogen to the air. The stable at the bottom has a cement floor to save the valuable liquid manure, and a cover to protect the manure from rain and leaching. This farmer also wets the manure occasionally, adds rock phosphate, and covers the pile with loam and straw to catch the nitrogen that is set free.—*After Dugger.*

110. How to Apply the Compost. "When applying two tons per acre or less, the best results can be obtained by putting the compost in the furrow and bedding out on it. Be careful not to bury too deep, especially on clay soils. When using more than two tons per acre, it is better to scatter broadcast.

"Bearing in mind the supplemental value of the cowpea, it is safe to say that by using compost at least 50 per cent can be added to the productiveness of the average one-hundred-acre farm, and that simply at the cost of a few tons of superphosphate and a little labor. With the barnyard manure and with the cowpea at his service to save and gather nitrogen for him, the average farmer is simply wasting his money when he continues to buy nitrogen in commercial fertilizer when he could easily produce all that his land needs upon his farm."

111. **Green Manures.** In addition to turning under stubble, it is sometimes advisable to turn under an entire crop. The green crop thus plowed under is called *green manure.* Green manuring provides a method of rapidly adding humus to the soil. Among the best crops for this purpose are cowpeas, velvet beans, soybeans, clover, and sorghum. Usually crops should be fairly mature before being turned under. Such green manuring should not take place immediately before the planting of a new crop, especially one of small grain. Cover crops are frequently sown in the fields at the last cultivation, grazed during the winter, and turned under in the spring. This is an especially valuable practice, as it furnishes grazing, saves the land from loss of fertility in winter, and adds valuable humus besides. All green manure should be turned under at least two to six weeks before the new crop is planted. Good green manure crops for the Southwest may be divided into legumes and non-legumes. We may also divide our legumes into summer and winter legumes. Good winter legume crops are: vetch, white sweet clover, yellow sweet clover, bur clover, and Austrian winter peas. Oats and rye may also be used for a winter green manure crop as they can be grazed in winter and plowed under for soil improvement in the early spring.

112. **Green Manure or Stock Feeding?** The question is often asked whether it pays better to plow under a crop or feed it to stock, put the manure on the land, and sell the stock. This depends upon so many circumstances that no general answer can be given. As over 80 per cent of the fertilizing elements of the crop is left in the manure after being fed to stock, it is usually wise to pass the crop through

stock before putting it into the soil. But, if the soil is very low in organic matter, the quickest way to replenish this is to plow under an entire crop, as more than 50 per cent of the organic matter is lost when fed. In each case one would have to consider the needs of the soil, the work involved in each method, the access to markets, and other factors before he could intelligently decide which procedure would pay best. This will be further discussed under Animal Husbandry.

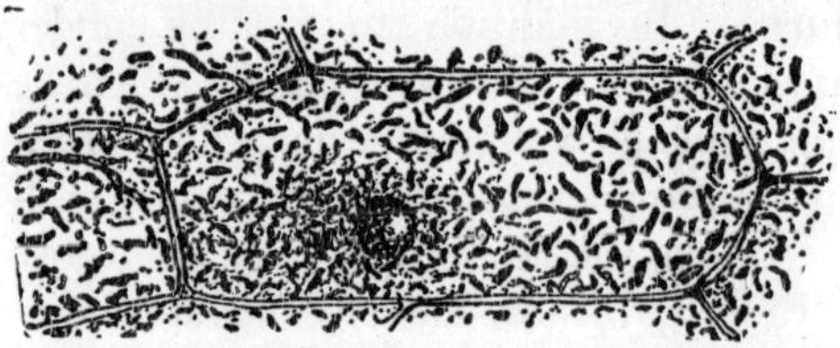

FIG. 78. This shows the nitrogen-fixing bacteria in the cells of the root tubercle of a legume.

113. Plants That Add Nitrogen to the Soil. Although there are millions on millions of tons of free nitrogen in the air and circulating in the soil, four-fifths of the air being nitrogen, plants cannot use this as food material. It must first be made into a soluble compound. You have learned that certain bacteria in the soil can take free nitrogen and help to make it into a soluble nitrogen compound. The scientists have found that there are certain plants upon the roots of which these nitrogen-fixing bacteria thrive. These plants are the *legumes* (lĕg′ūmz), such as peas, clovers, peanuts, alfalfa, bur clover, soybeans, velvet beans, and vetch. If you will examine the roots of these plants, you will see little wart-like nodules scattered over them. These are called *tubercles* (tū′bēr-klz), and contain millions of these bacteria. The plant feeds on the nitrogen compounds made by the bacteria on its roots, and deposits the nitrogen in its stem, leaves, roots, and fruit. If the whole plant is later turned under, all this soluble nitrogen is added to the soil. When the pea-vines that would produce a ton of hay are turned under, $6.00 worth of plant-food material is added to the soil. The roots alone, if left in the soil, add greatly to its fertility, as about 30 per cent of the plant-food material is in them. It has been estimated that in one ton of bur clover or sweet clover plowed under, there is added to the soil 40 to 50 pounds of nitrogen. A crop of two tons would add 80 to 100 pounds of nitrogen. This is equivalent to 400 to 500 pounds of commercial ammonium sulphate fertilizer which at current fertilizer prices is worth $8.00 to $12.00 an acre. The growing of legumes and the production of barnyard manure offer the most economic method by which the farmer may steadily improve his land and increase his income.

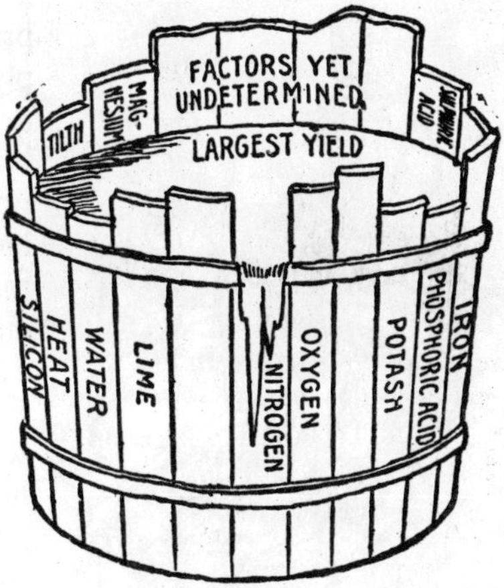

FIG. 79. Just as the tub can be filled no higher than the shortest stave, so the crop can grow no larger than is allowed by the most deficient necessary element in the soil. —*After Halligan.*

114. **The Most Deficient Food Element Sets the Limit of the Crop.** We know that one variety of crop uses more of one substance and another variety uses more of some other substance. We know also that some land is well supplied with one substance but lacking in some other. In such a case the material of which there is a plentiful supply cannot be used by the plant any longer than the supply of the deficient element holds out. For example, if a soil is deficient in nitrogen but well supplied with potash and phosphorus, the crop can use no more of the potash and phosphorus after the small supply of nitrogen has been used up, because the plant can make no new growth unless its food contains its proper proportion of nitrogen. There may be enough potash and phosphorus in a soil to produce one hundred bushels of corn to the acre, but if there is only enough nitrogen to produce twenty bushels, then that is all the field will yield. The most deficient element sets the limit of the crop.

115. **What Is a Commercial Fertilizer?** Materials rich in plant food may be prepared from mineral deposits, or otherwise. They contain much more plant-food materials than does manure. Such artificially prepared materials are called *commercial fertilizers.* While occasionally other elements need to be supplied, practically all fertilizers supply either nitrogen, potassium, or phosphorus, or some combination of these. We shall now give the names and a brief account of the chief materials used in commercial fertilizers, show how to calculate the value of mixed fertilizers, how to find out what fertilizers to use, and how to prepare them.

FIG. 80. This shows the tubercles on the roots of a soybean.

Courtesy of the U. S. Department of Agriculture.

116. **Fertilizers That Supply Nitrogen.** Along the coast of Chili there is a desert area which formerly supplied the world with nitrate of soda, a valuable source of nitrogen. While Chili still exports huge quantities of nitrates, the United States and other countries can now make this fertilizer. Since the World War, great progress has been made in learning how to change the nitrogen of the air into forms that can be used by plants. These changes are made to take place by the use of electricity. Some important nitrogen-carrying fertilizers made from the nitrogen of the air by electric processes are *calcium nitrate, ammonium nitrate, calcium cyanamid, sulphate of ammonia,* and *nitrate of soda.* Where electric current is cheap, these materials are being made at a lower cost than nitrate of soda can be obtained and shipped from Chili. It is estimated that nearly

two-thirds of the world's supply of nitrogen-carrying commercial fertilizers are now being made from the air. Sodium nitrate is the most available form of nitrogen for plants, but it is quite soluble in water and may be washed out of sandy soils easily by heavy rains. Ammonium sulphate contains more nitrogen than nitrate of soda and does not wash out of the soil as readily.

Fertilizers that are made from the air are called *synthetic fertilizers.* They can be made to carry higher amounts of nitrogen than the common nitrates found in nature. Recent synthetic fertilizers that carry a high content of nitrogen are *ammo-phos, diammo-phos,* and *urea.* Synthetic nitrogen fertilizers started on a large scale during the World War when nitrates were in demand for explosives. Since 1929, world production of synthetic nitrogen has increased over 50 per cent. Since 1913 there has been an increase in world nitrogen production of over one and a half million tons, of which synthetic nitrogen accounted for over one million tons. Increased nitrogen production, lowered nitrogen cost, and new synthetic fertilizers containing phosphoric acid and nitrogen are the outstanding developments of the fertilizer industry in recent years. *Cottonseed meal, dried blood,* and *tankage* are occasionally used to supply nitrogen in a fertilizer. These materials, however, are so useful for the feeding of livestock that their selling price is usually too high to permit them to be used profitably.

117. **Fertilizers Supplying Phosphoric Acid.** The principal source from which the phosphorus in commercial fertilizer is obtained is *rock phosphate.* Beds of this are found in Tennessee, South Carolina, Florida, and in several of the western states. Outside of the United States large deposits occur in Canada and northern Africa. Rock phosphate may be ground and spread on the land as raw phosphate. In this condition it is not soluble in water, and hence cannot furnish the plant food material; but in a soil well supplied with bacteria and humus the raw phosphate is slowly changed into

a soluble form and affords the cheapest supply of phosphorus for the crop. Most farmers do not like to use *raw phosphate*, because it must be placed in the soil some time before planting, and because results are somewhat uncertain. If the ground phosphate be treated with sulphuric acid, the phosphorus is changed into a soluble form and a product called *superphosphate* is the result. While superphosphate is soluble, it does not wash out of the soil as easily as nitrate. Superphosphate usually contains from 16 to 20 per cent phosphoric acid. Some manufacturers now manufacture *triple superphosphate* which has 45 per cent of phosphoric acid, and a higher superphosphate which runs as high as 75 per cent phosphoric acid. *Bones* were at one time of some importance as a source of phosphorus, either in the form of *bone meal* or *steamed bone*. The increasing use of bone meal in mineral mixtures for livestock now requires so much of the supply of bone that it is rarely used as a fertilizer.

118. **Fertilizers Supplying Potash.** The important materials supplying potash are *kainit, muriate of potash*, and *sulphate of potash*. Kainit contains 12 to 20 per cent of potash, and the other two about 50 per cent each. For potatoes and tobacco especially, *potassium sulphate* is preferable to potassium muriate or kainit. *Hardwood ashes* contain about one half as much potash as kainit. Ashes contain potash in the form of potassium carbonate, a very desirable form for use by any kind of crop.

119. **How Fertilizers Are Valued.** State laws now require that all commercial fertilizers be plainly labelled. The label must state what per cent of the different food materials the manufacturer guarantees to be in the fertilizer. Fertilizers are frequently named by numbers. For instance, a 4–8–4 fertilizer is one that carries 4 per cent total nitrogen, 8 per cent available phosphoric acid, and 4 per cent water-soluble potash. The state chemist each year finds what each of the fertilizer materials costs at retail in the large markets of the world, and publishes this price as the standard of

value for each year. For instance, the standard values set for 1936–37 were:

	CENTS PER LB.
Available phosphoric acid	6
Total phosphoric acid in tankage and bone	4
Nitrogen in mixed fertilizers	12
Potash	5½

With these prices known it is easy to tell the value of a mixed fertilizer. For example, if a ton of fertilizer contains 5 per cent available nitrogen, 15 per cent available phosphoric acid, and 5 per cent potash, its value can be found as follows:

1 ton=2000 lbs.		
5% of 2000 lbs.=100 lbs.	100 lbs. nitrogen at 12 cts. =	$12.00
15% of 2000 lbs.=300 lbs.	300 lbs. phos. acid at 6 cts. =	18.00
5% of 2000 lbs.=100 lbs.	100 lbs. potash at 5½ cts. =	5.50
	Total	$35.50

In this way we learn that the ton of fertilizer contains 100 pounds of nitrogen worth $12.00, 300 pounds of phosphoric acid worth $18.00 and 100 pounds of potash worth $5.50, which gives a total value of $35.50. This represents the value of the unmixed materials. A fair selling price would require that to this be added the cost of mixing, sacks, transportation, and a reasonable profit for the manufacturer. Before buying fertilizers one should write to the agricultural experiment station for the bulletin giving the fertilizer law, the valuations of materials for the year, and the analyses of the various brands sold in the state. Fertilizers are analyzed by the state chemist, to see that they contain what they are represented to contain.

The commercial value discussed above is no measure of the agricultural value of the fertilizer. It matters not what fertilizing materials may cost; if a ton of fertilizer caused an increase of forty bushels of wheat, and wheat sold at a dollar, the value of that fertilizer to the farmer would be forty dollars, less the additional expense of handling the fertilizer and the extra forty bushels of wheat.

120. **The Purchase of Fertilizers.** Fertilizers may be bought in the form of complete mixtures, containing nitrogen, phosphoric acid and potash, or the separate carriers of these plant foods can be bought separately. Usually it will be found more profitable to get high grade goods. By high grade goods is meant high analysis as shown on the guarantee tag. Experience shows that the higher the grade of the fertilizer the greater the amount of plant food secured per dollar spent. Low grade fertilizer costs more in freight charges and takes more labor in handling than the same amount of plant food in concentrated form. The United States uses annually about 7,000,000 tons of fertilizer. The largest fertilizer-consuming state, Georgia, uses about 1,000,000 tons, while Texas used 85,000 tons in 1937. The two most popular formulas are 4–8–4 and 4–12–4.

121. **How to Determine What Fertilizer Is Needed.** General recommendations for the use of fertilizers may be obtained from the state agricultural experiment station or from the county agricultural agent. The best way to find out for yourself what kind of plant food is needed on a particular one of your own fields is to study the effect of fertilizer on a crop. Sometimes the need of nitrogen is so great that the crop shows a sickly yellow color and a stunted growth. A lack of phosphorus may be indicated by stunted growth, a failure to develop a healthy root system, and a poor yield of grain or fruit. If potash is lacking, the leaves may die around the edges and the plants suffer from such diseases as rust or show white spots on leaves. However, the more reliable way is to check up on actual fertilizer needs by a row test. Stake off 22 rows about 100 yards long in a part of the field that allows each row to be in the same kind of

soil. Plant and cultivate exactly the same, but vary the fertilizer treatment as follows:

Rows Nos. 1 and 2........no fertilizer.
" " 3 and 4........8 pounds of nitrate of soda.
" " 5 and 6........10 pounds superphosphate.
" " 7 and 8........no fertilizer.
" " 9 and 10........5 pounds of muriate of potash.
" " 11 and 12.......8 pounds nitrate of soda, 10 pounds superphosphate.
" " 13 and 14.......no fertilizer.
" " 15 and 16.......8 pounds nitrate of soda, 5 pounds muriate of potash.
" " 17 and 18.......10 pounds superphosphate, 5 pounds muriate of potash.
" " 19 and 20.......no fertilizer.
" " 21 and 22.......8 pounds nitrate of soda, 10 pounds superphosphate, 5 pounds muriate of potash.

The plants to which nothing is added show what the soil will do unaided. The crops on each pair of rows will show just what effect is produced by each of the various amounts and kinds of fertilizers. If, for example, rows 3 and 4 showed no increase over the unfertilized land, what would that show as to whether the soil needed nitrogen? If the crop on rows 9 and 10 were better than that on rows 7 and 8, what would that show? If rows 17 and 18 were still better than rows 9 and 10, what would this show? By using various amounts and combinations of the materials, one can soon learn just what a soil needs to grow any special crop. Often a much simpler experiment than the one given here will tell you what you need to know.

122. **When to Home-Mix Fertilizers.** In many cases the farmer will save by buying the separate fertilizers and applying them separately. Nitrate of soda or ammonium sulphate is best applied in small quantities repeated at in-

tervals as a side dressing to the growing crop. Superphosphate is frequently applied separately previous to planting. A complete mixed fertilizer is usually better for vegetables. Factory mixed fertilizers are likely to be more uniform and in better physical condition. On the other hand, home mixing may save some money. Unless a rather large amount of fertilizer is to be used at once, it will hardly pay to go to the trouble of home mixing.

FIG. 81. This shows the effect of the absence of nitrogen, potassium, or phosphorus. The pot on the left lacks potash, the next lacks nitrate, the next lacks neither phosphate, potash, nor nitrate, the last lacks phosphate.

Courtesy of the Texas Experiment Station, College Station, Texas.

123. **Rotation of Crops.** Each crop takes certain kinds of matter out of the soil, some taking more of one element and some more of another. If any crop which took a great deal of a certain element out of the soil were grown year after year on the same land, you can easily see what would result. Seeing what would necessarily result, the wise farmer does not grow the same crop on his land year after year, but follows one crop with a different kind that draws largely from a different element in the soil, or encourages the growth of favorable bacteria which enrich the soil. The following of one kind of crop by a different one in a regularly arranged order is called *crop rotation* (rō-tā'shŭn). There is no one best rotation for all farms, as this depends on the nature of the soil, the climatic conditions, and the system of farming; that is, whether it is a stock farm, dairy farm, cane, grain,

or diversified farm. Under any system of farming, the wise farmer must practise rotation, but he himself must study out the best system of rotation for his situation. The following general principles should be followed: 1, the succeeding crops should take different elements from the soil; 2, each crop should leave the land in good condition for the following crop; 3, some of the crops should put back into the soil humus and needed food materials, such as nitrogen; 4, the land should never lie long bare to be blown or baked, or have the plant food in it washed out by heavy rains. Under most circumstances a complete rotation would include: (1) a clean cultivated crop; (2) a grain crop; (3) a legume crop; and (4) a sod crop. This does not necessarily take four years, as one or two of these crops may occupy the land in winter.

FIG. 82. Millet grown on equal and adjoining areas. That on the left has no fertilizer and produced one ton per acre. That on the right was given 200 pounds of potassium chloride per acre and produced three tons per acre. The cost of treatment was five dollars per acre. The increase in the crop was worth forty dollars per acre.

Courtesy of Mr. Louis H. Klaas.

124. **A Good Three-Year Rotation.** The following is

a good three-year rotation well suited to most Southwestern farms:

First year, cotton, followed by a cover crop, such as burclover or rescue-grass.

Second year, corn with cowpeas or soybeans sowed in the rows at the last cultivation. In very dry sections the legumes are sown sooner.

Third year, oats, followed by cowpeas or peanuts. If the farmer wishes to grow more cotton or more corn, this crop is grown two years instead of one, making a four-year rotation.

The above rotation can be modified to fit into the soil conservation program by planting oats after cotton in October, and allowing this crop to protect the soil in winter and also serve as a winter pasture. The oats may be plowed under in early spring. Then, the following fall, oats, wheat, or other small grain may be planted after the corn is harvested and be grazed during the winter and allowed to mature in spring. After this crop is harvested, a catch crop of sudan, or grain sorghum or peas may be planted.

125. **Advantages of Crop Rotation.** The practice of a well-planned rotation of crops has the nine following advantages:

1. It keeps the soil in better mechanical condition for plant growth. The continuous growth of one crop, such as cotton or corn, causes the soil finally to run together and get hard.

2. Crops differ as regards their food requirements. Certain crops would remove from the soil large amounts of one element, whereas other crops would remove large amounts of a different element. In growing different crops, no one element is so rapidly exhausted from the soil.

3. Some plants increase the supply of organic matter and nitrogen-fixing bacteria in the soil. This is true of legumes, such as clover, peanuts, and cowpeas. A rotation of crops will permit the growth of these soil-improving crops.

4. Some plants are deep-rooted while others are shallow-

rooted. By alternating the deep-rooted with the shallow-rooted plants, a portion of the food materials in the subsoil is used.

5. Where crops are grown in rotation, less commercial fertilizer is needed. That which is applied will bring more profitable returns than where no rotation is practised.

6. A good rotation provides for making more manure, because it grows crops for the feeding of live-stock. The wise farmer does not sell his hay and roughage, but feeds it to animals and sells the animals. The animals bring in the cash, and, in the manure they make, leave 80 per cent of the plant food to be put back into the soil.

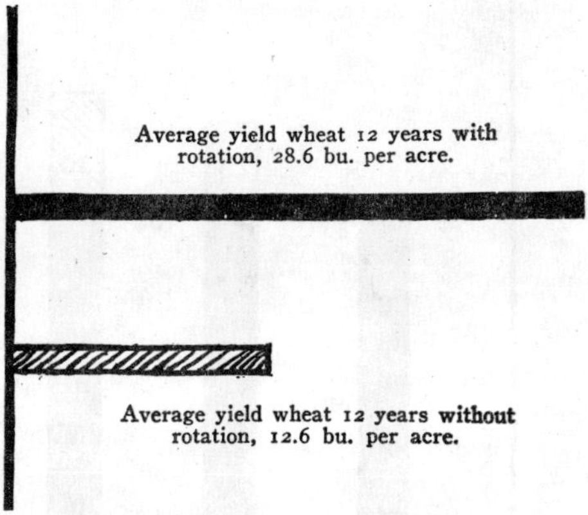

FIG. 83. This shows the yield of wheat, with and without rotation, at Rothamsted Experiment Station, in tests extending over forty-eight years.

7. Plant diseases and insect enemies do less damage when crops are grown in rotation.

8. When one crop has been grown on the same field for many years, it sometimes ceases to thrive in that soil. This fact is, in certain cases, due to a kind of soil sickness, which is now thought to be probably caused by poisons which that particular crop gives off to the soil. While these poisons hurt this particular crop, they do not necessarily harm others; therefore their injurious effects are avoided by rotation.

9. A good rotation gives the farmer an income during all seasons of the year, and also keeps his labor profitably employed all the year.

126. **Results of Experiments in Rotation.** At the

Louisiana Experiment Station· a three-year rotation, consisting of first year corn, second year oats, followed by cowpeas, third year cotton, was carried out for eleven years. At

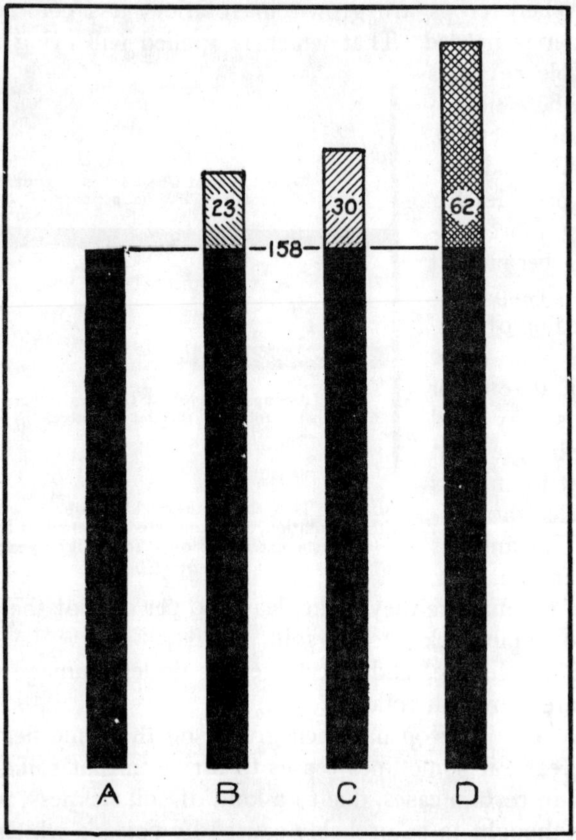

FIG. 84. The yield of cotton grown continuously on the same land without fertilization, 158 pounds of lint per acre, is represented by Column A. Column B represents the increased yield produced by fertilizer. Column C represents the increase produced by rotation; and Column D the increase due to rotation plus fertilizer.

Courtesy of the Texas Agricultural Experiment Station.

the end of this time the yield was from 12 to 25 per cent greater than at the beginning of the rotation. This was where no manure was added. In a part of the field in which manure was added, the increase was still greater. At the famous Rothamsted Experiment Station, in England, wheat was grown for forty-eight years without rotation in one part of a field, and in another part it was grown in a four-year rotation for the same time. The average yield per year for the twelve crops grown in the rotation was 28.6 bushels per acre, while the average yield for the same twelve years on that part of the field on which rotation was not practised was only 12.6 bushels per acre. In a similar experiment there, barley yielded in rotation an average of 2960 pounds per acre, but without rotation yielded only 1735 pounds per acre. The Texas Experiment Station found that cotton yielded 101 per cent more lint in a four-year rotation of cotton, cowpeas, corn, and wheat than cotton planted continuously on the same land. Corn in a two-year rotation with wheat produced 87 per cent more than corn grown continuously. These results were obtained at the Temple Sub-Station, which is located on heavy black prairie land.

The farmer who does not practise rotation, then, greatly lessens his yearly income from his farm, and gradually exhausts his soil, thus making the farm less and less valuable each year. On the other hand, the farmer who carries out a wisely planned rotation increases his yearly income, and, by building up the fertility of his soil, adds each year to the value of the farm itself.

QUESTIONS, PROBLEMS, AND EXERCISES

42. Find in your neighborhood an example of: 1, a rock worn down by the action of water; 2, pebbles rounded by river action; 3, rock surface with no soil; 4, rock broken by expansion from heat; 5, soil made by water deposit.

43. Bring in some soils from the neighborhood of your place, and try to divide them into sands, clays, sandy loams.

44. Find a soil in the river bottom and a soil on the upland, and see if you can tell any difference in the feel, the organic matter, and the color.

45. Which soils are more fertile, bottom soils or upland soils?

46. Which soils are more fertile, sandy soils or clayey soils? How do sandy loams compare with clay soils for cotton production?

47. Find a black soil. Place about 2 ounces in a can and let it heat in a furnace, hot stove or gas burner for a half hour, then notice if there is a color change. To what do you consider this due?

48. Make a rough outline map of your farm and show on this the location of the different kinds of soils. Dig down and measure the depth of the top soil and state what is the nature of the subsoil in each case.

49. Cut the top from a large can with a kitchen can cutter and measure the amount of water it will hold. Then press this can, open end down, into the soil. Now dig around the can and take it out full of soil in its natural position. Strike the soil off level with the top. Then slowly add water from your measuring glass until the can is filled. What per cent of the total space in the can was taken up by the dry soil? What per cent of the soil space was taken up by air?

50. Find a place where a fence separates cultivated land from pasture. Can you notice the difference in the depth of top soil on the sod side of the fence compared with the side that is cultivated?

51. Why does land with a crop lose less in winter by erosion than land that is bare? Why does it lose less by leaching?

52. Make holes in the bottom of a large can, lay a thin cloth on the bottom and fill the can two-thirds full of a mixture of good soil and nitrate of soda. Pour on water equal to a six-inch-rainfall, catch the water through the bottom, evaporate it and see how much has been leached from the soil.

53. With a clean spade throw out a shovelful of soil in a field; press a piece of litmus paper against the freshly cut surface soil. Use a piece of red litmus and also a piece of blue litmus. If the blue litmus turns red, the soil is acid. If the red litmus turns blue, it is alkaline. If no change occurs the soil is neutral.

54. Get a ten-cent bottle of muriatic acid from your druggist. Add a few drops of the acid to a handful of soil. If effervescence occurs, the soil has plenty of lime in it for the growing of legumes such as alfalfa and clover.

55. If four thousand pounds of manure after being exposed

for six months weigh only one thousand seven hundred and thirty pounds, what per cent of its weight has it lost?

56. Apply two tons of manure to one acre in your field. Plant and cultivate this exactly as you do the acre next to it. Keep a record of the amount produced on each acre, and find out how much the manure added to the crop the first year. The second year add no manure to either acre. Plant and cultivate both alike again. Note how much the manure affects the crop the second year.

57. In your field apply well-rotted manure liberally to an experiment plat, and turn under in winter to decompose still further before the drought begins. Apply the same amount of unrotted manure to a similar adjoining plat, but put it in late. Note which resists the drought best: the soil without humus, that with well-disintegrated humus, or that with the rough unrotted manure.

58. In the experimental garden, or at home, carry out the test shown on page 123. Make other similar experiments of your own.

59. Can you think of another good reason for practising rotation besides the reasons given in this chapter?

60. Is this a good rotation: first year, cotton; second year, wheat, followed by corn; third year, oats, followed by cane? If not, what is wrong with it?

61. Plan a good three-year rotation for a farm that raises hogs and cattle, and is in the moist belt. Explain why your rotation is good.

62. Plan a good three-year rotation for the same kind of farm in the dry belt. Explain why this rotation is good.

63. Make a list for each of your father's fields, showing what has grown on it during each year for the last four years. Study the list for each field, and state whether the rotation was wise or not, and why.

64. Ask your father to allow you to carry out your three-year rotation on a part of his land. Keep an account of the profits from the land each year. Keep an account also for an equal number of acres of the same kind of land adjoining this plat on which no rotation is practised and compare results. Also note the fertility of each plat at the end of the rotation, as is shown by the crop of the following year.

CHAPTER VI

TILLAGE AND FARM IMPLEMENTS

127. Introductory. In our study of the soil we have learned that the proper turning and pulverizing of the soil make it much more favorable to the growth of the crop. Such fitting of the land for the more favorable growth of plants is called *tillage* (tĭll'âj). The earliest form of tillage was the breaking of the soil in order to plant the crop. Crops were planted broadcast and not cultivated. Early implements used in tillage were hoes made of shells tied to poles and forked sticks used as plows, drawn by men and women. Later better plows were made of wood and beasts of burden were used to pull them. Man learned to put an iron tip on his wooden plow over a thousand years ago, but progress has been so slow that such plows were still used about a hundred years ago. These were finally surpassed by the iron and steel plows, which have been improved so rapidly during the past fifty years that we now have various forms of sulky plows and cultivators, and the many types of plows used with tractors. These improvements have been made very largely by Americans, who lead the world in the invention and manufacture of farm implements. It is said that it required four and six-tenths days for the old Roman farmer to produce a bushel of wheat. With his new implements and methods, the American of 1830 could do this with three hours' labor, but the present-day American farmer with still better tools and means of power can do it with nine minutes of work.

128. The Result of Improved Implements. The invention and use of improved farm implements has had some very interesting effects upon the American farmer and upon other American industries. First, these improved implements permit our farmers to cultivate and produce a greater number of acres of any crop and to do it more easily and better. Second, he and his family are happier and more contented because he has more leisure time and can get his work done quickly and at the right time. Third, as fewer people are needed on the farms, more of our people can devote their time to producing manufactured products, such as iron and steel, automobiles, clothing, houses, furniture, food products, and so on. About 1820, or a little over one hundred years ago, about 85 per cent of the people in the United States lived on farms. Today less than one fourth of our people live on farms; but they produce wheat, corn, cotton, meat, fruits, and vegetables to supply the other three fourths living in cities and to supply themselves as well, largely because of improved farm implements and machines.

129. **The Advantages of Tillage.** A great many of the advantages of tillage are already known to you, but let us bring these together and mention additional ones. Good tillage accomplishes the following things:

1. The soil particles are separated and broken and the soil made finer and deeper, so that the roots of the plant can penetrate more easily and the particles disintegrate and turn loose their food materials for the plant more rapidly.

2. It brings up new materials from below and turns vegetation and other matter under, in this way adding to the food material available for the crop.

3. When soil is cold and moist it aids the evaporation of water and the admission of air, which warms up the soil earlier in spring.

4. It enables the soil to hold more water and more air.

5. It destroys harmful insects, as well as their eggs, larvæ, and breeding places.

6. It prevents the growth of weeds that would rob the crop of food and moisture.

Just how tillage does all these things was made so plain in

FIG. 85. This shows how the soil is loosened and the air is caught in it by plowing.

the chapter on the soil that there is no need to repeat it here. It is necessary only to add a few general principles of tillage.

130. **When to Plow.** A large part of the labor of the crop should be done before planting. If roots are to have a bed that is easily penetrated, if there is to be a plentiful supply of water, air, and food material for the crop, the soil must be thoroughly broken before the crop is planted. As the freshly turned soil is full of large pores that interfere with the passage of capillary water in the soil, plowing should be done some time before the seeds are sown, in order to give the soil time to settle. When this cannot be done, the disc harrow and roller should follow the plow to press down the soil. If vegetation or manure is turned under, several weeks should be allowed for decomposition before seeds are planted. After all plowing, the harrow should be used at once, as the clods are much more easily broken then. After all rolling, the harrow should also follow.

131. **Plowing the Subsoil.** If any subsoil is to be turned up, this should usually be done in the fall, when it is drier than at other seasons, and when more time remains for it to be acted upon by bacteria, humus, and air before the crop is planted. While it is wise to deepen gradually the topsoil by turning up about an inch of subsoil each year, it is usually unwise to turn up two or three inches of hard subsoil at once, as it must be disintegrated and considerably changed before the plant can use it. When it is desirable to deepen the seed-bed at once in order that it may hold more

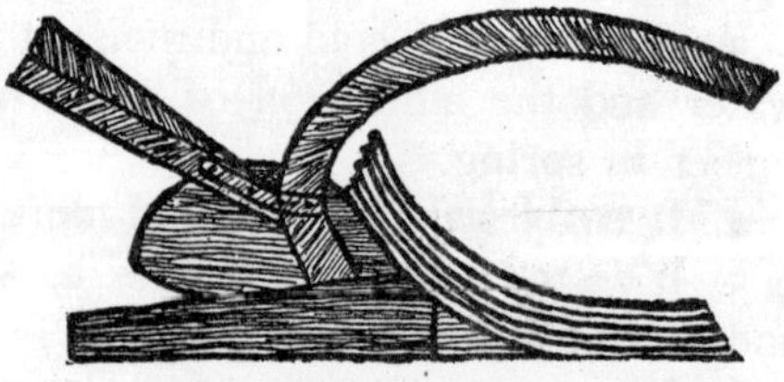

FIG. 86. The curve of the plow causes the layers of soil to slide by each other and thus separates the mass into smaller particles.

water, or for other purposes, it is best to use a subsoil plow that breaks up the subsoil but does not bring it to the surface.

132. **When Too Wet or Too Dry to Plow.** Plowing should never be done when the land is either too wet or too dry. Plowing land when too wet may injure it for many years. It is not easy to learn from a book how to to tell when land is right for plowing. If it sticks together and the furrow slice shows a shiny, polished surface it is too wet. If it breaks into hard clods that do not easily pulverize it is too dry. Sandy land does not need to be so dry as other land when plowed.

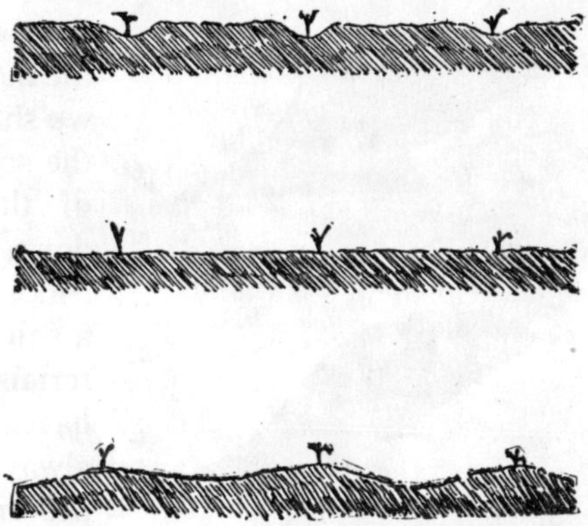

FIG. 87. In soil that is too wet or liable to become wet the crop is planted above the level of the ground; where the soil is dry or liable to become dry the crop is planted below the level of the ground. Can you see why this is done?

133. **Soil Should Not Be Long Left Bare.** It is usually unwise to plow up the soil and leave it long without a crop to cover it and prevent the loss of food materials to the air and through leaching. The loss from erosion is often very great when the soil is left bare.

134. **When and How to Cultivate.** While each crop has its own peculiarities which one must know in order to cultivate it properly, there are a few general principles that apply to all crops. All of these are easy to reason out now that we know what we do about the soil and about plant growth. Since it is through the tender root hairs in the new parts of the roots that food materials are taken in, then plainly we should not cultivate deep enough or near enough to the plant to disturb these unless it is our purpose to stop

the plant's growth. If we have to plant a crop on soil that is wet or liable to get soaked, it is plain that we should first bed up the land and plant on the bed so that the roots of the young plants will not be drowned by the spring rains. On the other hand, if the soil is dry or liable to such drought as will dry out the moisture in the upper soil, we should list, or plant below the surface, so that the roots of the plants will be deep down in the moist soil. Since weeds and grass rob a soil of water and food materials that the crop should have, obviously we should always cultivate in time to prevent or to kill while very young all grass and weeds. Other special directions with regard to tillage will be learned as we discuss the various farm implements and crops.

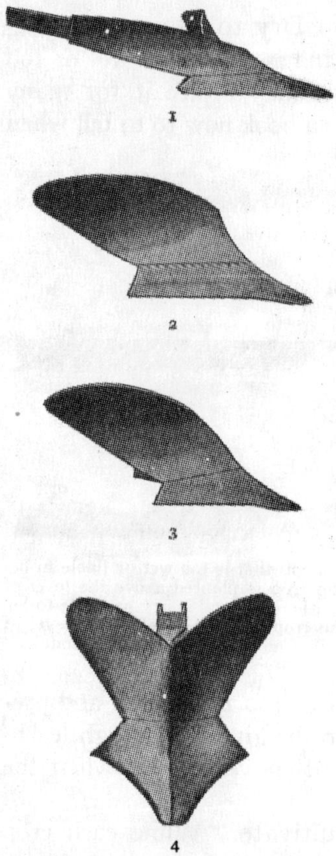

FIG. 88. Four types of plows: 1, breaker for prairie sod. 2, general purpose for light land. 3, black-land plow. 4, middle burster.

135. **Plows.** There are two general classes of plows, the mouldboard plow and the disc plow. The *mouldboard plow* is the oldest type. With this the soil is cut by the *share* and turned upside down and broken by the curved *mouldboard*. The amount of turning done depends upon the shape and size of the mouldboard. If there is fresh

sod to be broken and the land is heavy, the mouldboard is usually less curved, as in this way the plow turns the soil over less and pulls more easily. When the object is to turn materials under the soil the more curved mouldboard is used. When it is desired to throw the soil out on both sides a still different type called a *middle burster* is used. All of these types are easily understood from Figure 88. The very interesting way in

FIG. 89. Disc plow.

which the curved mouldboard tears the soil apart and crumbles it by sliding the layers of soil on each other is well shown in Figure 86. *Disc plows* are shaped very differently from mouldboard plows, the cutting and turning being done by a circular concave disc which rolls instead of sliding. They can be used in soil that is too dry and hard or too sticky to be worked with a mouldboard plow. The two types are about equally hard to pull under ordinary circumstances. Where fields are large and turns do not have to be made often, *sulky* and *gang plows,* which are made in both mouldboard and disc types, are usually to be preferred. If properly adjusted, sulky plows pull as easily as do walking plows, do somewhat better work, and yet allow the operator to ride.

136. **Harrows.** The implements used for pulverizing the clods, smoothing the surface, and putting on a dust mulch are numerous and varied, each with its peculiar advantages

FIG. 90. Tandem horse and tractor disc harrow.

and disadvantages. Where the broken land is rough and full of clods, or needs levelling and packing, the *disc harrow* is especially valuable. This harrow may be set to cut from one inch to three inches. In loose sandy land it is not en-

FIG. 91. Spike-tooth harrow.

tirely satisfactory because of sinking so deeply into the soil. The *spring-tooth harrow* has a number of strong curved steel springs forming the teeth, which are about two and one-

FIG. 92. Acme harrow.

half inches wide on the ends. These teeth are so arranged that they may be set to run two or three inches deep, thus breaking clods and fining the soil. The spring-tooth harrow is often used immediately after the disc harrow to further work down a rough soil. If much vegetation has been turned under, this harrow is unsatisfactory, as the teeth pull much of it out of the soil again. The fact, however, makes it a good tool for tearing up and destroying Bermuda grass. When used after the disc harrow or on plowed soil that is in good condition, the *spike-tooth harrow* pulverizes and levels the sur-

FIG. 93. Plank drag.

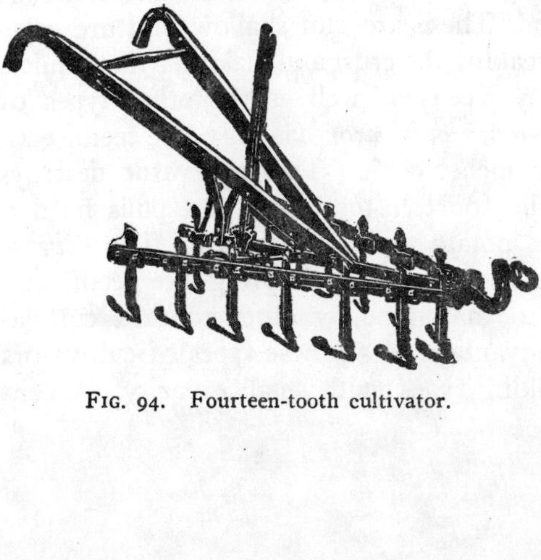

FIG. 94. Fourteen-tooth cultivator.

face soil very effectively. It is also useful for early cultivation of corn and cotton to break the crust and kill small weeds. The *Acme harrow* is also valuable in pulverizing the surface, but it is not satisfactory in stony soil or where heavy clods are to be broken. The *plank drag* serves very much the same purpose as the Acme harrow in smoothing and pulverizing the surface.

137. **Cultivators.** The implements used for tilling the soil after the crop is planted in order to prevent weeds and

FIG. 95. Five-tooth cultivator.

keep the surface broken are called *cultivators*. The *spike-tooth cultivator* has a number (usually fourteen) of teeth about six inches long by one and one-half wide flattened somewhat at the end. These are run shallow and are especially valuable for breaking the crust and making a dust mulch. They do not destroy weeds as well as the other types of cultivator. The *five-tooth cultivator* has only five teeth, each from three to four inches wide. This cultivator destroys weeds better than the fourteen-tooth one, but pulls harder. It is widely used in potato and corn culture. The *diverse cultivator* has from six to ten long spring-like teeth, and does work similar to that done by other toothed cultivators. The main disadvantage of all these types of cultivators is that they are walking types, with small capacity. A constantly increasing part of crop cultivation, especially in broad, level fields, is being done with *riding cultivators,* which do practically the same kind of work as do similar walking cultivators. There are usually four or six shovels on a one-row cultivator. On a two-row cultivator this equipment is simply doubled. These cultivators have made it pos-

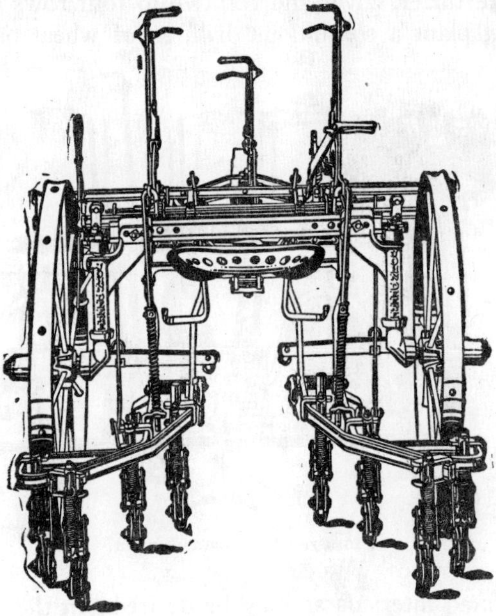

FIG. 96. One-row riding cultivator.

sible for one man to cultivate from five to seven acres a day with a single-row cultivator and nearly double that amount with a double-row cultivator. When the double-row cultivator is to be used, the crop should be sown with a double-row planter, to make it certain that each pair of furrows runs parallel. Tractor cultivators, designed to cultivate two, three, or four rows, are proving to be successful and economical. These can cover from twenty to forty-five acres a day.

138. **Planters and Reapers.** With a good planter one man can now not only plant as much in a day as eight or ten men can plant by hand, but he can do the work more uniformly and better. By this means a large crop can be got into the soil in a few days before a favorable season is gone. Some of these machines open the soil, plant the seeds, add the fertilizer, cover and roll two to four rows at a time. They can plant a continuous drill, as of wheat or oats, or

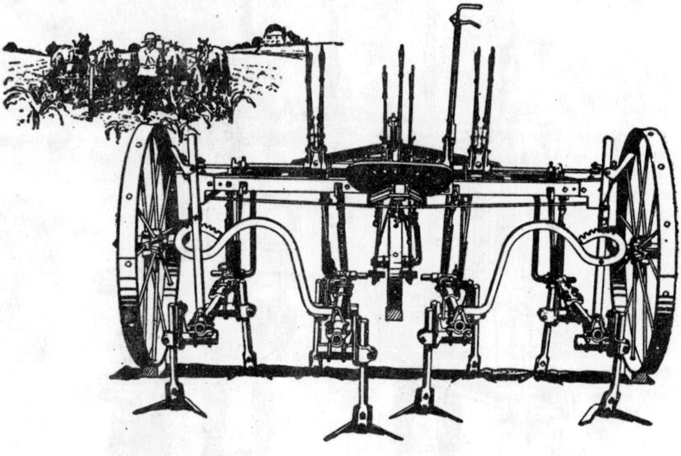

FIG. 97. Two-row cultivator.

only at fixed intervals as may be desired. In the same way one machine now may do the work of many men in reaping the crop, thus not only lessening labor, but enabling the farmer to use short favorable seasons to advantage. We cannot go into the various types of planting and harvesting machines here, but shall mention a few. You can learn about the others from any more advanced textbook on tillage. The *combined corn and cotton planter* is an implement that may be adjusted to plant either cotton or corn. The corn may be listed or planted level, and may be dropped at any distance apart that is desired. There are one-row and two-row types, the two-row type being especially desirable where

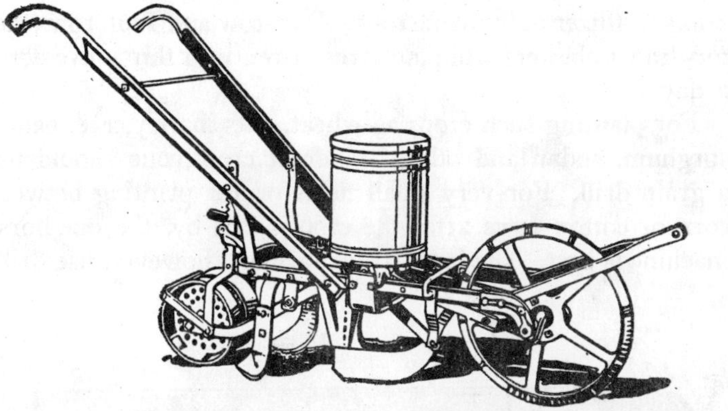

FIG. 98. Single-row cotton and corn planter.

it is intended later to use a two-row cultivator. Such a planter opens the furrow, selects the right number of grains from the seedbox, drops them into the right place, covers and rolls them. It may have also an attachment for distrib-

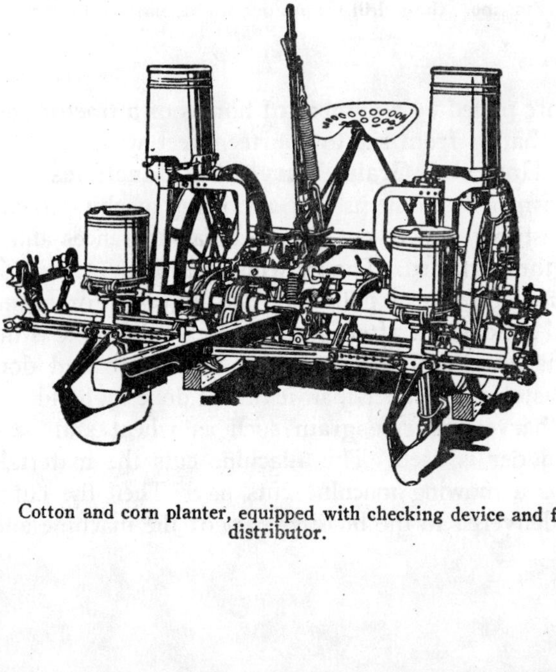

FIG. 99. Cotton and corn planter, equipped with checking device and fertilizer distributor.

uting fertilizer at the same time. Two-row and four-row tractor-drawn planters will plant from twenty to thirty-five acres a day.

For planting such crops as wheat, oats, barley, rye, beans, sorghum, sudan and other broadcast crops, one should use a grain drill. For very small fields or for planting between corn or cotton rows after the crop is laid by, the one-horse machine is best. The large grain farmers, however, use drills

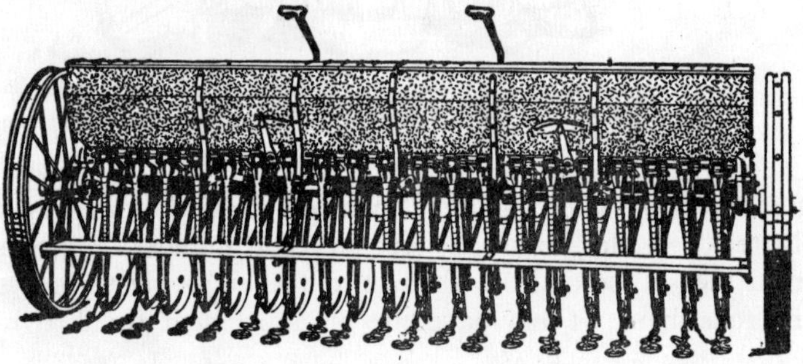

FIG. 100. Grain drill for planting wheat, oats, barley, etc.

which are pulled by a number of horses or a tractor and sow a strip that is from six to fourteen feet wide.

139. **Hay and Grain Harvesting Machines.** Nearly every farmer has some use for a mowing machine to cut hay, mow pastures, and to cut the weeds along fences and roads and in the barnyard. The mowing machine cuts a 4½ or 5 foot strip and can be pulled by two horses. For raking the hay and bunching it after it has cured, a sulky rake is needed. This machine is simple and easy to operate and does the work faster and cleaner than it can be done by hand.

For harvesting ripe grain such as wheat and oats, the grain binder is used. This machine cuts the material very much as a mowing machine cuts hay. Then the cut material is delivered to the binding part of the machine and tied

FIG. 101. Mower for cutting hay, weeds, etc.

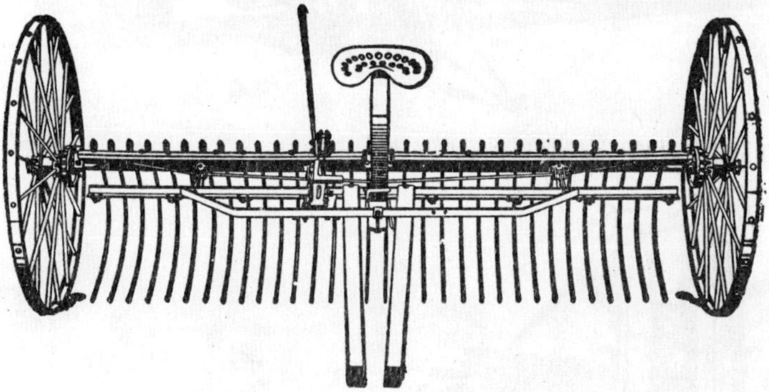

FIG. 102. Sulky hay rake.

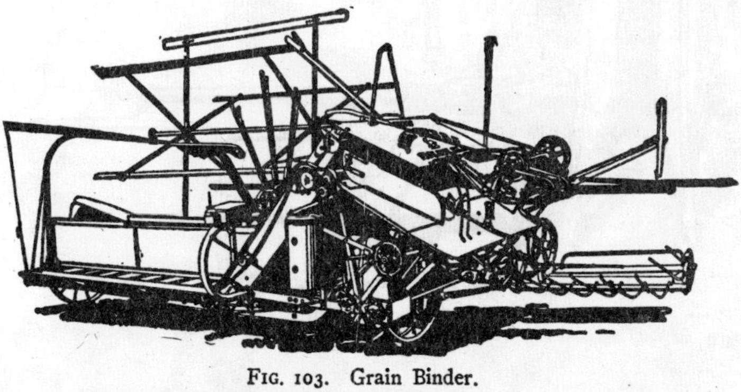

FIG. 103. Grain Binder.

FIG. 104. Combined harvester and thresher, or "Combine."

FIG. 105. Row Binder.

FIG. 106. Mechanical cotton picker.

into a neat bundle which is dropped on the ground. These bundles are placed in small stacks or shocks and allowed to dry out a few days and then are run through a threshing machine which separates the grain from the straw and cleans

FIG. 107. Manure spreader.

it. On the big wheat farms, machines called combines, Figure 104, pulled by powerful tractors, cut and thresh the wheat all in one operation.

Corn, grain sorghums, and similar row crops can be cut

and tied into bundles by means of a row binder, Figure 105. This makes it possible to save all of the crop and stack it or put it into the silo conveniently.

140. **Cotton Harvesting.** A great deal of time and

FIG. 108. Gas Engine pumping water.

money is being spent by various people who are trying to invent a machine which will pick cotton successfully. Many machines have been developed and tried out but none has proved completely satisfactory. Some of these machines are made to suck the cotton out of the boll while others are made with revolving drums having many spindles or fingers or hooks which catch onto the cotton and pull it from the bolls. Figure 106 shows one type of mechanical picker.

In Northwest Texas, where the cotton does not grow very tall and is not gathered until late in the season, often after the first frost, it is common practice to strip or snap the bolls from the stalk instead of picking the lint out of the boll. Machines have been developed for this and are called mechanical cotton strippers or sleds. Special ginning equipment is necessary to handle this cotton after the bolls are gathered.

141. **Manure and Fertilizer Distributors.** Manure and fertilizer distributors bring about a saving of labor on the farm in the same way that planters do, and should be in more general use. Planters are frequently made with a fertilizer distributor attached. The handling of manure is a very important matter on the farm, but it is usually done in a manner very wasteful of labor. The manure is handled once when thrown into a pile. Then it is thrown into a wagon. From the wagon it is thrown in piles on the field, then distributed from the piles. All this takes about twice the labor that should be used. By having the manure shed conveniently located, and by distributing the manure directly from the *manure spreader,* the labor is greatly lessened.

142. **The Gasoline Engine and Farm Machinery.** Gasoline engines are now made that are not expensive, and are simple enough to be run by any intelligent boy. It should not be long before every thoughtful farmer has such an engine to pump water, churn, cut and grind feed, saw wood, fill silos, and in other ways economize labor.

143. **The Care of Machinery.** Farmers have considerable money invested in machinery. For this reason they should take good care of it. This means keeping all parts tightened up and properly adjusted, greasing and oiling it often at all places where wear occurs, and then putting all machines in a shed or under cover when not in use. Many farmers make a practice of cleaning up, repairing and painting all of their farm machines and tools at least once a year.

QUESTIONS, PROBLEMS, AND EXERCISES

65. What does tillage add directly to the soil?

66. Plow one acre in the fall, turning up one inch of subsoil, and plant a cover crop. Plow under the cover crop three weeks before planting in spring. On an adjoining acre with the same

soil break the ground as usual in spring. Plant and cultivate both acres alike and compare the yields.

67. Cultivate ten rows of corn not over two inches deep. With five of the rows make the final cultivation much deeper, with a middle burster. Note the effect of each treatment on the yield of corn.

68. The teacher should take the class to an implement store, and to neighboring farms, and carefully explain each implement.

69. Make a list of such implements as it would be an economy for your father to purchase, stating reasons.

70. If a sulky cultivator enables a man to do in one day with two horses as much as he could do before in five days with one horse, how much is this cultivator worth to him per year?

71. State all the advantages that would come to the cotton-growing states if satisfactory cotton-picking machines were in use.

CHAPTER VII

FARM CROPS

144. Introductory. You have now learned the general principles that govern the nutrition, growth, and reproduction of plants, and the principles governing the cultivation of the soil and the conservation and increase of its fertility. These principles are universal; that is, they apply to all plants and all soils everywhere all the time. If some new plant that man had never cultivated were introduced, we should know that in cultivating it all these principles had to be observed: for instance, it would take its food materials through root hairs as do other plants, and it would manufacture its food out of the same elements. But, whether its roots would be deep or shallow, or in what proportion it took the different food elements from the soil, or whether the plant could stand drought or cold or shade, what was the best time to plant, the best method of cultivating and harvesting, and so on: all these we should not know. These special facts about each individual plant must be known before we can most wisely apply our general principles to the raising of that crop. If we have thoroughly mastered the important general principles, it will be easy to work out a wise plan of cultivation for any plant as soon as a few facts about its special characteristics and habits are learned.

To study the special characteristics and habits of all plants and apply to each of these the principles of growth and cultivation would take several lifetimes. In this course we shall study just a few of the most important farm crops of our section.

In connection with the growing of all plants farmers should consult their County Agricultural Agent and different divisions of their State Agricultural College frequently for information of the adaptability of different crops to their location and system of farming. New varieties of crops are continually being developed by research workers which are

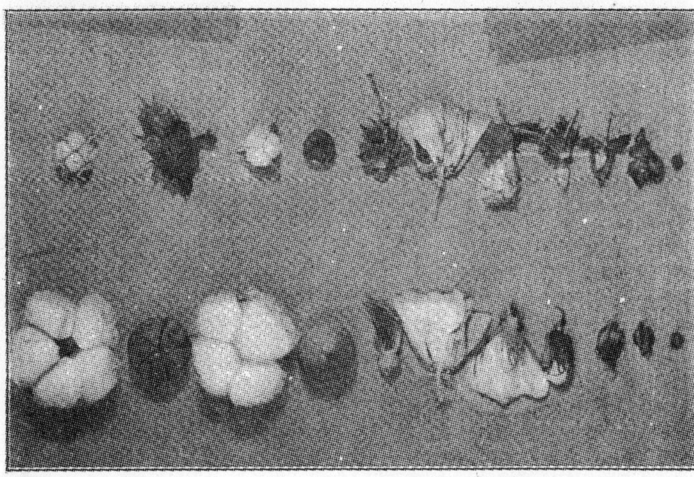

FIG. 109. Different stages in the development of the fruit of the cotton plant; Sea-Island above and Upland below.

outstanding in yield and quality due to their adaptability, resistance to disease, and insects. The purity, quality, and adaptability of seeds planted are very important in farming. Good seed are an investment, not a liability. In order to be informed on the new developments which are taking place consult the above agencies frequently.

COTTON

145. Cotton: Its Importance and Distribution. Cotton leads all other plants in the production of fibre. In many respects it is the most important plant on the globe. It furnishes the clothing for the larger portion of the world's inhabitants. The leading countries producing cotton are, in order of importance: United States, British India, China, Russia, Brazil, Egypt, Peru, and Argentina. The southern United States produces from two fifths to one half of the world's total supply of cotton, or from 11,000,000 to more than 17,000,000 bales per year. Of this amount Texas alone produces from one fourth to one third, or from 3,000,000 to 5,000,000 bales.

146. **Description.** The wild cotton as now found in tropical countries is a perennial bush or tree reaching a height

FIG. 110. Cotton leaves: *A*, upland; *B*, sea-island. Cotton bolls: *A*, upland; *B*, sea-island; *C*, Indian.

of fifteen or twenty feet. Cotton as grown in this country is an annual three to seven feet tall. It consists of a central solid woody stem called the *main stem.* From the joints of this main stem long ascending branches arise. These are called *primary branches.* Along the primary branches slender and shorter branches occur on which the bolls are attached. These slender branches are known as *fruiting limbs* and give direct rise to the bolls. Fruiting limbs are attached also to the main stem. The longest primary branches occur near the base of the main stem. These decrease in length toward the top of the stem, thus making the plant as a whole cone-shaped. In the cluster types of cotton, there are only a few long branches near the base of the plant. Above these the fruiting limbs are attached directly to the main stem. The *boll* is a pod containing the seed and lint. There are from three to five divisions of each boll, the contents of each division being called a *lock.* The number of bolls on a single plant may vary from a few to a hundred or more, depending upon the variety, soil, fertility, rainfall, and climate. The cotton plant under normal conditions develops a tap-root, which in a well-drained soil may go down to a depth of three to ten feet. However, if the soil is not well drained, or if the subsoil is very compact, this tap-root may go no deeper than nine inches or a foot. The lateral roots or feeding roots branch from the

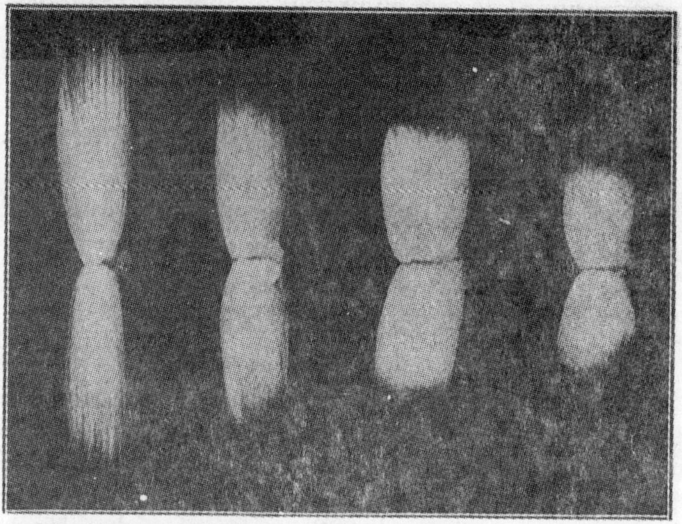

FIG. 111. Combings showing length of lint of four important types of cotton. In order from left to right these types are: (1) sea-island, (2) upland long staple, (3) upland standard staple, (4) upland short staple (Half and Half).

Courtesy of the Agricultural and Mechanical College of Texas.

tap-root at points from one and one-half to four inches below the surface of the soil. As the roots develop so near the surface of the soil, what kind of cultivation should be given cotton?

147. **Species.** There are five species of cotton that are important in the world's agriculture. These are:

1. *American Upland Cotton.* This species represents the common cotton grown in the southern United States. It includes both the short staple and long staple upland cotton.

2. *Sea-Island Cotton.* Until recently, this species was

produced chiefly on the islands and adjacent main lands of South Carolina and Georgia. Its profitable production in this area is no longer possible, due to the ravages of the cotton boll weevil. Sea-Island cotton is now grown successfully only on some of the smaller islands of the West Indies group. It differs from the American upland cotton in having longer and more slender limbs, dark seeds free from fuzz, longer and finer lint, and in yielding less per acre. On account of its long fibre, this brings a higher price per pound of lint.

FIG. 112. Stalk 33 inches high; 102 bolls.
Courtesy of the N. C. Department of Agriculture.

3. *Chinese Cotton.* This is cultivated largely in China and Central Asia. It is a short-staple, quick-maturing cotton.

4. *Indian Cotton.* This is cultivated largely in southern Asia. The lint is inferior to that of American cotton.

5. *Bengal or Tree Cotton.* This is grown in India, Ceylon, and Arabia.

Each of the above species contains a large number of varieties differing from each other as regards such characters as size of boll, shape of boll, length of lint, per cent of lint to seed, character of branching, climatic adaptation, soil adaptations, length of growing season, etc. A few of the most common varieties grown in the southern United States are Mebane, Cleveland, Lone Star, Express, Delfos, Half and Half, Mexican Big Boll, Cook, Acala, Rowden, and Misdel. New varieties are continually being developed.

148. **Improvement of Cotton.** The common practice of planting the ordinary grade of cotton-seed as it comes from the gin finally leads to an inferior grade of cotton, as well as a decreased yield. Neither can one depend always on buying improved seed from some other locality, for a variety may be excellent on one kind of soil or under one kind of climatic condition and yet produce poor crops on a different soil or in a different climate. The first step in the improvement of cotton is to find by trial a good variety that is well adapted to the locality in which it is to be grown. The second step is to look carefully after the selection of seed each year. The following method is simple and practical.

FIG. 113. A good stalk of cotton.
Courtesy of the U. S. Department of Agriculture.

149. **Selecting Seed.** When the crop is mature, go through the field and pick into a bag the seed cotton from the best plants. This seed should be selected from plants possessing the following qualities:

1. *Productiveness,* determined by number of bolls per plant and size of bolls.

2. *Earliness,* indicated by a short-jointed type with basal limbs near the ground.

3. *Freedom from disease,* such as boll rot, rust, and cotton wilt.

4. *Character of the lint.* Take seed cotton only from

FIG. 114. A prolific fruiting branch of Startex cotton, with bolls open and ready for picking.

plants that show a good heavy body of strong staple, which measures seven-eighths of an inch or more in length.

5. *Storm-proof quality.* Bolls that open back too wide allow the cotton to drop out in a storm, and those that stand straight up do not shed water well.

The seeds from each selected plant should be put into a separate paper sack, examined carefully for its good qualities, and ginned separately. If seeds are saved from a number of selected plants, the seeds from each plant should be planted in a separate row. These rows should be in good soil, all should be of the same length, the same number of plants should be left in each row, and all rows should receive the same cultivation and treatment. Harvesting should be delayed long enough to enable you to see which rows reproduced their good parent uniformly. These outstanding rows should each be harvested separately, examined for yield and quality of staple, ginned separately, and the seeds from each row planted separately the next year in a small block by itself. Continue multiplying the seeds of the best for two or three years, or until there are enough seeds to plant the entire crop. This process of locating superior individual plants should not stop then, but each year superior plants from your select blocks should be chosen and the seeds handled as directed above. By continuing this system of selection, of plant-to-the-row testing and separate block breeding year after year, one can constantly improve the quality and yield of his cotton. No farmer can afford to waste his labor and land on inferior quality seeds. If he will not take the time and care to produce his own pure-bred, select seeds, he should purchase them from a State Registered Seed Breeder.

FIG. 115. A poor stalk of cotton.
Courtesy of the U. S. Department of Agriculture.

150. **Soil.** Cotton will grow on almost any type of soil, from light sands to stiff clays, if it possesses fertility and is well drained. It usually grows best on clays or silty clays. In very sandy soils the plants have a tendency to rust. In

rich moist bottom land cotton very often produces a very rank growth with a small number of bolls.

151. **Rotations.** The usual custom in the South has been the growing of cotton year after year on the same land. However, experience has demonstrated that higher yields can be produced when cotton is rotated with other crops. The reasons for this have been learned by studying the principles governing the rotation of crops. There is no one rotation for cotton that is best for all conditions. Each farmer must decide for himself the rotation which fits his type of soil, climatic conditions, and system of farming. A good rotation applying to Southwestern farms in general would be: first year, cotton; second year, corn with cowpeas between the rows for seed; third year, wheat or oats, followed by cowpeas for hay. If more corn or more cotton is desired, that crop can be grown two years in succession, making a four-year rotation.

In addition to the above-named crops the farmer should grow also when possible a winter cover crop of clover, vetch, or small grain, to protect the soil from washing and leaching, and to add organic matter or nitrogen, or both, to the soil.

152. **Fertilizers.** Cotton responds readily to fertilizers. The kind and amount of fertilizer to use depends upon the character of the soil upon which the cotton is grown. Nitrogen and phosphorus are the elements most commonly needed. Potassium is present in most soils in sufficient amount. The most economical way to keep up the nitrogen supply in the soil is to grow the cotton in a rotation with leguminous crops, thus securing the nitrogen from the air. If it is not practicable to grow these legumes, nitrogen may be purchased as nitrate of soda, sulphate of ammonia, calcium nitrate, ammonium nitrate, or calcium cyanamid. In these materials the nitrogen costs from nine to fourteen cents per pound. At this rate a crop of peas that would yield one to two tons of hay per acre will, if turned under, put from four to fourteen dollars' worth of nitrogen in the soil, besides the valuable humus. Phosphorus is bought usually in the form of super-

phosphate, and is not nearly so expensive as the nitrogen. On average soils three hundred to four hundred pounds of fertilizer per acre, made of equal parts of cotton-seed meal and superphosphate, give good results. Commercial fertilizers for cotton now on the market, having such formulæ as 3–10–3 (nitrogen, phosphorus, potash), 4–10–2, 4–12–4, are applied at rates of two to six hundred pounds per acre. Some farmers, who know the deficiencies of their soil and the needs of the plant, get good results by applying a fertilizer that carries only one kind of plant food that may be all that is needed, such as, for instance, 18 per cent superphosphate. This is usually applied before or at the time of planting.

153. **Preparation of Soil for Planting.** Land intended for cotton should be broken from six to eight inches deep, except in the case of extreme sands, where deep plowing is not needed, as the soil is already open enough. Care must be taken not to turn up too much subsoil at one time, as this new soil can give little food materials to the plant until it has been exposed to the air and moisture for some time. Where the previous plowing has been shallow it is best to plow from a half inch to an inch deeper each year until the soil is of sufficient depth. Heavy clay soils are best plowed in the late fall or early winter. When plowed in the fall a cover crop should be planted to be turned under in the spring from three to six weeks before the cotton is planted. This necessitates plowing twice, but it will result in much higher yields. Loams and sands may be planted to a cover crop in the fall without deep plowing, and these crops turned under in the spring in time for the soil to settle and become rather firm before the crop is planted. The general practice among farmers in the Southwest when preparing their land for cotton is to use large plows and heavy teams or tractors, and throw their soil up into ridges or beds. Lands that are prepared early in the fall by flat breaking or bedding should be freshened by disking or harrowing before planting. These beds are partially harrowed down and the cotton planted on top of the bed.

154. **Planting.** Today practically all cotton is planted with one-row or two-row planters, either walking or riding planters. The seeds are put in much thicker than required for a stand. Planting is best begun as soon as the danger of frost is over. This is from the 15th of March in southern latitudes to the 15th of May in northern latitudes. About one bushel of seed is planted per acre.

155. **Cultivating and Harvesting.** Cotton should be cultivated as soon as practicable after the plants are up. This is best done with some form of fine-tooth cultivator, running as close to the plants as possible without covering them. To prevent the soil from being thrown on the young plants, a fender should be used. The cotton is then "chopped," or thinned, to the proper distance in the row usually by means of a hoe, though machines are now occasionally used. This distance will depend upon the fertility of the soil, varying from six inches on poor land to twenty-four inches on rich land. The subsequent cultivations should be such as to keep down weeds and conserve moisture. There is no need to cultivate cotton more than to control the weeds. Deep cultivation should be avoided, as cotton has many shallow roots, and cultivating deeper than two inches destroys many of them.

Cotton is still harvested almost entirely by hand labor. Picking machines that have been somewhat successful are now on the market, but these still are in the experimental stage. Two kinds of machines being used are sleds or strippers in the Panhandle of Texas and Oklahoma and mechanical pickers of various types in other parts of the cotton belt.

CORN

156. **Corn: Its Importance and Distribution.** Corn is a native of the New World. The early settlers found the Indians raising it when they landed in America, and learned from them its uses and how to cultivate it. Indeed, had it not been for this Indian corn many of the early settlers would have starved. The botanical name is *Zea mays,* and it is called maize, or Indian corn, to distinguish it from

small grain, such as wheat and barley, all of which are called corn by the people of Europe. The corn that we read about in the Bible was not Indian corn, but small grain.

The world's total production of corn varies from 3,500,000,000 to more than 4,000,000,000 bushels annually. Of this amount the United States produces about 2,000,000,000

FIG. 116. Boone County white corn and the corn from which it was developed by selection.

Courtesy of U. S. Department of Agriculture.

bushels, or over half of the total supply. Our corn crop of 1928 would fill a row of wagons stretching nine times around the world, each wagon holding 50 bushels and taking a space twenty feet long in the row. Corn is the most important crop grown in the United States. The acreage devoted to this crop is about two and one-half times that of cotton, and nearly twice that occupied by wheat. The leading corn-producing states in the United States are Iowa, Illinois, Nebraska, Missouri, Kansas, Texas, Indiana, and Ohio.[1]

[1]During the ten-year period from 1919–28 the average annual yield for the entire United States was 28.2 bushels per acre, for Illinois 35.5 bushels, for Iowa 42.6 bushels. During the same period the average yield for the states in the Southwest was approximately 20 bushels.

157. **Description.** Corn is a member of the grass family. It differs from most other grasses in having the male and female flowers borne separately. The male flowers are borne in a spreading *panicle* (păn'ĭ-kl) at the top of the plant, known as the *tassel.* These male flowers on maturing produce an immense quantity of yellowish pollen grains. It is

FIG. 117. Representative ears of pop, sweet, flour, flint, dent, and pod corn. *Courtesy W. A. Wallace, Earl N. Bressman, and John Wiley & Sons, Publishers.*

estimated that the tassel of each plant produces from 18,000,000 to 25,000,000 pollen grains. The female flowers from which the ears of corn develop are borne in the axils of the leaves. The young ear is surrounded by a covering of shucks, which are modified leaves, and serve to protect the ear. Growing out beyond the shucks are the fine, slender, thread-like *silks,* which are the elongated styles and stigmas on which the pollen is caught as it is carried about by the wind. The pollen grain, lodging on the silk, begins to grow, sending out a long tube which grows down the entire length of the silk, until at the base it reaches the ovule. It thus unites with and fertilizes the ovule, and the grain of corn develops. Unless the pollen grains come in contact with the silks, no corn grains will be produced. There are as many silks as spaces for grains of corn on the ear. Only one pollen grain is required for each silk. The pollen may come from a tassel on the same plant or one on some other plant.

The corn plant develops an extensive root system, a large part of which is shallow. There is no tap-root produced. The fine fibrous roots grow in a lateral direction and branch profusely. The greater portion of the feeding roots are found at a depth ranging from three to six inches below the surface of the soil, depending upon the character of the soil and the depth to which the seedbed has been prepared. It has been noticed that a large number of the roots, after growing in a horizontal direction for one or two feet, turn down more or less abruptly, presumably in search of moisture.

158. **Types of Corn.** There are at least six known types of corn. These are: 1, *dent corn;* 2, *flint corn;* 3, *sweet corn;* 4, *popcorn;* 5, *soft corn;* and 6, *pod corn.*

The bulk of the American corn crop is dent corn. It is the common race of corn grown in the Southern States, and is characterized by the presence of a small indentation, or dent, in the top of each grain. Varieties differ, but the usual tendency is to produce from one to two ears per stalk.

Flint corn is much harder than dent. The top of the grain is smooth or rounded, and the grains are shorter than those of dent corn. Plants vary in height from five to twelve feet, and have a tendency to produce two ears per stalk. Flint corn matures quickly, and is therefore grown near the northern limit of production.

Sweet corn presents a wrinkled, horny surface, and contains much more sugar than the other races. It matures early, the plants are small, and each plant bears a number of small ears. In the South it is grown mainly in gardens and is highly prized for table purposes.

Popcorn is characterized by its very small compact, horny grain. This compactness of the grain gives it its popping property. The plants are small, and several small ears are borne on each plant.

Soft corn bears a grain that is very soft and white. This corn was cultivated extensively by the Indians, because it is easily ground or crushed.

Pod corn is a curiosity and is not grown commercially. Each individual grain is inclosed in a small shuck, while the ear is covered by an outer shuck.

FIG. 118. These three stalks grew in the same hill. Differences in the seeds make the differences in the stalks.

Courtesy of Professor P. G. Holden.

159. **Varieties.** Each of the above races of corn contains a large number of varieties that have been produced as the result of crossing, selection, or growth under different conditions of soil and climate. Some of the most commonly grown Southern dent varieties are Surecropper, Ferguson Yellow Dent, Strawberry or Calico, Hasting's Prolific, Cocke's Prolific, Chisholm, Hickory King, Horton, Mexican June, Brazos White, Mosby, and Bloody Butcher.

160. **Improvement of Corn.** There are very few pure or uniform varieties of corn, due to the fact that it mixes so easily, because of the great distance to which the wind carries the pollen. Too little attention is given to improving our corn. Barren and poor stalks are allowed

to ripen and pollinate good stalks, and selection of seed is usually made after the corn is in the crib. The result is that the variety deteriorates from year to year. It is entirely possible to increase the yield of corn from 10 to 20 per cent by seed selection alone, an increase which comes with very little labor. It is probable that more is known about the breeding behavior of corn than about any other field crop. New practices in producing higher yields are being worked out, which makes it imperative that the wise farmer keep posted on these developments.

FIG. 119. These three ears grew on three different stalks in the same hill. Differences in the seeds make the differences in the yield.

Courtesy of Professor P. G. Holden.

161. Seed Selection. In selecting seed corn the entire plant as well as the ear should be considered. This makes it impossible to make a wise selection of seed corn from the crib, and necessitates selecting the seed in the field before the crop is harvested. A good plan is to go through the field at harvest time and select seed ears from the most productive plants, at the same time taking into consideration such points as the position of the ear on the stalk, the height of the ear from the ground, and the general healthfulness of the plant. Much improvement will be secured if nothing more is done than to select from productive plants enough seed to plant the next year's crop. However, even among ears that look equally good and come from equally good stalks, some will grow better and produce a great deal more corn than others. If the farmer wishes to have the very best seed, he must find out which of these good-looking ears are most vigorous and have the greatest producing power, and plant only from this stock. The vitality and growing power can to a great extent be determined by testing each ear in the germinator, taking six grains from different parts of the ear. Figure 120 shows the result of such a test. The difference in the producing

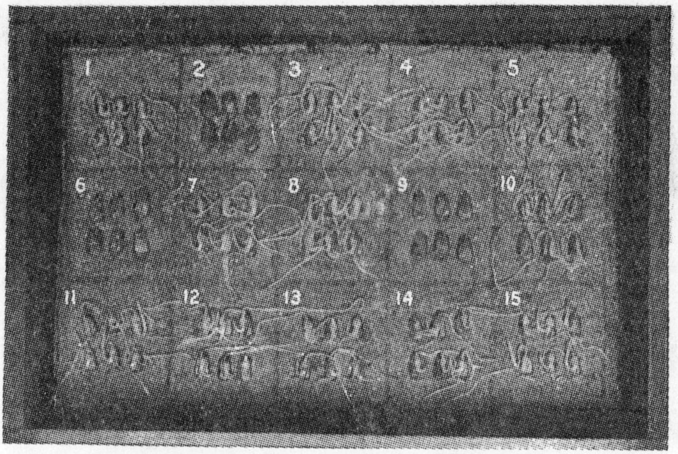

FIG. 120. The results of tests of fifteen ears of corn. Ears 2, 6, and 9 are entirely dead; ears 3, 4, 7, and 8 are particularly vigorous.

Courtesy of Professor P. G. Holden.

power of different ears can be found out by what is called the ear-to-row test.

162. **Ear-to-Row Test.** At least 25 of the best ears selected from the field should be planted on a uniform plot of soil according to the following plan. Number the ears from 1 to 25 consecutively. Lay off 25 rows of equal length on soil of uniform productiveness. These rows should be at least long enough for 150 hills of corn. Plant row No. 1 from ear No. 1, row No. 2 from ear No. 2, and so on until the 25 rows have been planted from the 25 ears. All rows should contain an equal number of hills with an equal number of plants in the hill. The cultivation and general treat-

ment should be the same for all rows. As soon as the tassels show and before any pollen has been shed, carefully pull the tassels out from one half of every odd-numbered row, say the north half, and from the south half of every even-numbered row. This insures cross-fertilization, which has been shown to increase the production in corn. Seed should be

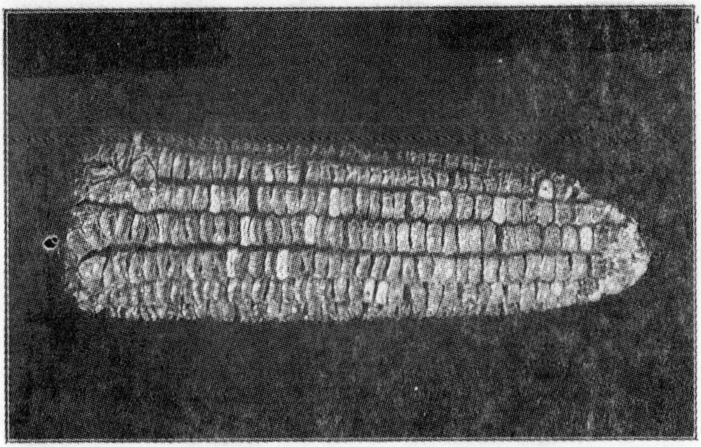

FIG. 121. Corn is pollinated by the wind and must be isolated to keep it pure. This ear of sweet corn has several seeds of field corn resulting from wind pollination.

saved only from the detasselled stalks, as the ears on these stalks were of necessity fertilized by the pollen from some other stalk or stalks. At harvest time the ears from each row should be husked and weighed, keeping the ears from the detasselled stalks separate from those of stalks producing tassels. By weighing separately the corn from each row the best-yielding strain can easily be determined. Select the best ears from the detasselled portions of the eight or ten best rows for planting the general crop. If this does not furnish enough seed for the general crop, these selected ears may be planted on a half acre of good land and seed grown for the main crop. There should always be enough good ears selected from the best-yielding rows to plant a new ear-to-row plot the following year. All seed corn should be hung in a well-ventilated place where it is not exposed to sudden changes of temperature or to attacks of mice or weevils.

163. **Results of Ear-to-Row Test.** Ninety farmers in Iowa sent Professor Holden ears of their seed corn to test by the ear-to-row method. He tested them and found that

FIG. 122. Comparative yield of five highest and five lowest yielding ears at Story County, Iowa Station. The average of the five highest was 80.3 bushels per acre; of the five lowest 40.8 bushels. The seeds were all secured from seed corn being used as seed by the farmers. How much is your state losing each year by planting inferior untested seed?

Courtesy of Professor P. G. Holden.

the yield per acre from different ears varied all the way from 31.5 bushels to 80.5 bushels per acre. The six best producing ears averaged 77.5 bushels, and the six lowest 35.6 bushels, showing that by planting seed of the best-yielding ears only the farmers would have added to their yield on the average 41.9 bushels per acre, or more than double what they would have made by planting the six poorest-yielding ears. Professor Holden also tested 102 ears of the selected seed corn at the station, and these fine-looking ears varied in yielding power all the way from 90.5 bushels from ear No. 75 to 36 bushels from ear No. 93. Ear No. 19 gave 79 barren stalks, while ear No. 83 gave only 6; ear No. 54 had 258 broken stalks, while ear No. 85 had only 41. Not all of these qualities can be found out by testing in the germinator before planting, but the vitality and growing strength can be found out in this way. By testing in the germinator and throwing out all weak growing ears, and then by ear-to-row test discovering the other good and bad qualities, the farmer can now breed and improve his corn easily and rapidly.

164. **Soil.** Corn will grow on a wide range of soil types. It makes its best yield on a deep, fertile, moist loam. A large per cent of the roots go down deep to supply the plant with water during dry periods. This necessitates deep preparation, especially if the soil contains considerable clay. Upland clay soils should be deepened gradually each year and at the same time vegetable matter added by the proper rotation of crops or by the addition of barnyard manures. The seedbed should be at least six inches deep, preferably eight. Bottom lands are better adapted to corn than upland soils, because they contain more moisture, but these must be well drained.

165. **Rotation.** Corn is adapted to the same kind of rotation that was given for cotton. When both corn and cotton are included in the rotation, the corn usually follows the cotton.

166. **Fertilizers.** Corn makes a very profitable use of rough manures containing organic matter, such as barnyard manure or green manures. When the supply of organic matter is maintained by the above methods, nitrogen fertilizers need not be used. The same is generally true of potash, as the decaying vegetable matter will make sufficient potash soluble in the soil to supply the needs of the crop. Phosphorus is usually more deficient in the soils than potash, and where the above system of maintaining the organic content of the soil is practised, phosphatic fertilizers will generally be found profitable for corn. If the soil is deficient in organic matter, then both nitrogenous and phosphatic fertilizers should be used. A common application of fertilizer for corn is 200 to 400 pounds of a 3–8–3 or 4–10–4 complete

FIG. 123. The root system of corn.
Courtesy of Kansas Experiment Station.

commercial fertilizer per acre. The amount of fertilizer used depends mainly upon the fertility of the soil.

167. **Planting.** Where possible the corn planter should be used, as it insures a more uniform planting and germination, and also saves much labor. One-horse planters with fertilizer distributors are quite generally used, and are very satisfactory. The use of the two-horse check-row planter is restricted to level lands only. Where the conditions permit its use it may be very profitably employed.

168. **Time of Planting.** Corn should be planted as soon as the soil becomes warm in the spring and the danger of frost is over. This varies from February 20 to April 15, depending upon the locality. If planted before the ground is warm, or while the soil is still too wet, corn is liable to rot before germination. Certain varieties of early corn, such as Mexican June, are often planted in June with good results.

169. **Depth of Planting.** Corn is planted from one to four inches deep, depending on the soil and the season. It should be planted deep enough to insure the presence of enough moisture for good germination. On sandy lands deep planting is generally the rule, while shallow planting is better on wet clay soils. Early-planted corn should not be covered as deeply as late-planted.

170. **Cultivation.** Too often deep-tilling implements are used in cultivation of corn. These large shovel plows pulverize the soil very ineffectively, and also destroy many of the feeding roots. Small-tooth cultivators or some form of sweeps make the most satisfactory implements for cultivating corn. These should not go deeper than from one and one-half to two and one-half inches, unless it be at the first cultivation on clay land that has a tendency to run together and bake.

171. **Harvesting.** The usual custom of pulling the corn blades for feed is expensive, and also decreases the yield of the grain. Cutting the tops just above the ears at the time when the outer shucks have turned brown is also unprofitable. This practice does not materially reduce the yield of grain,

if done when the ears are fairly mature, but it is a very expensive way to get feed. The most profitable way to harvest the corn crop is to cut and shock the whole plant. This should be done when practically all of the shucks have turned brown and the grain has become hard. This practice does not decrease the yield of grain. In this way all the forage is saved, and the use of the land is obtained for fall cover crops. In some localities the corn thus harvested is run through the shredder, which shucks and separates the ears from the stalk, at the same time tearing the stalk to pieces. The shredded leaves and stalks are known as *stover* (stō′ver), and are readily eaten by stock.

172. **Corn Judging.** How to judge corn will be treated when Four-H Clubs are discussed in Appendix III.

SMALL GRAINS

173. **Wheat.** Wheat is the national bread crop and is grown more or less in every important agricultural country in the world. It is probable that the acreage planted to wheat is greater than that for any other crop grown.

The origin of wheat is not known. When man first began to keep records, wheat was an important crop. Wheat was not grown in the New World before its discovery by Columbus. It was grown in the Virginia Colony as early as 1618. Production moved westward as the country developed. The average production for the years 1925–26 to 1929–30 totalled 822,512,000 bushels with an average yield of 14.1 bushels per acre.

174. **Types of Wheat.** Wheat is an annual grass which bears its seed in a spike or head. Each spike is made up of a number of spikelets, that is, flower clusters which usually develop into two or more grains. There are eight types of wheat, based upon head characters: (1) Common Wheat, (2) Club Wheat, (3) Durum Wheat, (4) Poulard Wheat, (5) Polish Wheat, (6) Spelt Wheat, (7) Emmer Wheat, and (8) Einkorn Wheat. Of these different types only the first three are grown extensively in this country.

Common Wheat. About 85 per cent of the wheat produced in the United States is of this type. The plants have long, broad, somewhat dense, flattened, tapering spikes. This is our main wheat for making flour for bread. It is extensively grown in Kansas, North Dakota, Nebraska, Oklahoma, Ohio, Illinois, Montana, Texas, and other states. Practically all of the wheat grown in Texas is of this type. The leading varieties grown are Turkey, Kharkov, Kanred, Marquis, Fultz, and Mediterranean.

Club Wheat. This type is grown on about 7 per cent of the acreage planted to wheat in this country. It has a short, dense, four-sided spike, shaped somewhat like a policeman's billy. This wheat is used for the making of crackers, breakfast foods, and in some places for flour. It is grown mostly in the states of Washington, Oregon, California, Idaho, and to some extent in Michigan and New York. The leading varieties of Club Wheat are Hybrid 128, Sun, Red Club, California Club, and Little Club.

Durum Wheat. About 6 per cent of the wheat acreage of the country is of this type. The spikes of durum are compact, usually rather slender and always bearded. It is used for the making of macaroni, vermicelli, spaghetti, and other edible pastes. Durum wheat is grown principally in North Dakota, South Dakota, Minnesota, and Wisconsin.

175. **Adaptation.** Wheat is exceeded by few plants in its range of adaptation to climate and soil. It is grown in the far north and where the winters are cold and its growing season short, and in the hottest tropics where the atmosphere is not too humid. Our finest flour-making wheats are grown where the plants experience cold winters, a cool moist growing season, and dry, bright, and hot maturing conditions.

Wheat grows on a wide range of soil types, but it prefers a deep fertile clay loam. Most suitable of all is a lime soil containing considerable clay. Wheat needs a fairly rich soil, which contains plenty of organic matter. For the highest yields, the soil should be prepared in time for all crop residue to decompose and to store up abundant moisture. Wheat should be grown in a rotation with other crops in order to keep up the organic matter of the soil and to keep down insects and diseases. Where moisture is sufficient, commercial fertilizers can be used to a good advantage.

176. **Seed.** The breeding and developing of new varieties of wheat is seldom a profitable venture for the farmer. Usually a farmer would do well to secure a sufficient supply of seed each year, or at least every few years, from a state experiment station or a reliable seed breeder to plant a seed block on his farm. In growing wheat, only pure clean seed of an adapted variety should be used. Wheat is a self-fertilized plant and the seeds will stay pure if they do not become mixed through mechanical means. Since practically all farmers sell most of their wheat, and since different market classes are produced in different sections, it is doubly important that the right variety be planted.

177. **Planting Dates.** The kinds of wheat grown in the high plains region of Texas should be sown in the fall. Most of it is planted between October 1 and November 15. Temperature and moisture are important factors in determining the time and rate of planting. In the central and northern part of the state most of the wheat is planted between the above dates, but some is often planted between January 25 and February 27. In either period early planting is advisable.

178. **Rate of Seeding.** When good clean seed of an adapted variety is available, from 2 to 4 pecks per acre are often sufficient. The range for the rate of planting is from 1 to 6 pecks per acre. Where winter killing is likely to be serious, the rate is often increased beyond this.

179. **Planting.** Wheat and other small grains are either planted with seed drills or sown broadcast. There are advantages to each method. In sections where large acreages are planted, the grain drill is usually used. The advantages in drilling wheat are: (1) larger yield per acre; (2) more uniform distribution of seed in the row, even depth, uniform growth and ripening; (3) less seed required to plant an acre.

When a drill is used the seeds are put in rows, 6, 8, or 10 inches apart.

180. **Harvesting.** Most of the wheat grown in the Great Plains region is now allowed to ripen in the field and is harvested with a *combine.* This machine heads and threshes the grain all in one operation. Where *binders* are used the wheat is cut when in the hard dough stage, tied into bundles, put into small shocks, and when dried the grain is threshed out.

181. **Yield.** The yield of wheat for the entire United States for the five-year period ending in 1930 averaged 14.1 bushels per acre. In farm trade, a bushel of wheat is considered to weigh 60 pounds. A measured bushel may weigh a few pounds above or below this weight. When the grain passes through a central market it is sampled, weighed, and graded. Weight per bushel is a very important grade factor. The heavier a bushel of wheat, the better the quality.

182. **Oats.** The oat is another member of the family of grasses. It came into use much later than wheat and barley, but is today widely grown throughout the world. Oats are often thought of as a feed for horses, but large quantities are made into oatmeal and other cereal dishes for human food. The world production of this crop is often between 4 and 5 billions of bushels. Our yield per acre here in the United States averaged 29.6 bushels for the five-year period ending 1930. For the period from 1928 to 1932 Texas planted an average of 1,455,000 acres of oats which gave an average yield of 23 bushels per acre.

The flowering part of an oat plant is called a head or panicle. It is made up of spikelets or groups of grains which are borne on seed branches. The grains of most varieties on being threshed retain their hulls.

183. **Adaptation.** Oats prefer a cool moist climate. They are adapted to a wider range of soil types than wheat and produce good yields on fertile land, if the climatic conditions are favorable. Many farmers plant oats in the fall of the year to protect the soil and provide winter pasture. In a rotation oats generally follow corn.

184. **Varieties.** The most important varieties of oats grown in the Southwest are: (1) Red Rustproof, (2) Ferguson 71, (3) Nortex, (4) Fulghum, (5) Burt, (6) Fraizer, and (7) Winter Turf. The first three are planted in the fall from October 1 to November 10 and the last three from January 25 to February 27.

185. **Soil Preparation.** Oats seldom receive a very well prepared seedbed. Since oats usually follow corn, the land is seldom plowed. The cornstalks are often dragged down and the oats disked or drilled in. Moisture and soil fertility are more important factors in the production of oats than is seedbed preparation.

186. **Seed.** Every grower of oats should pay close attention to the seed he plants. Only pure, clean, adapted seeds, which have been treated for seed-borne diseases, should be used. Oat varieties do not run out, but often become badly mixed in threshing, which increases their susceptibility to diseases and insect attacks. The rust-resistant varieties are best for Southern planters.

187. **Planting.** Most oats are planted now with a seed drill. Drilled oats usually make larger yields than oats sown broadcast, due to uniform distribution of seed and depth of planting. If the seed can be put into moist soil, two inches is deep enough to plant. If the seeds are drilled in, the young plants will be down in a furrow, which protects them somewhat from the cold, and, in cold climates, from what is known as heaving or soil lifting.

188. **Rate of Planting.** Oats are strong feeders and often when moisture is sufficient and the soil is fertile one kernel will give rise to a large number of stems or culms. The rate of planting depends upon the variety, the rainfall, and the fertility of the soil. From 4 to 16 pecks per acre may be taken as the extreme range. Most farmers in Texas, if using a drill, will plant from 4 to 10 pecks, or will broadcast 6 to 12 pecks, per acre.

189. **Harvesting.** The seeds in oats mature from the top of the panicle downward. If the crop is grown for grain, it

should be cut when the seeds are in the late dough stage. The harvesting of oats is best done with a self-binder. For the best results, the bundles should be put up in small shocks for curing, and should not be handled much before threshing. Many acres of oats are grown for hay. When this is practised, the crop should be cut when in the milk stage.

190. **Barley and Rye.** Both of these plants are grasses which grow their seed in spikes. In Texas they are used chiefly for winter pasture and soil building. When conditions are favorable and the plants give evidence of maturing grain, they are left for harvesting. For the production of high quality grain, barley prefers fertile soil and almost ideal growing conditions. The 1928–32 production of barley in the United States was 282,241,000 bushels a year, but of this Texas only accounted for 3,646,000 bushels. Farmers classify barley by the number of rows of spikelets in the spike, as two-row and six-row. Each kernel in barley is borne in a separate spikelet. The hull remains on most barley when it is threshed.

191. **Rye Is a Very Hardy Plant.** Farmers often say that "when land gets too poor to produce rye, it should be turned out." Production in the United States from 1928–32 averaged 38,655,000 bushels, or an average of 12.2 bushels an acre. The grains of rye may be yellow, gray, green, or graying brown. They are generally borne two to a spikelet, in long cylindrical spikes. In threshing, the kernels are freed from the chaff.

THE FORAGE CROPS

192. **Sorghums.** There are four distinct groups of cultivated sorghums: grain sorghums, grown for grain and forage; sweet sorgos, grown for forage and syrup; broom corn, grown for the making of brooms; and grass sorghums, grown for forage and pasturage. Examples of grain sorghums are Yellow Milo, Black-hull Kafir, Feterita, and Hegari; of sweet sorgos, Sumac or Red Top, Honey, Amber, Orange, Gooseneck and Sugar Drop; of broom corn, Standard and Dwarf Broom Corn; of grass sorghums, Sudan and Johnson Grass.

193. **Grain Sorghums.** Sorghums used for grain are usually classified under four different groups: the kafir group, the milo-durra group, the kaoliangs, and shallu. Only the first two groups are of importance in the Southwest. In the kafir group, only three varieties will be mentioned, the red, the white, and the black-hull. These varieties are easily distinguished on sight. The red and black-hull varieties generally yield a little more grain and forage than white kafir, and are considered more desirable. The important varieties of the milo-durra group are Dwarf Yellow Milo, Dwarf White Milo, Feterita, Brown Durra, and White Durra. The Yellow Milo is grown more extensively than any of the other varieties, especially in western Oklahoma and the Panhandle of Texas. It matures in a shorter time than kafir, and is especially valuable for arid and

FIG. 124. Typical heads of Dawn Kafir (top), Dwarf Yellow Milo, and Spur Feterita (below), three of the best varieties of grain sorghums for the Southwest.

Courtesy of the Texas Agricultural Experiment Station.

semi-arid regions. The grain of this variety is large, brittle, and easily masticated by stock.

White Milo differs from Yellow Milo mainly in color of seeds, and it should be grown under the same or similar conditions. Feterita, another grain sorghum of this group, is of considerable importance in the Southwest plains country. It has large chalky-white kernels and a short dense head similar to the head of the milos. Feterita is a little earlier maturing than Yellow Milo, but has a slender, weak stem which permits it to be blown down easily. An improved strain called Spur Feterita, developed by the Texas Experiment Station, is being more and more widely used. Brown Durra is grown extensively in certain sections of the country, especially in California, where it is often called "Egyptian Corn." As the grain shatters easily, it is not considered as valuable as Yellow Milo. White Durra is not grown extensively in this country. Owing to its liability to injury by insects and fungus diseases, and to the fact that the seeds shatter easily, it is not a very satisfactory crop.

Another grain sorghum which has recently come into importance in the Southwest plains sections is known as Hegari. The stalk, seed, and head characteristics resemble closely those of Black-hull Kafir, the principal difference being that the heads are perhaps a bit more compact, the kernels more chalky in appearance, and the entire plant a little smaller. Under Federal standards the threshed seeds of Hegari are classed with the White-seeded Kafirs.

Kafirs, milos, and durras are especially adapted to semi-arid regions, owing to their drought-resisting properties, and should usually be depended on there for grain rather than corn. They vary from four to eight feet in height.

194. **Soil and Seed.** These crops are best adapted to a good clay or loam soil. They make a fair growth even on a poor soil, but of course do best on a soil of high fertility.

Care should be taken to plant only reliable seed that are not mixed. The Texas Experiment Stations have developed and made available pure line seed of all the leading varieties.

Tests of these seed have been made to determine disease resistance, productivity, and other desirable qualities.

195. **Preparation of Soil and Seeding.** The seedbed is prepared in very much the same way as for corn. Where the crop is grown for grain, plowing is often done in the late winter or very early spring, so that the land will hold more moisture and the seedbed get firm before planting

FIG. 125. Field of improved Black-hull Kafir developed by the Texas Experiment Station.

season. The land should be thoroughly pulverized and the seed planted in drills three and one-half to four feet apart, and three to five, and in very poor land eight, inches apart in the drills. It is planted best with a corn planter with a drilling attachment. In regions of little rainfall, listing is a very common practice. This is done by throwing out furrows at planting time with a lister, the seed being drilled in the bottom of these furrows. This insures deeper rooting and better enables the plants to withstand drought.

196. **Cultivation.** The cultivation of these crops is the

same as for corn. The sorghums are shallow-rooted, a fact which necessitates shallow cultivation. Small-tooth cultivators are generally used.

197. **Harvesting.** If grown primarily for grain, the crops should not be harvested until the seeds are fairly mature. The entire stalks may then be cut with a corn harvester and shocked like ordinary corn, or the heads may be first cut off with a knife or header and the stalks afterward harvested for stover. The heads should not be stored in large piles until they are thoroughly dry, as they heat easily. The grain may be threshed by running the heads through an ordinary grain thresher, or, if the entire plant has been cut, the heads may be inserted into the thresher until the seeds are removed, the stalks being then withdrawn. While much higher yields can be produced, the usual yield varies from twenty to forty bushels per acre.

SORGOS

198. The sweet sorghums differ from grain sorghums in that they have a great abundance of sweet juice in their stem and produce but small amounts of grain. For many years after their introduction into the United States they were used chiefly for making syrup, but now they are largely used as dry forage, silage, and as a green feed. During 1936 there were grown in the United States 2,565,000 acres of the sorgos for forage, and 215,000 acres for syrup. Of this Texas planted 606,000 acres for forage and 30,000 acres for syrup.

199. **Varieties.** There are many varieties differing chiefly in size of plant, earliness, shape of panicle, color of seed, and seed covering. Black and Red Amber mature in from ninety to one hundred days. They have slender stems and loose, spreading panicles. Sumac and Red Top mature in from one hundred and ten to one hundred and twenty-five days, have a compact panicle, and are well suited to the making of syrup. Orange matures in from one hundred and five

to one hundred and twenty days, has a medium compact panicle, and heavy stalk. Honey and Gooseneck are late-maturing varieties, often requiring from one hundred and twenty-five to one hundred and forty days.

200. **Planting.** Sorgos find a place on every farm where there is any livestock. When planted on land of average fertility, they produce good yields of appetizing forage. They may be sown either broadcast, drilled or in rows for intertillage. The use to be made of the crop often determines the method of planting. Sorghums are sensitive to cold and should not be planted until the soil is warm.

201. **Seeding.** Pure seeds are best and any farmer can secure them for his plantings at very reasonable prices. Rates of planting vary according to the use to which the crop is to be put. For syrup four to six pounds, for forage drilled forty-five to sixty pounds, for forage broadcast one to two bushels per acre are the rates of planting often used.

202. **Harvesting.** Sorghum for forage should be cut when the first heads reach the dough stage. For good silage the plants should be mature. It is not advisable to harvest sorghums by pasturing cattle. If the plants have gone through a drought period, there is danger of prussic-acid poisoning. Sorghum for syrup is cut with a binder or knives after the juice becomes sweet and the seeds have reached the hard dough stage.

GRASS SORGHUMS

203. **Sudan.** Sudan and Johnson Grass are our two most important grass sorghums. Sudan grass was introduced into this country in 1909 from the Sudan region of Africa. It is an annual grass, which is a near relative of the sorghums. It resembles Johnson grass somewhat in its above-ground appearance, but doesn't possess rootstocks. It is grown for pasturage, soiling, hay, and silage. Sudan is not exacting in its soil requirement, but produces most on fertile well-drained soil. It is not unusual for as many as four cuttings

of hay to be harvested in one season from the same planting. After each cutting new tillers arise and in the course of a season one plant may give rise to as many as fifty stems. The rate of seeding will vary with the method of seeding: (1) drilled or broadcast, twenty to twenty-five pounds; (2) in rows thirty-six to forty-four inches apart, two to four pounds; (3) in rows eighteen to twenty-four inches apart, four to six pounds per acre.

FIG. 126. Single plant of Sudan grass, illustrating its habits of growth when planted in rows.

Courtesy of the U. S. Department of Agriculture.

204. **Johnson Grass.** This is a perennial grass which has some very desirable as well as seriously objectionable qualities. It has a very wide range of soil adaptation. Growth is made in accordance with the fertility of the soil. If cut for hay when the first heads are appearing, it makes a fine quality of hay. In many places it serves as a valuable plant in meadows when rotated, and in temporary pastures for both hay and grazing. In cultivated land it is a weed. The rootstocks it possesses make it

very hard to eradicate. It can be eradicated by: (1) heavy pasturing for one or two years; or (2) by preventing seeding and plowing the land thoroughly at frequent intervals for one or two summers.

LEGUMES, OR NITROGEN-GATHERING PLANTS

205. **What Legumes Are.** By legume is meant that class of plants the members of which bear their seeds in pods, and increase the supply of nitrogen in the soil because of certain bacteria which live on their roots and have the power of taking in the insoluble nitrogen from the air and working it into the soluble nitrogen compounds called nitrates, on which plants can feed. The common representatives of this group are cowpeas, soybeans, peanuts, the clovers, vetches, and alfalfa. These are especially valuable plants to agriculture, because they improve the soil as well as produce the best of foodstuffs and forage.

206. **Nodules.** If one will carefully examine the roots of cowpeas, clover, or any of these legumes, he will find a large number of small galls, knots, or nodules growing on them. These galls are known as *nodules*. They are produced by bacteria that live on the roots of the plant. Each nodule contains thousands of these bacteria, which are alive and must have food. Part of this food they get from the plant on which they grow and part of it from the air which circulates through the soil. The food that these bacteria get from the air is pure nitrogen, which makes up four fifths of the air. This insoluble nitrogen unites with other elements in the bacteria and makes a soluble nitrogen compound. The bacteria finally die and the legume takes the nitrate left by the bacteria and uses it in its growth. Finally when the legume dies and decomposes, the nitrogen which was taken out of the air by the bacteria is left in the soil for the use of other plants.

207. **Inoculation for Legumes.** The bacteria that produce tubercles on clover are different from those that grow

on alfalfa. In fact, nearly all the different kinds of legumes have their own particular bacteria, and if the right kind of bacteria is not in the soil, it must be added before the legume can be grown successfully. This addition to the soil of material containing the proper bacteria is called *inoculation*

Fig. 127. Red clover in both pots; no nitrogen in soil of either, but bacteria in the pot on the right.

Courtesy of the University of Illinois Agricultural Experiment Station.

(ĭn-ŏc-ū-lā'shŭn). The best way to inoculate a field is to add to it soil that contains the proper bacteria. For example, if one wished to inoculate an acre of land for alfafa, he would apply to it about two hundred and fifty pounds of soil from a field that was growing alfalfa successfully. This would be sprinkled on as would a fertilizer and harrowed into the land. The different varieties of helpful bacteria are now cultivated in laboratories. These cultures may be procured from commercial concerns, or in some cases from the United

States Department of Agriculture. By inoculating the seed with the proper culture before planting the same result may be obtained as with the soil inoculation. Inoculation does not have to be repeated every year, as the bacteria live from year to year in the soil. All legumes do not need inoculation. For example, the bacteria that grow on the roots of

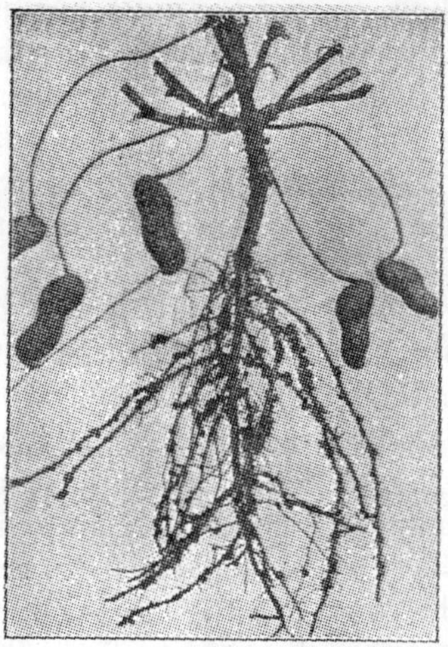

FIG. 128. Peanut plants.

Cut on the left, courtesy of the United States Department of Agriculture; on the right, courtesy of "Farm and Ranch."

cowpeas are present in nearly all soils and would not have to be added.

We shall speak of only a few of the most common legumes.

208. **Peanuts.** The peanut plant is an annual, growing from one to two feet high, depending upon the variety grown and the soil. The fruit, or seed, which is not a nut at all, is borne in pods underneath the surface of the soil, on tips of stems. These stems start out from the axils of leaves above ground and, after blooming, push their way into the soil and there develop the seed.

209. **Varieties.** Two well-defined types of peanuts are

recognized: those with large pods and those producing small pods. A common representative of the former group is the Virginia peanut, used for roasting. The Spanish peanut is the small-podded variety. These are used mostly for making confectionery and feeding hogs.

210. **Soil and Fertilizers.** The peanut does best on a loam soil containing plenty of lime and not too much humus. If barnyard manure is used, it should be applied to the preceding crop, so as to give it ample time to decompose thoroughly before the nuts are planted. Nitrogenous fertilizers are seldom applied, as the peanut can secure its own nitrogen from the air. Potassic and phosphatic fertilizers are largely used.

211. **Planting and Cultivation.** The land should be plowed and prepared as for corn. The large-podded peanuts are usually planted about the time corn is planted, while the Spanish peanuts may be planted considerably later, at any date from the time that cotton comes up until about July 1. The small-podded peanuts are generally planted in rows about twenty-four to thirty inches apart, and from four to eight inches apart in the row. For the large-podded varieties the rows should be from thirty to thirty-six inches apart. The land should be cultivated often enough to control weeds and to keep it finely pulverized.

212. **Harvesting.** Peanuts intended for seed or market should be harvested before frost. A common method of harvesting is to run under the row on each side with a turning plow from which the mouldboard has been removed. This plow should be run at sufficient depth not to tear the pods from the branches. The plants are then lifted by hand or with a fork and stacked, usually on the same day that they are dug. The plants should be stacked with the tops turned outward and the stacks made as slender as possible. They are capped with grass or straw.

213. **Cowpeas.** The cowpea is the most important Southern legume. It is grown on the widest variety of soil of any Southwestern hay crop. It fits into almost any system

FIG. 129. Cowpeas in corn rows.
Courtesy of Texas Experiment Station.

of crop rotation that the farmer wishes to practise, and is valuable either as hay, pasture, or seed crop. A crop of cowpeas may be grown after small grain comes off in the spring, before small grain is seeded in the fall, or between two crops of small grain. They are generally grown as a secondary crop, being sown at the last cultivation of corn,

FIG. 130. Soybean field. A good legume for hay and for building up the soil. *From Halligan's Fundamentals of Agriculture. Courtesy of Messrs. D. C. Heath & Co.*

except in regions of very dry summers, where they must be sown earlier. They may be either pastured off, used for seed production or for hay. They are often planted in drills between the corn rows, or between the hills of corn in the same row. In either case they are allowed to mature and the seed is harvested. Twenty bushels per acre is a good average yield. Cowpeas should never be planted until the soil gets thoroughly warm. Deep preparation of the soil is not essential to the successful growth of cowpeas, though on heavy clay soil it is very profitably employed. When sown broadcast, from one to one and one-half bushels of seed per acre are required. When sown in drills, one peck of seed is usually enough. Superphosphate makes up the bulk of the fertilizer used, although considerable amounts of kainit are under certain conditions desirable. No nitrogen fertilizer is necessary. Cowpeas should be harvested for hay when the most mature pods are beginning to turn yellow. One and one-half tons of hay per acre is a good average yield.

FIG. 131. Typical soybean plant.
Courtesy Iowa College of Agriculture, Agricultural Experiment Station.

214. **Soybeans.** The soybean is rapidly growing in favor throughout the Gulf States and the Southwest. It is planted for hay, silage, pasturage, beans, and for soil building. The climatic adaptations are about the same as those for corn. It grows on nearly all types of soil, though the best results are obtained in mellow fertile loam or sandy loam. The seed should be inoculated before planting. Some of the varieties being used successfully are: Laredo, Biloxi, Mamoloxi Mammoth Yellow, Barchet, and Otootan. The rate of planting varies with the variety used, the fertility of the soil, and the method of planting. For best results the beans should be planted in rows and cultivated.

The method of harvesting depends upon the purpose for

which the crop is to be used. When seeds are to be produced the mower, grain binder, or combine is used.

215. **Alfalfa.** Where adapted, alfalfa is one of the most valuable perennial legumes. It is grown primarily for hay, but is sometimes used for pasturage, soiling, or silage. When planted on well-drained, fertile lime soil it produces large amounts of very nutritious and palatable hay. Wet weather seems to be more injurious than dry weather. In humid sections alfalfa does best on well-drained, deep soils high in lime, but under these conditions it is hard to cure for hay. Much alfalfa is grown in irrigated sections of the West. High-grade alfalfa hay must be bright green in color, have a high percentage of leaves, and fine stems. To produce such hay requires a semi-arid climate where there is good sunlight, but not too much dew.

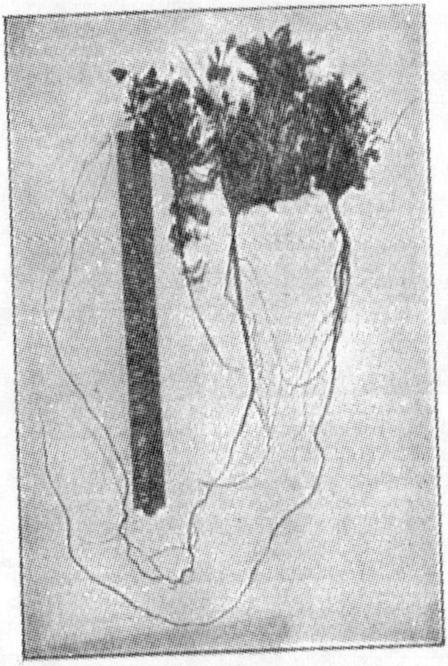

FIG. 132. An alfalfa plant only a few months old. The roots are three feet long. Under very favorable conditions alfalfa roots are known to have run over forty feet.

Since alfalfa is only grown in localized areas one would do well to investigate the suitability of his location, adaptability of soil type, and presence of root rot, before planting extensively.

216. **The Clovers.** To this group of plants belong *red clover, white clover, crimson clover, alsike clover,* and *mammoth clover.* These are known as the true clovers. Japan clover, or lespedeza (lĕs-pē-dē′zȧ), bur clover, and sweet clover, while commonly classed as clovers, are not closely related to the above plants, and are not true clovers. Bur clover is

closely related to alfalfa. The true clovers most commonly grown in the South are the crimson, red, and white. Crimson clover is an annual, and therefore has to be seeded every year. It is seeded in the fall and will produce a crop of hay in time for corn to be planted on the land the next summer. About twenty pounds of seed per acre are sown. Red clover is a perennial. It is primarily a hay plant, but is sometimes used for pasture. It grows best on fertile land containing considerable lime. In the South it is best sown in the fall at the rate of ten or twelve pounds of seed per acre. White clover is primarily a pasture plant, and is seldom grown for hay, owing to its prostrate, or creeping, habit of growth. It is usually sown in mixtures of grass seed for pasture at the rate of from two to six pounds of seed per acre. Bur clover is an annual, but reseeds itself readily. It makes its growth in the late fall and early spring, and hence is a good supplement to pasture grass mixtures, giving good grazing at a season of the year when the grasses are dead. Japan clover is an annual, making its growth during the summer months. It is used primarily for pasture, but some hay is produced from it. Japan clover often covers waste land that has been abandoned because of its poverty, greatly aiding in restoring this land to productiveness. Its value is too little appreciated by Southern farmers. Sweet clover is common throughout most of the South and Southwest, but its importance has been generally recognized only during the last few years. There are both annual and biennial varieties. The biennial varieties are of greater importance and wider distribution. Sweet clover is most useful as a pasture or a green manure crop. While best results are obtained on a rich soil with high lime content, it will produce a moderate crop even on poor, worn-out soil, if properly inoculated and if lime is added. Sweet clover makes good pasture in a large part of the Plains section, where it is too dry for other clovers.

217. **Sugar Cane.** Sugar cane was probably the first plant used in the manufacture of sugar and is still one of the

most important crops for this purpose. It is a coarse perennial grass, grown in tropical and semi-tropical countries for its stems, the juices of which are used for the making of sugar and syrup. The growing of sugar cane in the Southwest is a comparatively small farm enterprise.

FIG. 133. Stem of sugar cane, showing the "eyes" at the joints from which the plants grow.

218. **Soils.** The plant makes a large amount of growth and requires a deep, fertile, well-drained alluvial soil. In hilly or rolling country the alluvial bottoms that are well drained generally produce the best yields. The land is thrown up in beds 5 to 7 feet wide with water furrows between them.

219. **Planting.** At planting time a furrow is opened in the center of each bed and a double row of canes is planted in the bottom of each furrow. In this country sugar cane does not produce seed. It is propagated by planting the stripped stalks or nodes cut from stalks. The buds or eyes located at the joints of the cane grow and produce plants. Planting is best done in the fall, although some cane is planted in February or March.

220. **Varieties.** Varieties which are popular in the syrup-making section include: 1. Louisiana Purple or Red Cane; 2. Red Ribbon or Louisiana Striped; 3. Crystalina or White Transparent; 4. Honey Green or Green; 5. Green Ribbon; 6. D-74; and 7. Cuban Red.

221. **Culture.** Frequent shallow cultivations to control weeds and improve growing conditions during the early stages of the crop are important.

222. **Harvesting.** Most farmers try to judge their harvesting of sugar cane by frost dates. The longer the plant grows in the fall the higher the sugar content is developed. There is also a tendency for the leaves to loosen as the season advances. Harvesting consists of three operations: strip-

ping the leaves, cutting off the tops, and cutting the stalks. A special implement has been developed for use in stripping the leaves while a heavy hoe or stout knife is used to cut off the plants. As the stalks are cut they are laid in piles to await the wagon, cart or truck which is to haul them to the mill.

223. **Yield.** Twenty to twenty-five tons of cane per acre

FIG. 134. Cutting sugar cane with a corn knife.

are a fair yield. Often more than this is produced. A ton of cane yields from 150 to 160 pounds of sugar. This gives more than 3000 pounds of sugar per acre. As much as 4500 pounds of sugar per acre has been produced.

RICE

224. **Rice and Its Distribution.** Rice is an annual belonging to the grass family. It is grown for its grain, which is borne in a spreading panicle somewhat resembling that of oats. This grain is one of the most important cereals used for food. It forms the principal article of diet for more than one half of the world's inhabitants. Asia produces more rice than any other continent. Next to Asia comes Europe, followed by North America. The leading rice-

producing states in the United States are Louisiana, Texas, California, and Arkansas, producing a total of from thirty to forty million bushels of rough rice. Of this amount, Louisiana produces about eighteen million bushels and Texas about eight million bushels.

225. **Types and Varieties.** There are two types of rice grown in this country. These are upland rice and lowland rice. The upland rice is grown on relatively dry soils without irrigation. The lowland rice is the more important type. There are few varieties of rice grown in the United States. In the Eastern States white rice and gold seed rice are grown in considerable quantities. In Louisiana and Texas the most important varieties are Honduras, Caloro, Blue Rose, Texas Fortuna, Rexoro, and Japan rice. The Honduras rice produces a rather large grain, and is not so easily blown down because of the stiff straw produced. The Japan rice produces a short thick grain, and the plants do not grow as tall as Honduras rice. It is said to yield more grain than Honduras. Blue Rose is intermediate in size between Honduras and Japanese.

226. **Rice Soils in Texas and Louisiana.** The rapid development of the rice industry in Texas and Louisiana has been due to the opening up of large areas of prairie land in southeast Texas and southwest Louisiana. These rich soils have shown a remarkable adaptation to rice. They have heavy clay subsoil, and for that reason are very retentive of moisture, and, being practically level, are especially adapted to irrigation. They are sufficiently far from the coast to be free from storms and the attacks of birds.

227. **Preparing the Ground.** Rice land is usually plowed in the spring. The better the soil is pulverized the greater is the yield. Deep plowing is more satisfactory than shallow plowing, although rice does best in a rather compact soil. This compact condition can easily be produced by the use of a heavy roller after the land has been plowed. The plow should be followed in a short time by the disc harrow and then by the smoothing harrow.

228. **Sowing.** Rice should usually be sown from March 17 to April 20 for best results. Drilling rather than broadcast sowing is preferred, as a more uniform stand can be attained. Broadcast sowing is still very common, but this

FIG. 135. Panicles of Caloro, Blue Rose, Texas Fortuna, and Rexoro rice, representative of short, medium, long, and long-slender-grain types, respectively (left to right).

Courtesy of the Texas Agricultural Experiment Station.

practice should be discarded. One to two bushels per acre are sown.

229. **Germination.** Very often the seed germinates poorly because of too little moisture in the soil. Some farmers let on enough water to saturate the ground immediately after sowing, drawing off at once any surplus water. A few sprout the seeds before planting by placing bags of rice in

water. However, if the soil is dry when these germinated seeds are sown, failure is sure to follow.

230. **Irrigation.** Rice is best produced on land which can be kept flooded from the time the plants are six to eight inches high until near the time of maturing. Therefore land must be chosen that has some convenient supply of water for irrigation, has a retentive subsoil, and is practically level. In Louisiana and Texas the water used for irrigation is pumped from bayous and rivers, or from underground wells. By means of pumps and a system of canals the water is brought to the highest part of the fields. Low levees, or embankments, are constructed throughout the fields, chiefly with the plow, so that the water can be maintained at a uniform depth through the different portions of the field. This depth should be from three to six inches. The water is applied when the plants are about eight inches high, and a constant circulation of the water is maintained by a continuous inflow at the highest portion of the field and an outflow at the lowest portion. The water should all be drawn off in time for the soil to get firm before harvest time, as this allows the use of improved machinery in harvesting the crop. The irrigation takes the place of cultivation in keeping down the weeds.

231. **Harvesting.** Where the water can be drained off the land, rice is best harvested with the self-binder. It is first put up in shocks in the field and capped in such a way as to shelter the heads from sun and rain. It remains in the shock until the straw is cured and the grain is hard. Threshing is done in the same manner as with other grains.

QUESTIONS, PROBLEMS, AND EXERCISES

72. Bring in a cotton plant and point out the main stem, primary limbs, and fruiting limbs.

73. Examine five cotton bolls, each from a different stalk, and make a record of the number of locks in each boll, number of seeds in each lock, and any points in which the bolls differ.

74. Find three different varieties of cotton in your neighborhood and describe each.

75. Dig carefully around a cotton stalk standing in the field and see what effect would be produced by cultivation two inches, three inches, four inches, and five inches deep.

76. Select the best stalk of cotton in your father's field. Gather the cotton, pick the seeds by hand, and plant these away from all other cotton. Cut out all poor stalks before they bloom, save seeds of the one best stalk again, and pick by hand and plant as

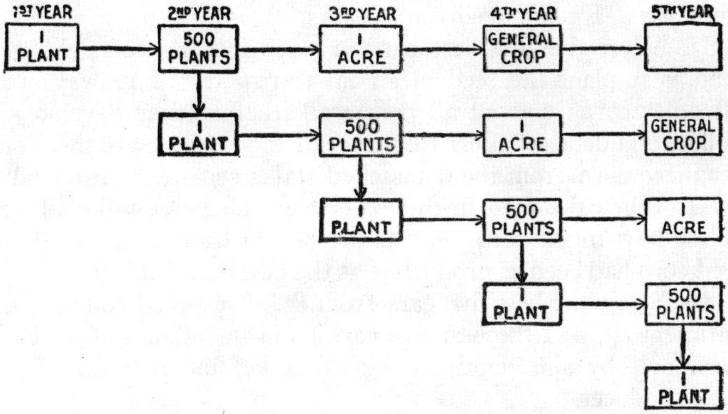

Fig. 135A. Five-year breeding plan for cotton or other crop.
After Webber, "Yearbook U. S. Department of Agriculture, 1902."

before. Use the seeds of the other stalks to plant a large seed patch, and continue this selection for five years in accordance with the system shown in the diagram in Figure 135A.

77. Select six of the best stalks in your father's field, and enough of the poorest stalks to furnish a quantity of cotton when gathered equal to that obtained from the six best stalks. When ready to plant, take an equal quantity of the ordinary gin-run seed planted on the farm and plant side by side, in separate rows, first, the mixed gin-run seeds; second, the seeds from the selected plants; third, the seeds from the very poor plants. Cultivate all alike, and keep a record of the amount produced by each variety of seeds. Then calculate how much cotton would have been produced on the entire farm by planting altogether from each kind of seed.

78. Draw and label the parts of an actual corn plant.

79. Bring in corn shucks that show that shucks are modified leaves.

80. Dig down in the row of a growing cornfield, and make note of how the feeding roots are distributed.

81. Plant corn one, two, three, and four inches deep. After four weeks, dig up the plants and note the character of each, especially the character of the roots, and tell which is the best depth to plant on that soil.

82. Plant rows of corn from grains taken from tips and butts, and parallel to these plant rows from the middle parts of the same ears. Keep a record of results.

83. Select the five best ears in your father's cornfield, and next year plant the seed in an ear-to-row test, far away from all other corn; cut out all poor stalks before they develop any pollen, and detassel alternate halves of each row. Save the cross-fertilized corn from the detasselled stalks separately for seed.

84. When the corn in the experiment above is gathered, calculate how much corn your father would have raised if all his seed corn had been as productive as the best ear of this lot.

85. Select the best five ears from the detasselled stalks in experiment 83, and the best five ears from the other stalks. Plant these side by side, cultivate exactly alike, and note how much each produces.

86. Select fifty ears of your father's seed corn and test for germinating power.

87. Collect and describe as many varieties of kafir and durra as you can find in your community.

88. Make a selection of especially fine stalks of either kafir or durra growing on your farm, and breed up a finer variety by the same methods given for corn and cotton.

89. Find plants of Japan and bur clover. Draw and describe each and bring the plants to school.

90. Select seed from especially fine plants of peas or other legumes and breed an improved variety.

91. Get your father to help you make the following experiment. Plant peas in the rows of half the corn in one field. Also sow rescue grass and bur clover seed, about fifteen pounds per acre, at the last cultivation of this same half. Gather the peas for seed and graze the clover and grass till the spring plowing, when all sod is turned under. Plant cotton, or corn and peas,

again on both halves of this field. Keep account of cost of seed and labor, and of value of all crops raised on each half of the field, and of the value of the grazing. Find out whether the legumes and grass paid, and if so, how much. (A great deal of the value of the legumes and green manure is still in the soil after the first year, and will add to whatever crop is grown on the land for several years.)

92. In a very dry climate, would it be better to plant peas broadcast or to plant them in rows and cultivate? Why do you think so?

93. Plant test rows of six varieties of peas and two varieties of soybeans on your corn land. Give all equally good land and the same cultivation, and see which is best suited to your needs. Try this in a dry season and in a wet season.

CHAPTER VIII

THE GARDEN

232. **Home Gardening Differs from Truck Growing.** Raising a home vegetable garden is quite different from trucking, or raising vegetables in large quantities for the market. Trucking is practicable only where the soil and climate are favorable and the transportation and marketing of the produce are easy. In trucking, large fields and special equipment for tillage and marketing should be used. In the home garden the aim is to furnish the family a supply of fresh wholesome food at all seasons and add an attractive feature to farm living. The surprises and delights in growing the variety of plants found in a garden are many. Only a few farmers can profitably be truck growers, but practically every farmer should have a good home garden. The growing of truck is very important in many states, but in an elementary textbook treatment of this must be omitted and the space given to the more generally needed home garden.

233. **Value of the Garden.** The farmer probably gets a larger return from the time, money, and land devoted to a vegetable garden than from any other expenditure on the farm, provided it is intelligently managed. At the University of Illinois a careful account was kept of a half-acre vegetable garden for five years. During that time the garden produced an average of one hundred and five dollars' worth of vegetables per year at a cost for seeds, labor, and insecticides of thirty dollars per year. At present prices the value of the garden products would be nearly double the Illinois figure. A vegetable garden is valuable not merely

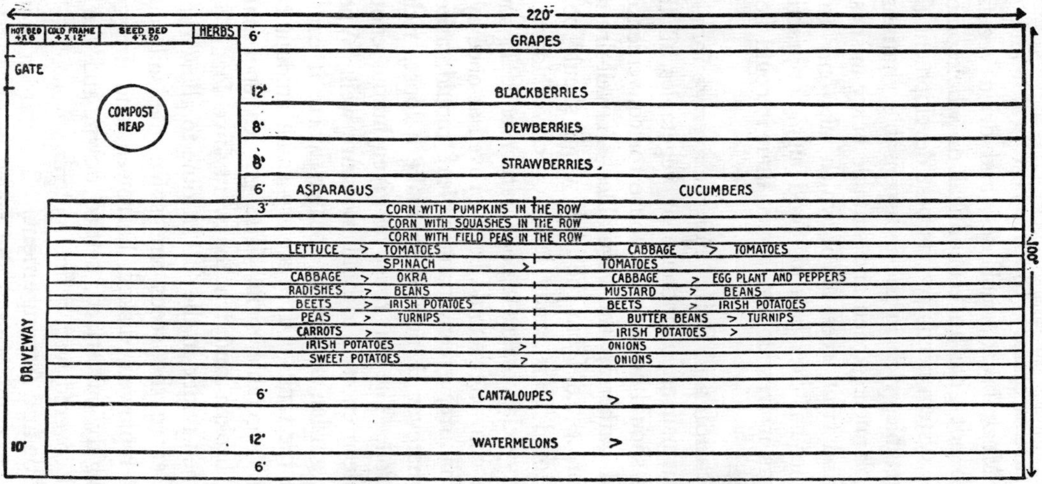

FIG. 136. Plan for a half-acre garden that is easy to cultivate. Lettuce > tomatoes means that tomatoes are planted after the lettuce is harvested.

because it produces foodstuff worth so much money, but also because it furnishes at all seasons of the year the fresh green foods that are necessary for the best health and working efficiency. Meat, bread, molasses, and dried vegetables all the time do not give an economical or wholesome ration. The human system needs for its best development the fresh garden vegetables rich in the indispensable vitamins. With a properly planned garden and suitable berry bushes, grapevines, nut and fruit trees, all of which take only an acre or so, the farmer has over half his food supply at practically no cost, and has it fresher and better than it could be bought at any price.

234. **Location and Soil.** For the average family a half acre will furnish an abundance of vegetables all the year. The garden should be near the house for convenience in caring for and gathering the vegetables. A well-drained spot somewhat protected from the high winds should be chosen. The soil should be a sandy loam or clayey loam. Coarse sand or heavy clay makes a poor garden soil. If such must be used, the character should be improved at once by the addition of manure, green manure, leaf mould, ashes, lime, sand, or whatever is needed to make a loose, rich, finely pulverized soil. The soil must be given humus enough and be broken deep enough to hold moisture well. When practicable, the garden should be located where it can be irrigated from the tank. Often a very small amount of water will save a vegetable crop. The garden spot should be thoroughly broken and ten to twenty-five loads of stable manure turned under in the fall in time to allow for decomposition. Before the seeds are sown the soil should be plowed and replowed, disked, harrowed, and dragged until it is thoroughly pulverized, settled down, and the surface levelled and covered with a fine mulch.

235. **Shape and Arrangement.** The garden should have a wide gate to admit wagon and team, should be oblong, so that the rows may be long, and should be so planted that the tillage can be done largely with teams. The rows should ex-

tend the entire length of the plat, and should not be less than thirty inches apart for the use of the horse cultivator, and fifteen inches for the hand wheel cultivator. Small square patches worked by hand make gardening needlessly burdensome and expensive. The grapevines and berries are usually planted on one side of the vegetable garden, the grapes in rows about twelve feet apart, and the berries in six or eight-foot rows. Blackberries and dewberries should be in almost every garden in the Southwest, and in almost every section some of the numerous varieties of bunch grapes, such as Carman, Black Spanish, Brilliant, and others adapted to the soil and climate. Where no other grape can be planted a few of the wild grapes for making jelly, jam, and grape juice should be placed where they may be gathered easily. Occasionally a few of the vegetables which cannot stand the hot sun may be grown under the arbors.

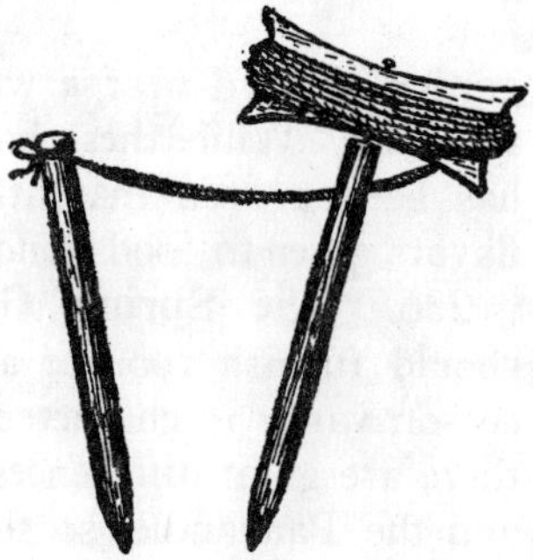

FIG. 137. A homemade garden reel.
Courtesy of the U. S. Department of Agriculture.

Also, certain vegetables that are especially sensitive to heat may be protected by inexpensive frames made of slats or laths that admit only a small amount of the sunshine. Mint and parsley beds should be planted somewhere in the

FIG. 138. A homemade sled marker.
Courtesy of the University of Illinois.

garden or yard near a water supply, as they need frequent watering. While these have no food value in themselves, it has been proved that attractive decoration and appetizing flavors given to foods tend to stimulate appetite and digestion.

236. **The Spring Garden.** A well-managed garden should furnish food at all seasons of the year. The same season varies in character from year to year and, of course, there are great differences in the climates of the Gulf Coast and the Panhandle, so that no statement would fit all sections; but a few general suggestions will help to guide the beginner. It is not possible to give exact planting dates for different sections of the country. The beginner is urged to get that information from those who have the reputation in the community of being good gardeners. The following vegetables should be included in the garden to supply the family during the spring and early summer: (*a*) Cabbage will stand a light frost and thrives best when the weather is cool. The seeds are usually planted in a specially prepared seedbed in time for them to be large enough to transplant to the field three weeks or a month before the time of year when the last cold weather is expected. Handled in this way, they mature before hot weather. Hot weather is objectionable for cabbage in that it causes them to become tough and strong in flavor. Collards, cauliflower, and kale are closely related to cabbage and should be planted and handled in the same manner. (*b*) Spinach, Swiss chard, mustard, and turnips are crops which grow best during the cool part of the year. They will stand a light frost and should be planted as soon as the hard freezing weather is past. These crops are started differently from cabbage in that they are always planted directly in the field. These vegetables will provide tender greens within a month or six weeks after planting. Turnips will produce fleshy roots for use somewhat later in the season. (*c*) Lettuce is likewise a cool season crop and should be planted in the very early spring. In many sections of the South, it may be grown throughout the winter. In most parts of Texas it can be safely planted by the middle of

February. (*d*) Radishes, carrots, and beets are root crops which should be grown in every home garden in the early spring. They are hardy crops and can be planted early. Radishes mature in about a month and quickly become pithy. It is important therefore that only a row or two of this crop be planted at one time, but that other rows be planted at weekly intervals. Beets and carrots come on later but are available for a much longer period. (*e*) Onions, sweet corn, and garden peas are other vegetables which may be planted in the very early spring. Onions may be started from seedlings or from dry sets.

237. **The Summer Garden.** Only a few vegetables grow well in both cool and warm weather. Hence gardeners must plant certain kinds for the cool spring season and other kinds for the hot summer. Of the crops discussed for the spring garden, only Swiss chard, onions, carrots, collards, and perhaps a few cabbage will last very long after hot weather sets in, and the quality of even these will be lowered by the heat. Fortunately, there are others, which thrive during long hot days: (*a*) Tomatoes, eggplant, and peppers are plants of a group known as the vegetable fruits. They are warm season crops and are injured by a light frost. It is customary to start the young plants in a protected seed bed and move them to the field as soon as danger of frost is over. They will continue to grow throughout the summer if there is sufficient rainfall. (*b*) The cantaloupe is likewise a warm season crop and should not be planted until danger of frost is past. A temperature of 40 degrees F. is likely to dwarf and retard the growth of the young plants. The seeds are often planted directly in pots and shifted to the field without disturbing the root system. Another practice is to plant the seed in a piece of sod about four inches square and two inches deep. After the young plants have made considerable growth, the sod and plant are transferred to the field. Cucumbers, squashes, and watermelons may be planted the same as cantaloupes and at the same season. (*c*) Other crops which should be planted with the advent of

warm weather are beans, peas, and okra. These are planted directly in the field and are not transplanted. (*d*) New Zealand spinach is a kind of vegetable which thrives during the long hot days of the summer months if moisture is abundant. It differs from ordinary spinach in this respect, and also in that it is a vine plant. The plants may be started from seed in late spring. It can be propagated from cuttings also. New Zealand spinach has the taste and flavor of spinach. The young growing tips of vines are the edible part, and cuttings may be made from the same plant on successive dates.

238. **The Fall and Winter Garden.** In the North, gardeners plant special crops for storage and use throughout the season when vegetables cannot be grown in the open. This should be done more generally in every section. In the South, sweet potatoes are stored on many farms. There are many other crops which can be kept for winter use, such as pumpkin, cashaw, Fordhook squash, and Irish potato. In some localities, conditions are more favorable for fall than spring gardening. In places where hard freezes seldom occur before Christmas, the following vegetables may be grown in a fall garden: cabbage, broccoli, collards, kale, turnips, mustard, Swiss chard, carrots, beets, and spinach. The time for starting the fall garden varies, but generally it is planted late in August or early in September.

239. **Permanent Garden Crops.** Most vegetables grow only a short while before they are ready for harvest, and usually the plants die or are plowed under shortly afterwards. Some garden crops, of which the following are examples, live and produce from year to year. (*a*) Asparagus. The young shoots of this delightful vegetable are the edible part. They grow from the "crown" of the plant and are cut when they are about 6 inches high. The crowns are produced by planting seed. After one year they are planted at about one or two feet intervals in wide rows. For the first three years young shoots should be allowed to grow and should not be

cut. After this, they may be cut as fast as they appear for a month or six weeks each spring. Asparagus plants live and produce for fifteen years and longer. (*b*) Rhubarb. This is normally a cool season crop and is grown generally in northern home gardens. It is grown also with a measure of success by many gardeners in the South. Certainly rhubarb pie in the early spring justifies almost any amount of trouble in growing it. Once established, a rhubarb bed will last many years. (*c*) Shallots. These are onions that grow from year to year. They multiply so rapidly that they never produce a very large bulb. This makes them ideal for green onions and they will supply the family nearly every month in the year. (*d*) Berries. Blackberries, dewberries, and strawberries are grown so commonly in home gardens that they must be considered in the permanent planting plan. These and the permanent vegetables should be planted in adjacent rows on one side of the garden.

240. **Sources of Seed.** The seed of many vegetable crops must be purchased. Cabbage, for example, produces good heads when grown in the South, but seldom produces good seed. Onion plants produce good seed heads but oftentimes rain causes them to mould and decay before the seed matures properly. Most of our onion seeds are grown in places where it is very dry when the seeds are maturing. The seed of many crops, however, can be saved. This is true of cantaloupes, watermelons, sweet corn, okra, and all kinds of beans and peas. If seeds are saved, it is important to have only one variety growing in a place. This is especially true if the vegetable is likely to be cross-pollinated, as is the case with sweet corn or cantaloupe. Beans, on the other hand, are seldom cross-pollinated and come true even though another kind of bean may be growing nearby. In saving seed, select only the best specimens from the type of plant that you wish. Allow the fruit to ripen fully and then dry the seeds thoroughly before putting them away. They should be stored in a cool, dry place. It is more important that they be dry than

cool, though both are important. It is advisable to guard against damage by mice, rats, and insects while they are in storage.

241. **Cultivation.** Wherever water can be secured for irrigation the crops are of course made more certain and the vegetables more tender. It is useless to plant most vegetables

FIG. 139. An inexpensive wheel hoe for cultivation of the garden.

unless the soil is very fertile and well supplied with moisture. Good tillage and repeated additions of humus and fertilizing material make a good garden possible even in dry sections and without irrigation. At times it is necessary in addition to careful cultivation to cover the soil with a mulch of leaves, chopped straw, hay, or other material that will hold in the moisture.

The methods of planting and cultivating each vegetable are easily learned from the directions on the seed packages and from the bulletins of the State and of the United States Department of Agriculture, and from standard books on gardening. The general principles of plant growth, tillage, and fertilization which you have learned will enable you to apply or to modify intelligently these directions to meet your needs. In planting, the seeds should usually be covered to a depth about equal to three or four times their own thickness. The soil should be pressed down closely upon them either by rolling or tramping, and then loose soil raked over the packed soil to hold the moisture. The soil must be kept

FIG. 140. Combination garden tool.

constantly stirred and no weeds allowed to grow and rob the vegetables of moisture and plant food.

All fence rows and corners should be kept clean even in winter months, to prevent as far as possible the harboring of insects. By planting in long rows after some such plan as is shown in Figure 136, it is possible to do nearly all garden cultivation with the horse cultivators. Where rows are too narrow for this, the type of wheel hoe shown in Figure 139 does excellent work with far less labor than when hand hoes and forks are used.

242. **Garden Tools.** For work in gardens hand-power tools are now made with which the greater part of the work of the garden can be done with from one half to one tenth the labor required with the old-time tools. Such hand-power combination wheel tools are shown in Figure 139 and Figure

FIG. 141. A horse cultivator for garden use.

140. With such tools one can turn light soil, plant, distribute fertilizer, and cultivate with ease several times as much as with ordinary spades, hoes, and rakes. A horse cultivator suitable for garden use is shown in Figure 141.

243. **Hotbed and Cold Frames.** In order to have very early vegetables it is often best to plant during cold weather in the house in boxes, or in a specially prepared bed and frame out of doors. Figure 142 illustrates a convenient form of out-door arrangement, called a *hotbed*. A hole is dug about a foot deep and as large as the hotbed is to be or larger. This is filled with damp horse manure that is beginning to heat. On this the wooden frame is set. About six inches of good garden soil is placed inside the frame and soil is piled up outside all around the base. The decomposition going on in the manure serves to keep the soil warm. Such frames may be of any size. They should be from eighteen to twenty-four inches deep on one side, sloping down to

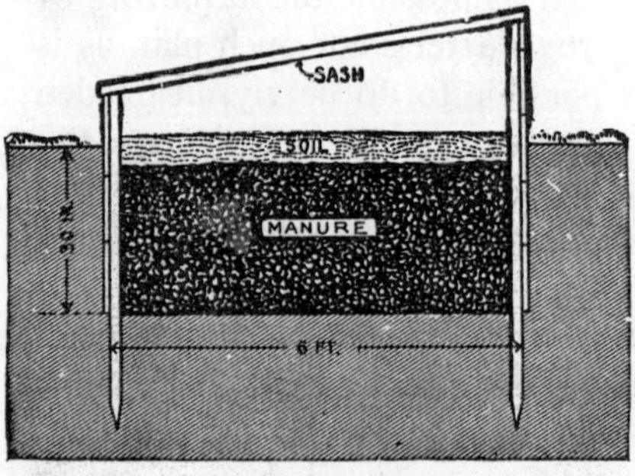

FIG. 142. A hotbed.

FIG. 143. Tomato plants ready for setting in the field. Note the large amount of soil carried with the roots.

twelve or eighteen inches on the opposite side. As they must be covered with sash, it is well to have a shape that some cheap stock size of sash will fit. Three by six and four by eight feet are convenient sizes. They should be narrow enough to enable one to reach across easily. Where manure is not used and no bottom heat is provided, such a frame is called a cold frame. At times these are covered with cloth instead of glass. The hotbed and cold frame offer protection not only against cold but to some extent against insect pests. Plants that are started in such frames are tender and must be gradually hardened by first raising and later taking off the covering on mild days, thus by degrees exposing the plants to the weather. Any one can make a hotbed, and the expense is so small that every family can have one.

FIG. 144. Handy box for use in seeding or when plants are transplanted while in hotbed in order to increase their size.

244. **Transplanting.** You have learned that in transplanting the delicate root hairs are usually torn from the roots as they are taken from the soil, and because of this the plant is unable to make any headway after being transplanted until new root hairs are developed. In many cases

the plant never recovers. This injury may be largely avoided by planting the seeds in small pots or cans and transplanting the plant and soil together. A similar result can be secured by planting in a shallow box like that shown in Figure 144. By fastening one side of this box with screws or nailing it lightly so that it may be easily removed, it is possible to cut around the plant with a trowel and remove it and the soil together, so that the roots and root hairs are undisturbed. Frequently the plants are transplanted once in the hotbed while very small, being reset about four inches apart. When this is done they grow more vigorously, and a larger mass of soil and root can be taken up with each plant when it is carried to the garden. Transplanting is best done on damp days or late in the afternoon. If the soil is not thoroughly moist, water should be poured into the hole and the loose soil drawn in and lightly pressed upon the roots. After this, more loose soil should be drawn around the plant over the wet spot to hold in the moisture. It is important when transplanting to trim carefully all bruised roots and to take off an amount of the top to correspond to the amount of the root lost. Unless the soil and plant are moved together, with the roots left undisturbed, the roots should be carefully spread out before being covered with soil. Figures 145 and 146 show clearly the right and wrong ways to transplant.

FIG. 145. The right and the wrong way to set out plants with a dibber.

245. **Watering Plants.** Plants should not be watered while the sun is shining on them. The water should be put on late in the afternoon or early enough in the morning to soak in well before the sun gets hot. Frequent shallow waterings are not as good as occasional thorough soakings of the soil, because in surface wetting most of the water is lost by

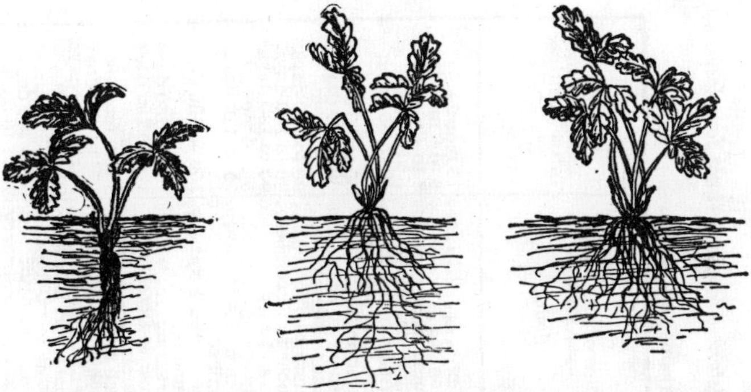

FIG. 146. Transplanting. The roots of the plant on the left will never grow well. The plant in the center is set too high. The one on the right is correctly planted.

evaporation, and the growth of a very shallow root system is encouraged. When water is poured around individual plants or run down a trench, the wet places should be covered with dry soil as soon as the water has soaked in, in order to prevent baking.

246. **Garden Pests.** Insects and diseases are discussed in another chapter, but a few additional practical suggestions here for young gardeners will be helpful. Where there are only a few or especially valuable plants to protect, the cutworm, which destroys so many fine young vegetables, may be kept out by wrapping the stems of plants with a thin soft paper when they are transplanted. The paper should be loosely rolled around the stem and the plant so set in the soil that the paper extends below the surface far enough to hold it in place. The paper soon disintegrates or may be removed after the plant is large enough to protect itself. Another successful plan is to surround the

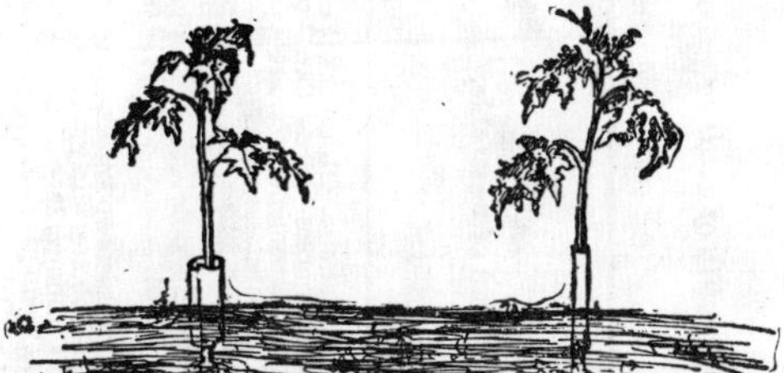

FIG. 147. Paper and tin shields for young plants in a small garden.

GARDENER'S PLANTING TABLE

Quantity of seeds or number of plants required for a row 100 feet in length, with distances to plant, times for planting, and period required for production of crop

Brackets indicate that a late or second crop may be planted the same season.

Kind of vegetable.	Seeds or plants required for 100 feet of row.	Distance for plants to stand— Rows apart. Horse cultivation.	Distance for plants to stand— Rows apart. Hand cultivation.	Distance for plants to stand— Plants apart in rows.	Depth of planting.	Time of planting in open ground. South.	Ready for use after planting.
Artichoke, Globe....	½ ounce........	3 to 4 ft..	2 to 3 ft..	2 to 3 ft..	1 to 2 in...	Spring..............	15 months.
Artichoke, Jerusalem	2 qts. tubers..	3 to 4 ft..	1 to 2 ft..	1 to 2 ft..	2 to 3 in...	Spring..............	6 to 8 months.
Asparagus, seed.....	1 ounce........	30 to 36 in	1 to 2 ft..	3 to 5 in..	1 to 2 in...	Autumn or early spring	3 to 4 years.
Asparagus, plants...	60 to 80 plants	3 to 5 ft..	12 to 24 in	15 to 20 in	3 to 5 in...	Autumn or early spring	1 to 3 years.
Beans, bush........	1 pint........	30 to 36 in	18 to 24 in	5 or 8 to ft	½ to 2 in...	February to April. [August to September.]	40 to 65 days.
Beans, pole.........	½ pint.........	3 to 4 ft..	3 to 4 ft..	3 to 4 ft..	1 to 2 in...	Late spring..........	50 to 80 days.
Beets..............	2 ounces......	24 to 36 in	12 to 18 in	5 or 6 to ft	1 to 2 in...	February to April. [August to September.]	60 to 80 days
Brussels sprouts.....	¼ ounce........	30 to 36 in	24 to 30 in	16 to 24 in	½ in.........	January to July.......	90 to 120 days.
Cabbage, early......	¼ ounce........	30 to 36 in	24 to 30 in	12 to 18 in	½ in.........	October to December..	90 to 130 days.
Cabbage, late.......	¼ ounce........	30 to 40 in	24 to 36 in	16 to 24 in	½ in.........	June and July.........	90 to 130 days.
Cardoon...........	½ ounce........	3 ft......	2 ft......	12 to 18 in	1 to 2 in...	Early spring..........	5 to 6 months.
Carrot............	1 ounce........	30 to 36 in	18 to 24 in	6 or 7 to ft	½ in.........	March and April. [September.]	75 to 110 days.
Cauliflower.........	¼ ounce........	30 to 36 in	24 to 30 in	14 to 18 in	½ in.........	January and February. [June.]	100 to 130 days.
Celeriac...........	¼ ounce........	30 to 36 in	18 to 24 in	4 or 5 to ft	⅛ in.........	Late spring...........	100 to 150 days.
Celery............	¼ ounce........	3 to 6 ft..	18 to 36 in	4 to 8 in..	⅛ in.........	August to October.....	120 to 150 days.
Chervil...........	1 ounce........	30 to 36 in	18 to 24 in	3 or 4 to ft	1 in.........	Autumn.............	1 year.
Chicory...........	¼ ounce........	30 to 36 in	18 to 24 in	4 or 5 to ft	½ in.........	March and April......	5 to 6 months.
Citron............	1 ounce........	8 to 10 ft.	8 to 10 ft.	8 to 10 ft.	1 to 2 in...	March and April......	100 to 130 days.
Collards...........	¼ ounce........	30 to 36 in	24 to 30 in	14 to 18 in	½ in.........	May and June........	100 to 120 days.
Corn salad.........	2 ounces......	30 in.....	12 to 18 in	5 or 6 to ft	½ to 1 in....	January and February. [Sept. and October.]	60 days.
Corn, sweet........	¼ pint.........	36 to 42 in	30 to 36 in	30 to 36 in	1 to 2 in...	February to April.....	60 to 100 days.
Cress, upland.......	½ ounce........	30 in.....	12 to 18 in	4 or 5 to ft	½ to 1 in....	January and February. [Autumn.]	30 to 40 days.
Cress, water........	½ ounce........	Broadcast	....	....	On surface.	Early spring..........	60 to 70 days.

Cucumber	½ ounce	4 to 6 ft.	4 to 6 ft.	4 to 6 ft.	1 to 2 in.	February and March. [September.]	60 to 80 days.
Dandelion	¼ ounce	30 in.	18 to 24 in	8 to 12 in.	½ in.	Early spring or autumn	6 to 12 months.
Egg-plant	⅓ ounce	30 to 36 in	24 to 30 in	18 to 24 in	½ to 1 in.	February to April	100 to 140 days.
Endive	1 ounce	30 in.	18 in.	8 to 12 in.	½ to 1 in.	February to April	90 to 180 days.
Horseradish	70 roots	30 to 40 in	24 to 30 in	14 to 20 in	3 to 4 in.	Early spring	1 to 2 years.
Kale, or borecole	¼ ounce	30 to 36 in	18 to 24 in	18 to 24 in	½ in.	October to February	90 to 120 days.
Kohlrabi	¼ ounce	30 to 36 in	18 to 24 in	4 to 8 in.	½ in.	September to March	60 to 80 days.
Leek	½ ounce	30 to 36 in	14 to 20 in	4 to 8 in.	1 in.	May to September	120 to 180 days.
Lettuce	½ ounce	30 in.	12 to 18 in	4 to 6 in.	½ in.	September to March	60 to 90 days.
Melon, muskmelon	½ ounce	6 to 8 ft.	6 to 8 ft.	Hills 6 ft.	1 to 2 in.	February to April	120 to 150 days.
Melon, watermelon	1 ounce	8 to 12 ft.	8 to 12 ft.	Hills 10 ft.	1 to 2 in.	March to May	100 to 120 days.
Mustard	¼ ounce	30 to 36 in	12 to 18 in	4 or 5 to ft	¼ in.	Autumn or early spring	60 to 90 days.
New Zealand spinach	1 ounce	36 in.	24 to 36 in	12 to 18 in	1 to 2 in.	Early spring	60 to 100 days.
Okra, or gumbo	2 ounces	4 to 5 ft.	3 to 4 ft.	24 to 30 in	1 to 2 in.	February to April	90 to 140 days.
Onion, seed	1 ounce	24 to 36 in	12 to 18 in	4 or 5 to ft	½ to 1 in.	October to March	130 to 150 days.
Onion, sets	1 quart of sets.	24 to 36 in	12 to 18 in	4 or 5 to ft	1 to 2 in.	Early spring	90 to 120 days.
Parsley	¼ ounce	24 to 36 in	12 to 18 in	3 to 6 in.	⅛ in.	September to May	90 to 120 days.
Parsnip	½ ounce	30 to 36 in	18 to 24 in	5 or 6 to ft	½ to 1 in.		125 to 160 days.
Peas	1 to 2 pints	3 to 4 ft.	30 to 36 in	4 or 5 to ft	2 to 3 in.	September to April	40 to 80 days.
Pepper	⅛ ounce	30 to 36 in	18 to 24 in	15 to 18 in	½ in.	Early spring	100 to 140 days.
Physalis	¼ ounce	30 to 36 in	18 to 24 in	18 to 24 in	½ in.	March to May	130 to 160 days.
Potato, Irish	5 lbs. (or 9 bu. per acre).	30 to 36 in	24 to 36 in	14 to 18 in	4 in.	January to April	80 to 140 days.
Potato, sweet	3 lbs. (or 75 slips).	3 to 5 ft.	3 to 5 ft.	14 in.	3 in.	April and May	140 to 160 days.
Pumpkin	½ ounce	8 to 12 ft.	8 to 12 ft.	Hills 8 to 12 ft	1 to 2 in.	April and May	100 to 140 days.
Radish	1 ounce	24 to 36 in	12 to 18 in	8 to 12 to ft.	½ to 1 in.	September to April	20 to 40 days.
Rhubarb, seed	½ ounce	36 in.	30 to 36 in	6 to 8 in.	½ to 1 in.		2 to 4 years.
Rhubarb, plants	33 plants	3 to 5 ft.	3 to 5 ft.	3 ft.	2 to 3 in.		1 to 3 years.
Rutabaga	¼ ounce	30 to 36 in	18 to 24 in	6 to 8 in.	½ to 1 in.	August and September.	60 to 80 days.
Salsify	1 ounce	30 to 36 in	18 to 24 in	2 to 4 in.	½ to 1 in.		120 to 180 days.
Spinach	1 ounce	30 to 36 in	12 to 18 in	7 or 8 to ft	1 to 2 in.	September to February	30 to 60 days.
Squash, bush	½ ounce	3 to 4 ft.	3 to 4 ft.	Hills 3 to 4 ft.	1 to 2 in.	Spring	60 to 80 days.
Squash, late	½ ounce	7 to 10 ft.	7 to 10 ft.	Hills 7 to 9 ft.	1 to 2 in.	Spring	120 to 160 days.
Tomato	⅛ ounce	3 to 5 ft.	3 to 4 ft.	3 ft.	½ to 1 in.	December to March	100 to 140 days.
Turnip	½ ounce	24 to 36 in	18 to 24 in	6 or 7 to ft	¼ to ½ in.	August to October	60 to 80 days.
Vegetable marrow	½ ounce	8 to 12 ft.	8 to 12 ft.	Hills 8 to 9 ft.	1 to 2 in.	Spring	110 to 140 days.

Courtesy of the U. S. Department of Agriculture

young plant with a tin ring which extends about an inch above and an inch below the surface of the soil, and stands about an inch from the plant all around. By melting the tops and bottoms from old tin cans and cutting the cans in circular strips about two inches wide, these rings can be made at no expense except the labor. They will last for years. In order to make it easy to remove the rings from the plants they should be slit, so that they are merely bent around the plant and held in place by the soil. Figure 147 shows such protectors in place.

In addition to protecting the plants from cutworms, all means for destroying these worms should be employed without ceasing. They should be destroyed by hand and by poison, and no remnants of crops should be left ungathered for them to feed on, so that they thrive and multiply. A few worms left will develop into moths that will lay eggs enough to fill the garden with worms again. All weeds and grass should be kept out of the garden, and the fields immediately adjoining it should be cultivated and kept clean. Spasmodic fighting of such pests accomplishes little, but persistent intelligent work brings them under control.

The insects which destroy so many young melon, cucumber, and other similar plants can often be guarded against economically by the use of wire-netting cages, which can be made for about two cents apiece and which last for years. Such a cage is easily made by cutting strips from thirty-six-inch wire netting. Ravel out one or two wires on one side and on one end of a piece of netting six by eighteen inches in size. Bend this strip and make a hoop about six inches in diameter by running the loose ends of the wire at one end of the strip through the meshes at the other end, and bending them back so as to hold the two ends together. Then cut a piece seven inches square and press it down on top of the ravelled end of the hoop, and work the loose ends of the wires of the hoop through the meshes of the top piece, and bend these down so as to hold the top on. Such a cage keeps out insects and shades the young plants slightly. When the plants are tough enough to withstand attacks, the cages are lifted off and put away for future use.

Toads, horned lizards, and birds should be encouraged in the garden, as they are great natural insect destroyers. Toads should be kept in the hotbeds, cold frames, and seedbeds. Young fowls, before they are old enough to do harm by scratching or pecking the vegetables, will destroy great numbers of insects.

The QUESTIONS, PROBLEMS, AND EXERCISES on the garden will be given at the end of the next chapter.

CHAPTER IX

SCHOOL GARDEN

247. **Every School Should Have a School Garden.** Each pupil should have a home garden and put into practice there the things that he learns in school. The work is more interesting and instructive if there is also a school garden. With even as little as a quarter of an acre forty-two pupils may each have an individual plat five by fifteen feet, and still half the land be left for trees, shrubs, flowers, experimental tests, and general observation plats. It is better, however, if the garden for this number covers a half acre, as shown in Figure 148. Patrons and friends gladly contribute the needed manure, fertilizers, and work of teams when these cannot be secured out of the school funds. The work in the garden is a pleasant diversion for both teacher and pupils and in no wise interferes with the other work of the school.

248. **How the School Garden Is Laid Out.** The school garden should have around it attractive borders of flowers set against a background of shrubs. There should also be one plat set aside for the cultivation of flowers. One should be devoted to a nursery for forest and nut trees. In this the seedlings are raised, with which part of the practice in grafting, budding, pruning, and transplanting is obtained. Another plat devoted to fruit trees serves in part the same purpose, and gives opportunity for the study of insects and diseases, and for learning how to care for fruit trees. Students will have opportunity to raise seedlings, bud, graft, and transplant; and can take some of the plants home to help establish planting at their homes. Another plat is devoted

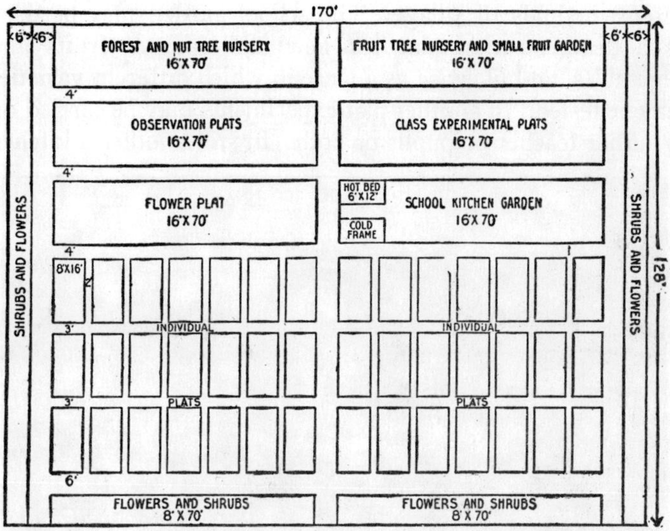

FIG. 148. A good layout for a half-acre school garden where there are only a few pupils.

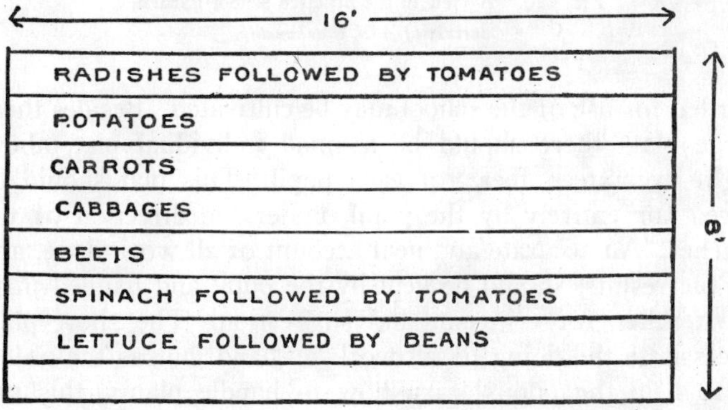

FIG. 148a. A good planting plan for the pupil's individual plat.

to demonstrations for observation by the entire class of such things as the effect of different varieties of seeds and different methods of tillage. The school garden may be used as a demonstration plat for new and unusual fruits and vegetables, and likewise as a place in which different varieties may be tested. In another plat experiments may be carried on by either teacher or pupils or both. In still another a kitchen-

FIG. 149. A view in the practice school garden.
Courtesy of A. S. Blankenship.

garden for use of the school may be cultivated. Besides these large plats there should be a small individual plat, about eight by sixteen feet, for each pupil. This plat should be cared for entirely by the pupil, under the direction of the teacher. An accurate and neat account of all work done, and of the results, should be kept by the pupil and handed in to the teacher for criticism and suggestion. The short plats necessitate the doing of all work by hand, but as the object is to help the pupils learn how to handle plants, this uneconomical plan is justified.

249. **The Inexperienced Teacher May Learn with the Pupils.** The school garden will not only facilitate and encourage the instruction of students, but it will enable the

teacher to become more familiar with new varieties of fruits and vegetables adapted to the location, and new practices as they are introduced from time to time. Every school library should have a set of good reference books on horticulture, such as *Vegetable Growing* by James Edward Knott; *Vegetable Gardening* by Watts; *Vegetable Crops* by Thompson. Bulletins from the United States Department of Agriculture and various state experiment stations and agricultural colleges are free and valuable. Leading farm and horticulture magazines will be found also helpful and inspiring in planning and maintaining the school garden.

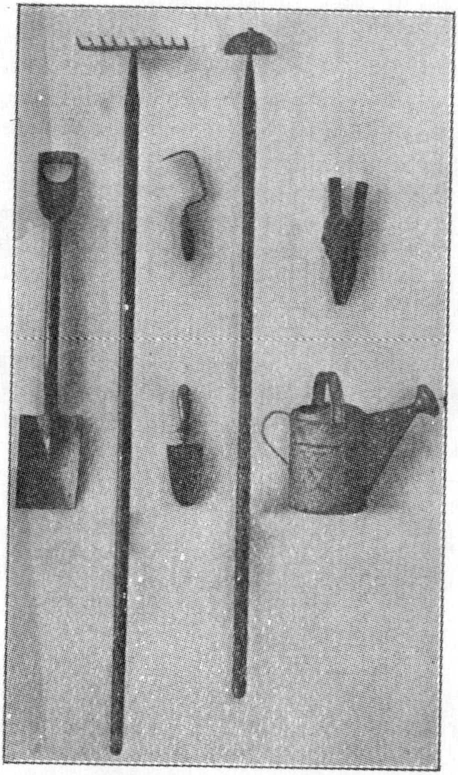

FIG. 150. A good set of tools.
Courtesy of Edward Mahoney.

QUESTIONS, PROBLEMS, AND EXERCISES

94. How many days in the year do you eat green vegetables?

95. Are all of your dried and canned vegetables grown at home?

96. How large is your home garden?

97. What varieties of vegetables are grown in it?

98. Write out a practical plan for improving your home garden.

99. State where you would locate a garden on your place, and give your reasons for choosing this spot.

100. Tell the character of soil and subsoil on the spot chosen in the problem above, and state fully what you would do to bring this particular soil into the right condition.

101. Lay out a garden plan for a quarter-acre garden, similar to that in Figure 148, including especially the vegetables liked by your family. Be careful to have a succession of vegetables covering the year, and the land occupied all the year with some crop.

102. Get your father to allow you to carry this plan out next year. Keep an account of all expenditures and labor, and of the value of the vegetables used and sold. Report this to the school.

103. Watch for plants that are especially able to resist certain insects.

CHAPTER X

FRUITGROWING AND SHADE TREES

250. **The Home Fruit Garden and Commercial Orchards.** With fruits as with vegetables we must first learn about the home fruit garden. The growing of fruits in large quantities for the market or commercial orcharding must be left for later study in advanced courses in *horticulture* (hôr′tĭ-kŭl-tūr), the branch of agriculture which deals with garden and orchard crops. Horticulture comes from the Latin words *hortus,* a garden, and *cultura,* cultivation. Fruitgrowing for the market is a very profitable business in many parts of the Southwest, and as soon as more growers learn the science of horticulture it will be more so. A study of the home fruit garden, or home orchard, will be the best beginning in this subject.

251. **Value of Home-Grown Fruits.** For thousands and thousands of years before man learned to plant field crops and vegetables, or to cook his food, fruit made up a large part of his diet. Sound ripe fruit is still one of the most wholesome and delicious of foods. We need such food in both summer and winter to keep ourselves at the highest point of physical and mental power. At present a large part of the market fruit is picked when hard-ripe or even green, is ripened unnaturally, and is frequently stored for long seasons in great refrigerators. Some fruits, such as pears, persimmons, and bananas, may attain better quality under these conditions, but others, such as citrus fruits and peaches, have better flavor when ripened on the tree. Many farmers at very small expense can produce at home wholesome fruits of

excellent quality, often far better and more wholesome fruit than they can buy; for the tenderest and most delicious varieties of fruit are not usually raised for the market, as, with a few exceptions, they do not keep well nor stand the rough handling in shipping. Let us therefore learn how to grow fruits at home.

252. **Fruits at All Seasons.** It is possible to have a succession of fruits ripening during almost the entire grow-

FIG. 151. A Southwestern Home Demonstration girl and some of her home-canned products.

ing season, and to finish out the year with stored fruits, grape juice, canned and dried fruits, marmalade, jams, and fruit butters, all prepared at home at small expense. In some sections, climatic conditions, such as long summer drouths, may make it impossible to do this without some kind of irrigation. With fruits as with vegetables, no one list will suit all sections. Oranges and lemons which grow on the Gulf coast will freeze further north. Apples that make splendid crops in north Texas do not succeed as well in south Texas. Cherries and gooseberries which are popular in the North cannot stand the hot, dry summer of the Southwest.

Each one must learn by inquiry and experiment just what fruits grow well in his locality. If possible, every home should have some of each of the following: strawberries, raspberries, blackberries, dewberries, plums, apricots, peaches, pears, apples, cherries, persimmons, figs, grapes, and, in the semi-tropical districts, oranges, lemons, and pomegranates. The strawberries give the earliest fruit, followed by the other berries, the plums, apricots, apples, peaches, and pears. There are so many varieties of peaches which ripen at such different seasons, and grow well over such a wide area, that a well-selected orchard will furnish fresh peaches from June until late fall. Apples have even a wider distribution. Every wise farmer should test out new fruits and new varieties occasionally, as only in this way can he learn just what is suited to his soil and climate. However, in most cases, it is best to plant such fruits as neighbors and near-by nurserymen have tested and found suitable.

253. **Where to Locate the Orchard.** The orchard should be located on a hillside, where the drainage is good and where the trees are somewhat protected from the cold winds. If planted in a bottom, the trees are apt to bloom too early and cause the crop to be destroyed by late frosts. Fungus diseases also are more troublesome in valleys. No one soil suits all fruits equally well. All demand good drainage. The plum, quince, and pear do best on a heavy soil, peaches on a rich sandy loam. Some varieties of grapes do well on a heavy soil, some on either a heavy or a light soil, and some only on a light soil. The soil should be rich in nitrogen as this element is the one most likely to be deficient. If there is no soil perfectly suited to a mixed orchard, it is possible to improve it greatly before the trees are planted by adding sand, ashes, humus, or whatever is needed by each fruit to the particular spot on which it is to be planted. Barnyard manure and cover crops, such as peas, should be turned under to build up the soil before the trees are planted.

254. **How to Plant an Orchard.** Orchards are usually planted in regular rows according to the plan shown in

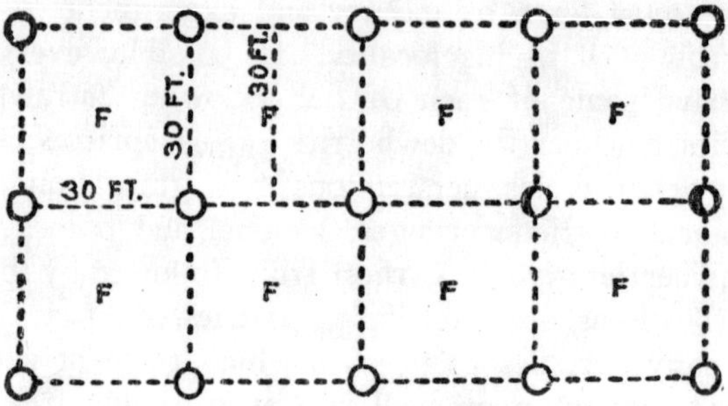

Fig. 152. Planting in squares

Figure 152. The equilateral triangle method shown in Figure 153 gives a more even distribution on the land and enables one to put more trees on the same amount of ground without crowding. The rows should be carefully laid off and trees so planted that the straight rows show plainly from all directions. Peach trees are usually set about sixteen to twenty feet apart, apples from thirty to forty feet, bunch grapes in rows eight, and blackberries in rows six feet apart.

You have already learned about transplanting, and therefore know before being told that fruit trees should be moved when dormant, that part of the top should be cut off, that the roots should be disturbed as little as possible, should never be allowed to dry, should have all bruised and broken

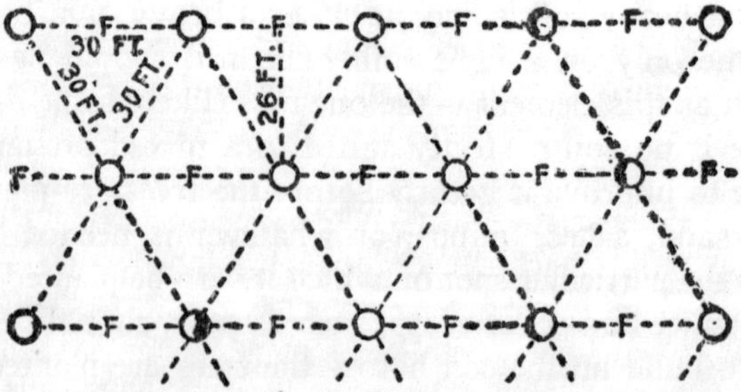

Fig. 153. The equilateral triangle planting plan, which, by giving a better distribution over the land, allows more trees per acre without crowding than does the square planting plan.

parts trimmed off smooth, should be spread out in natural order in the ground, should be set in moist soil, and should have the soil packed closely around them. While it is usually not practicable to move soil and root together, it is well to take as many healthy roots as possible. As soon as dug, the roots of the young tree should be covered with a moist wrapping, carried to the orchard, and never uncovered until the hole is dug and all is ready, so that the roots can at once be covered with soil.

The hole for a tree should be dug large enough to receive the roots in natural order. If the roots are too long to do this economically, they should be cut back somewhat. They should not be doubled up. When the hole is being dug the topsoil should be thrown on one side, as it is usually the best soil, and should be put back into the hole immediately touching the roots. Manure should not be placed in contact with the roots, though it is sometimes advisable in poor land to put some well-rotted manure in the hole away from the roots and in the soil that is used for filling in above them. The trees should be set in the orchard as deep as they grew in the nursery or about two inches deeper. The soil should be tramped well around the roots and loose soil raked over the surface of the packed soil.

255. **How to Handle Bought Trees.** When trees are bought from a nurseryman they should be planted out as soon as received. Each should be taken from the wrapping only after the hole is prepared and when it can be immediately covered with moist earth. Many transplanted trees die because the roots were allowed to lie exposed to the air until they were dried out. In case it is not practicable to plant the trees as soon as received, open up in a well-drained spot a sloping trench deep enough to admit all the roots and a bit of the stems of the young trees. Place the trees in this close together and cover with moist earth, packing it in carefully so that the roots are in close contact with the moist soil. If the soil is not moist, water should be poured into the hole before the final layer of loose dirt is drawn around

the trees. This temporary placing of plants in the soil for protection is called *"heeling in."* If the soil is well drained and is kept moist, heeled-in plants will keep perfectly until a favorable transplanting season.

256. **Pruning.** Fruit trees, bushes, and vines grown under cultivation are pruned for several reasons: first, the young plants, especially trees, need to be shaped properly; second, the bearing plants must be kept in condition to produce good crops of high quality fruit; and third, the old trees, bushes and vines must be pruned to remove some of the old wood and make new wood.

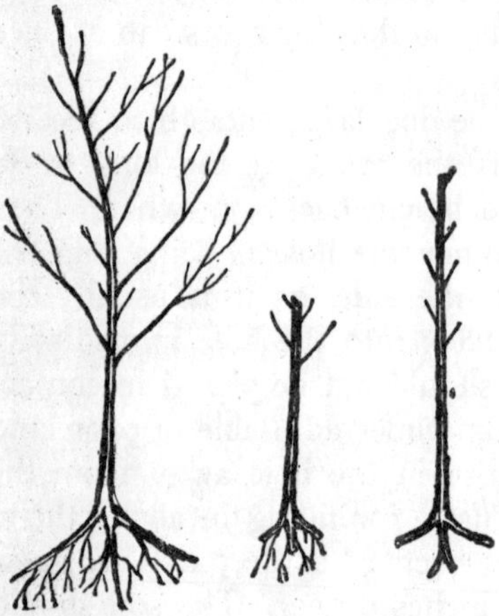

Fig. 154. Pruning nursery trees. On the right the tree is improperly pruned, not enough being taken off. The one in the center is correctly pruned.—*After Halligan.*

Fruit trees, when planted, should be cut back only moderately, so as to leave a main trunk about thirty inches tall. If the tree is small, this trunk will have to be formed the second year. On this trunk, three to five side branches are chosen, on different sides, and at different levels. These are selected as soon as growth begins in the spring, and all other shoots are rubbed off. If you will follow up every two weeks through spring and summer, and rub off the shoots that you do not need, you will have a fine tree by fall, and you will not need to prune it much in the winter. By following the same method for two seasons, and not cutting the tree heavily, you can have peaches and plums bearing in the third season.

Bearing trees are pruned differently, according to the ways in which they grow and bear fruit, but the principal object is to make orchard operations easier, and to cause the tree to produce regular crops of good fruit. Pruning does not increase yields, as many people believe, but when it is done carefully, it does cause the tree to produce a better grade of fruit, and therefore a better money return. The general practices of pruning should be to take out awkwardly shaped limbs, thin out the lateral branches so that sunlight can get in to the fruit, cut back the long branches so that they will not break with fruit, promote the growth of fruiting branches, and so direct the growth that the tree will be well proportioned and symmetrical. Trees should be pruned when dormant, though at times additional summer pruning is advisable.

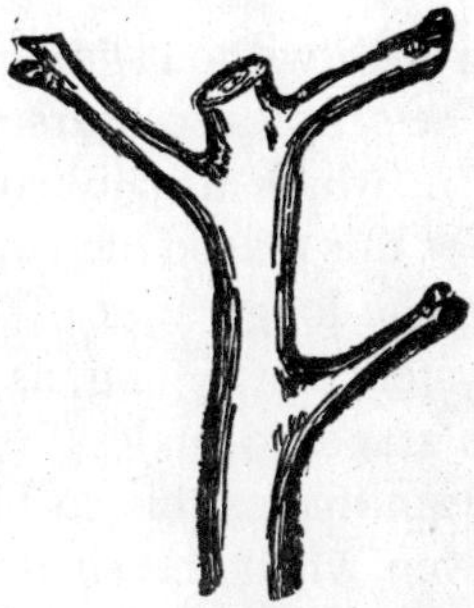

FIG. 155. Pruned to direct growth. The growth will be in the direction taken by the topmost bud left when the branch is cut off, as this bud grows most rapidly. This being true, in what direction will each limb in the cut above grow?

The apple and the pear bear their fruits upon short branches of the previous year's growth, called *fruiting spurs,* which grow out from limbs that are one year or more old. The bearing shoots are not usually the long ones near the ends of the branches. In pruning care must be taken, therefore, not to cut off too many of the short fruiting spurs. The peach bears on wood of the last season's growth, but directly on the branches instead of on spurs. With the peach, cutting back the long branches is necessary in order to limit the crop and prevent the tree's breaking. The Japanese plum fruits on both spurs and year-old wood, and may well be cut back similarly to the peach. The quince bears its fruit at the end of new shoots of the present season's growth, so that the pruning must be such as will stimulate new

FIG. 156. The right way to cut off the old stem after the new budded branch has got started.

growth without going so far as to limit too greatly the crop. The grape also bears fruit upon the shoots of the present season, which usually come out from canes of the past season. For this reason the vines should be cut back severely each year, as the long canes of old wood bear no fruit. The Munson system of training, as illustrated in Figure 157, is recognized as standard, unless it is desired to make an arbor. The diagram makes this so plain that explanations are unnecessary. Even when the shade of an arbor is desired, better results

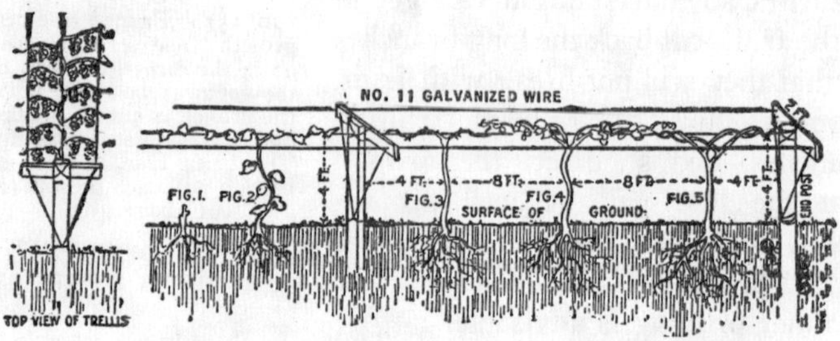

FIG. 157. The Munson system of grape culture.
From "Foundations of American Grape Culture."

will be secured if the vines are planted close together, and the lateral branches trained out in regular order, and the canes cut back each year in a manner similar to that illustrated in Figure 157. Under this system the new growth will soon cover over an entire arbor each year, if the old canes have been properly trained and trimmed. In pruning the grape, care must be taken not to cut the vines just as the sap is beginning to flow in the spring. They may be safely cut later in the season, but the proper time for pruning is when the vine is dormant. Blackberries and raspberries also need severe pruning, as they bear their fruit on short shoots growing out from canes of the previous season's growth. Strawberries bear best the first year, and after two years should be taken out and room given to young plants.

257. **Cultivation.** Fruit trees need cultivation for the same reasons that other plants need it. If weeds are left to

absorb the food materials and water, and the soil allowed to crust over and crack open and the water to evaporate, a rapid growth of the trees cannot be expected. When trees are young and the roots short, it does no harm to plant vegetables or other shallow growing crops among them, but after the trees have been planted a few years, the soil should be

FIG. 158. A young vine that shows how grapes flourish in the Southwest.

cultivated clean during the growing season of the trees, so that no weeds, grass, or crops will use up the water and food materials needed by the trees. After the middle of the summer a fall crop of clover, oats, bur clover, or other winter cover crop should be planted, as this protects the soil from winter washing and leaching and supplies vegetable matter to turn under in the spring. If a legume is planted in this way, or in the middles earlier in the season, the amount of manure or fertilizer that should be added is greatly lessened. Trees cannot be expected to bear heavy crops each year unless food materials are supplied. It is especially necessary that an abundant supply be given when a large crop is being

FIG. 159. Clean cultivation.

FIG. 160. Peas in the middles.

borne, as the tree must during this season lay aside enough reserve food to mature all its fruit, and also enough to start the new crop the next year, and sustain the tree until its new leaves are developed. It is usually unwise to allow the trees to bear very heavy crops. Peaches especially should be thinned, so that they are about five inches apart on the stem. This increases the size of the fruit and, by lessening the drain on the tree, makes it more likely that a crop will be produced the next year.

258. **Protection from Cold.** In our changeable climate the loss of an entire fruit crop from early blooming and a

FIG. 161. Pecans interplanted with plums, and with potatoes planted in the middles.

late frost has been a serious drawback to the growing of fruit. Often the entire crop could be saved by protecting the orchard one night. It has been found possible to do this economically, even in our windy country, by the use of orchard heaters. These fires are usually made by burning crude oil in the orchard in vessels which hold about ten gallons of oil, and are so constructed as to keep a fire burning for many hours on one filling. The best heaters are those with tall stacks, which get red hot and radiate much heat. These have been known to raise the temperature of an orchard six to eight degrees. Where there is much wind the change in temperature is not so great. Another protection for the orchard is a row of evergreen trees on the side from which the cold winds usually come. Now that the use of these

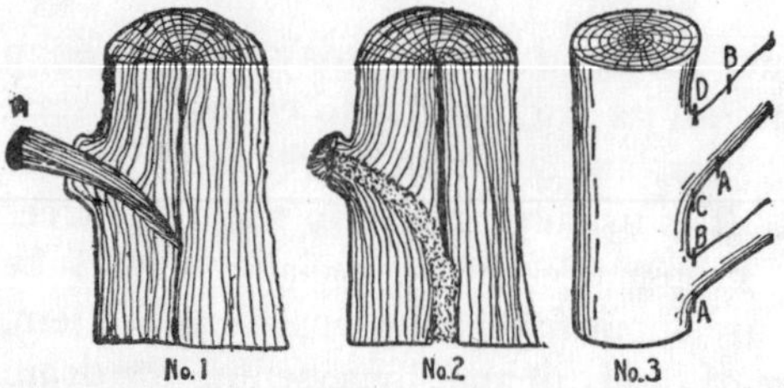

FIG. 162. No. 1. Limb cut off too far from the tree and cannot heal. No. 2. The same limb with the heart decayed and the decay carried into the heart of the tree (after Davis). No. 3. On the lower branch the right point at which to cut off a limb is shown. A cut should first be made one third through on the under side of limb at *A*, in order to prevent splitting the tree. Then saw through from above at *B*. The upper branch illustrates the best method to follow when a very large limb is to be cut off and the danger of splitting is very great. Saw first at *A*, and then saw above a little further out on the limb until it breaks off. Then the limb may be easily cut off properly at *C D*.

protections is understood, it is possible to save the fruit crop practically every year.

259. **Nut Trees.** Every country home should have, if possible, a few nut trees. The native walnuts, hickories, and pecans grow in almost all sections when properly cared for. It seems probable that the budded and grafted Persian walnut, or "English walnut" as it is usually called, will also grow in many sections. Certainly the fine thin-shelled pecans, the most delicious of all nuts, grow to perfection over a very large part of the Southwest. These trees grow and bear on the plains and on high hills, but do best along the river bottoms. For many years it was thought that pecans would not bear till ten or twelve years old, but varieties are now found and propagated which bear within two years of the

FIG. 163. A native pecan tree cut back and grafted and budded with a fine "paper-shell" variety.
Courtesy of the North Texas Agricultural College.

time they are budded. Occasionally nuts are borne the first year after the tree is budded. With these early fruiting varieties, such as Halbert and Texas Prolific, which are most delicious "paper-shell" varieties, it is now possible to have a young pecan orchard bearing fair crops almost as soon as a peach orchard, if one cultivates and fertilizes properly the young trees. Wherever there is a native pecan tree that is not giving a good annual crop of nuts, it should be cut back and grafted, or budded with a standard variety in the manner explained in Chapter III. When budded on large trees, the new buds grow much more rapidly than when set on nursery stocks, and hence a large fruiting is secured earlier by budding on the old trees than by setting out a young orchard.

260. **Shade Trees.** The comfort and beauty of shade trees are so much enjoyed by all that it is surprising to see so many homes and towns without shade. The fact that it takes so long for a tree to grow large enough to give shade is undoubtedly the principal cause of this neglect. Let us remember the joy and comfort given us by the trees planted by those who went before us, and prepare for our descendants and for our own middle life and old age by planting the splendid, long-lived trees, such as oak, elm, hickory, and pecan. Even in ten or fifteen years these trees will give considerable shade. It is best when planting these slow-growing, long-lived trees to plant in between them the quick-growing, short-lived ones, such as the umbrella chinaberry, the cottonwood, and sycamore. The hackberry is a tree of fairly rapid growth that makes a fine shade, will grow in almost any soil, stands drought well, and is comparatively free from insect attack. An objection to it is that it is very difficult to grow grass or flowers under it or near its roots. The proper methods of planting and caring for trees have already been given, and the methods of protecting fruit, nut, and shade trees from the ravages of insects and diseases will be given in the next chapter.

Warning should be given against the bad habit so often practised of sawing off short the stems of large trees, six inches or more in diameter, when transplanting, and of severely cutting back large shade trees every few years. When a tree is transplanted, the cambium layer is rarely able to heal over the wound if the stem is cut off at a point at which the diameter is four or more inches. Decay will therefore soon enter the tree. The transplanted tree, when large, should

FIG. 164. The long-lived, stately live oak is the most valuable shade tree for the Southwest.

be cut off higher up where the diameter is not over three or four inches, and the lateral branches thus left on the stem should be cut back, leaving stubs one or more feet long. Care, however, must be taken not to allow more leaves to grow the first year than the crippled roots can supply with water. After the first year or so, unsightly branches should be cut out, the growth balanced and directed by pruning, and long limbs that are in danger of being broken by the wind cut back, but wholesale severe cutting spoils the natural gracefulness of the branches and retards the growth.

261. **Filling Decayed Trees.** Through unwise pruning or through some mishap, many fine trees get decay in the

heart wood. Unless arrested this will soon eat out all the heart wood, so that the tree will break in the first severe wind. Such decay may be arrested, and the life of the tree indefinitely prolonged, by proper treatment. All of the de-

A B C

FIG. 165. Tree surgery. At *A* is shown a decayed tree, with all decayed wood properly cut away and the cavity painted with creosote and asphalt. The same tree is shown in *B* properly filled with cement, with joints of asphalted paper to permit slight bending. At *C* is shown a similar tree photographed a year after being treated and showing the growth of healthy new wood beginning to cover the cement.

Courtesy of J. Y. Chadick, tree surgeon.

cayed material should be carefully cut away and cleaned out, the entire cavity painted thoroughly with an *antiseptic*[1] mixture of creosote oil and warm asphalt, and then filled with concrete. The concrete is usually made with one part cement and two parts sand, or with equal amounts of each. The cavity is completely filled up to the edge of the growing

[1]An antiseptic (ăn-tĭ-sĕp′tĭk) is something that destroys the germs which produce disease.

wood, as shown in Figure 165. If the cavity is large, the concrete is put in in joints, with heavy asphalt roofing paper in between the joints, as is shown in the picture. This makes possible slight bending in the wind, and thus prevents the concrete from being so easily broken. When the cement is set, the edges are carefully covered over with warm asphalt, to make sure of filling all cracks. If properly done, this will prevent further decay, and if the opening is not too large, it will be covered slowly by new tissue thrown out by the cambium layer.

QUESTIONS, PROBLEMS, AND EXERCISES

104. How much land is devoted to fruit on your farm? Give the number of trees or vines of each kind.

105. Select a spot for an orchard on your place and state why you select this spot, describing soil, subsoil, drainage, and protection.

106. Make out a planting plan for a home orchard for your family, giving a diagram of the proposed orchard and the location of each tree, bush, and vine, with names of varieties. State why each variety is selected.

107. State what you would do to the soil around each of the kinds of trees or vines that you plant.

108. Working in pairs, let each two pupils set out under direction of the teacher either in the school orchard or at home at least two kinds of fruit trees and vines, getting actual experience in root and top pruning, and in correct planting.

109. Heel in correctly some young trees.

110. By use of the school orchard and of neighboring orchards, let each pupil, under the direction of the teacher, practise in pruning: (1) to direct growth, (2) to prevent breaking, (3) to regulate fruiting, and (4) to improve the appearance.

111. Prune and train two grapevines according to the Munson system, and leave two equally vigorous vines of the same variety to run freely. Make an accurate measure of the fruit produced by each of these for three years.

112. Give three reasons why it is better to have a tree bear 100 peaches that fill a bushel measure rather than 200 peaches that only fill the same measure.

113. Make careful records each year of the number of hours it would be necessary to protect the orchards from cold in your locality.

114. How many nut trees are there on your place? If there are any, cut off and top graft and top bud some of these. If there are none, plant nuts, and bud or graft the seedlings with fine varieties. Buds can be bought usually for a cent or two each from neighboring nurserymen.

115. Let each class plant one or more shade trees on the school grounds, planting some slow-growing trees, such as the live oak, walnut, or pecan, and some rapid-growing ones.

116. Find edible wild fruits in your locality, pick out especially desirable specimens, transplant these, and see what improvement can be made in them by cultivation and by variation and selection.

CHAPTER XI

PLANT ENEMIES

262. **Varieties of Plant Enemies.** The enemies of the farm, garden, and orchard are usually grouped into five classes: (1) weeds, which injure crops by depriving them of light, water, and food materials; (2) animals and birds (while most birds are very helpful to crops through destroying harmful insects, a few do considerable damage); (3) parasitic plants such as mistletoe; (4) insects; (5) diseases. In this course we shall consider only the insects and diseases, leaving the others for later study.

Fig. 166. An inexpensive cage in which to keep insects for study.

263. **Losses from Insects and Diseases.** Insect damage and losses are factors which the farmer must consider in the economical production of crops of all kinds. Independent of any direct field damage to crops, insects may attack stock or carry diseases of livestock on the farms, or may become carriers of plant diseases.

Insects as a group are more beneficial than they are destructive. They are necessary in the pollenation of many kinds of plants. Some kinds of fruits will produce a much greater amount if a sufficient number of insects are present to pollenate all of the flowers. Fruit growers, for instance, have learned that a few hives of honey bees in an orchard may often increase production, and accordingly rent colonies of bees which are placed in their orchards while the fruit is in bloom.

Crop losses due to insect damage each year in the United States is estimated to be more than one billion five hundred million dollars. The boll weevil alone, for instance, is known to cause an annual loss of about eighteen million dollars. If to this were added the damages from the bollworm, cotton-

FIG. 167. This picture shows cotton and corn growing in soil infected with cotton root rot. Note that the corn is immune, but the cotton is dying.

flea hopper and cotton-leaf worm the annual losses would run up to many more millions of dollars. Other insects entail just as much damage. The insects which affect the grain, cereal, vegetable and fruit crops take off annually many more millions from the profits of the farmers.

264. **Losses from Plant Diseases.** Plant diseases, like insect pests, are annually responsible for heavy monetary losses, and for making crop production less profitable. The organisms which cause plant diseases may be introduced into new territories, in a number of ways, but mainly by infected seeds; infected plants; infected soil particles carried by farm implements, farm animals, or man himself; or

by wind, blowing rains, or by soil erosion. Plant diseases may also be carried about from place to place by various insects, an example of which is cotton wilt which may be spread about by insects that have fed upon infected plants. Many of the virus diseases of plants are also spread about by insects. Certain plant diseases are able to attack a large number of crops. For example, cotton-root rot affects over one thousand different species of plants. In other cases, a disease may attack only one or a very limited number of other crops. Examples of this are cotton wilt which attacks only cotton and okra, and watermelon wilt which attacks only this crop. Taking advantage of these facts, farmers are frequently able to control a certain plant disease by rotating with non-susceptible crops.

FIG. 168. This shows the root of a cotton plant in an advanced stage of the cotton-root rot disease.

The annual losses from plant diseases are staggering. It has been estimated that during 1933 the losses to the United States from wheat diseases were about twenty-seven million bushels, corn diseases four million bushels, Irish potato diseases forty million bushels, tomato diseases thirty-one million bushels, peach diseases four million bushels, and cotton diseases two million bales. In Texas alone the losses from the cotton-root rot disease are estimated at fifty to one hundred million dollars per year. The annual losses in Texas from plant diseases may be conservatively estimated at 15 per cent of all the crops grown in the state. If half of the losses from plant diseases in Texas could be prevented the increased income annually would be sufficient to support the state government and handsomely take care of all the educational and charitable institutions.

Fig. 169. A tomato leaf infected with blight, or downy mildew, caused by a fungus.

265. **Why Insect Pests Have Increased.** There are many reasons why insects injurious to cultivated plants have increased in recent years. For these same reasons they will increase still more in the future if proper precautions are not taken. In the first place the wild trees and plants have been cut down and the land put under cultivation, so that the insects which formerly fed on wild plants must now feed on cul-

Fig. 170. An onion infected with anthracnose or smudge, caused by a fungus.

tivated crops. These crops are grown with more certainty and regularity than the wild plants were, and hence support the insects better. Then, too, the cultivated places, which used to be more or less separated from one another, are coming more and more to be contiguous, so that the pests can pass directly from one field to another. Again, as more kinds of new plants are cultivated, the varieties of insects that attack these are multiplied and brought to our attention. Probably most effective of all in scattering these plant enemies have been the improved means of transportation. Both insects and diseases are shipped into new districts along with foodstuffs, seeds, or nursery plants. It is for this reason that the transportation of seed, nursery stock, or other material likely to spread disease or insects should be strictly regulated by law.

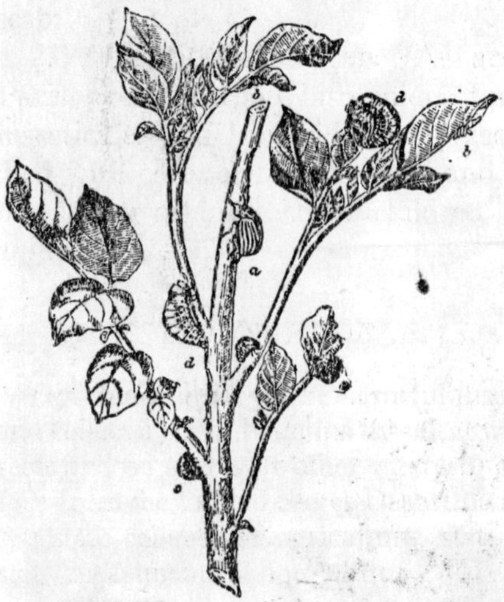

FIG. 171. The Colorado potato beetle: *a*, beetle; *b*, masses of eggs; *c*, half-grown larvæ; *d*, mature larvæ.

Courtesy of U. S. Department of Agriculture.

266. **The Spread of Black Rot, Boll Weevils, and Colorado Beetles.** There are many remarkable examples of the spread of plant diseases. One of the most notable is the spread of the black rot of grapes. When the early settlers came to America they found the wild grapes here afflicted with this disease, which was then unknown in Europe. They sent some of these native vines back to Europe, with the result that this disease soon broke out in

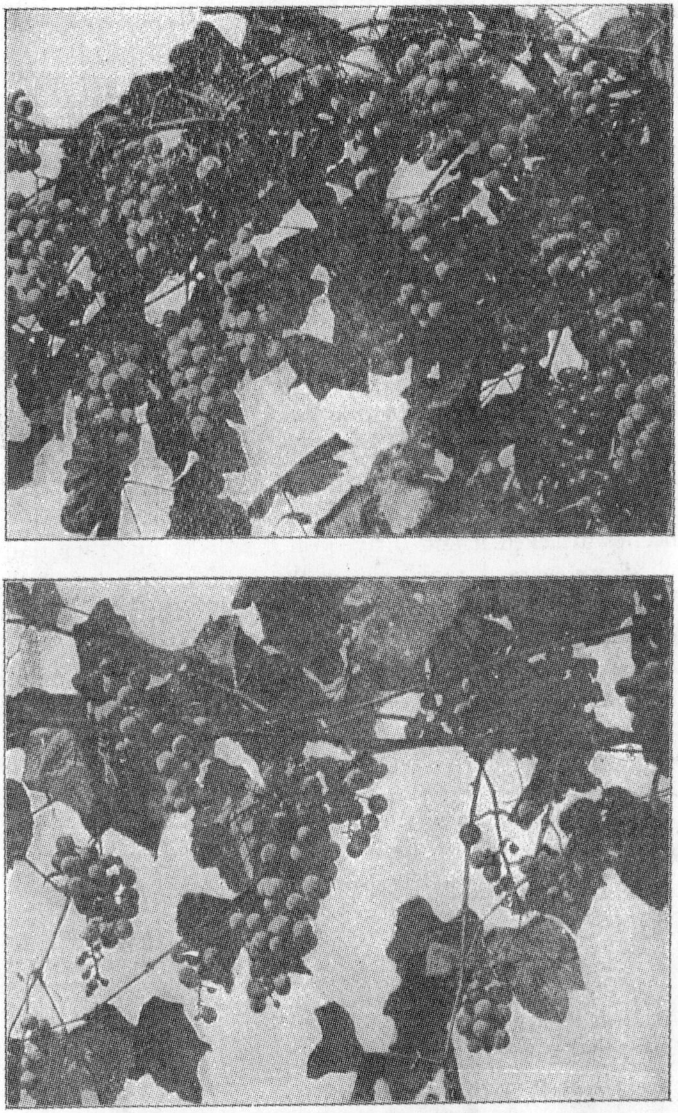

FIG. 172. Black rot on grapes: above, sprayed; below, unsprayed.
Courtesy of U. S. Department of Agriculture.

FIG. 173. The cotton boll weevil: *A*, as seen from above; *B*, as viewed from the side; *C*, larva; *D*, pupa. All about two and one-half times the natural size.

Courtesy of U. S. Department of Agriculture.

the European vineyards. Ever since that time this disease has caused great losses in the vineyards there, which must even yet be carefully sprayed to prevent very serious damage to the crop.

We have in America two recent instances of the rapid spread of a new insect pest. The ordinary potato beetle, commonly called the "potato bug," first appeared in the potato fields of Colorado about 1855. It had been living in that state on wild plants related to the potato, and when the cultivated potato was brought to Colorado by settlers the beetles attacked it and thrived on it so well that they multiplied and spread rapidly over the country. By 1864 they had extended to the Mississippi, and in 1874 reached the Atlantic States. The Mexican cotton boll weevil crossed the Rio Grande about 1892. It had infested the cotton of Mexico for years, and in some districts had forced the abandon-

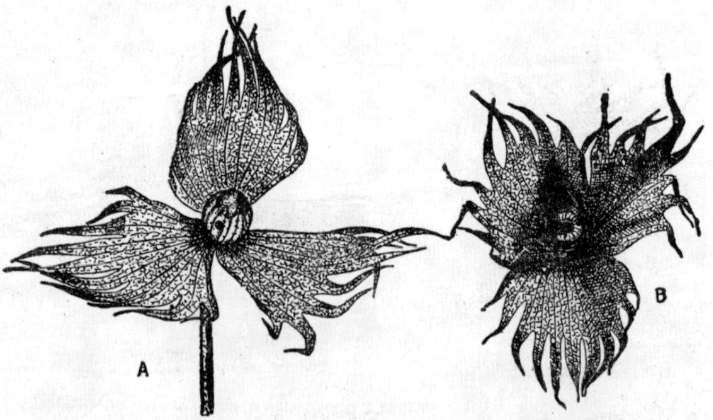

FIG. 174. *A*, square punctured by boll weevil, showing the flaring back of the bracts; *B*, the weevil maturing within the boll.

Courtesy of U. S. Department of Agriculture.

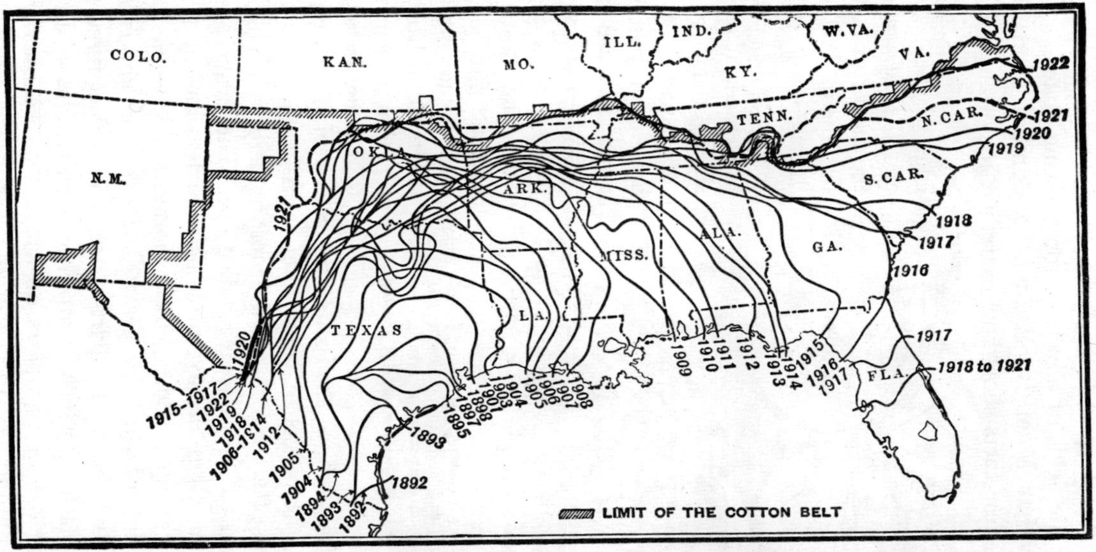

FIG. 175. Chart showing the spread of the Mexican cotton boll weevil.
Courtesy of U. S. Department of Agriculture.

ment of cotton cultivation altogether. In less than twenty years this pest spread nearly all over Texas, and is now ravaging the fields of all the cotton-growing states.

267. **What Must Be Known to Combat Plant Enemies.** These facts show how extremely important it is that we obtain a knowledge of these insects and diseases, and of the means of controlling them. While the damages of only a few can be entirely prevented, it is possible to reduce greatly the damage of all, and to prevent their rapid increase and spread. Let us then first see what insects are, and how they live and multiply, for it is by knowing their habits and life history that we learn how to destroy or prevent them. After this we shall study the causes of plant diseases and learn the means of controlling them.

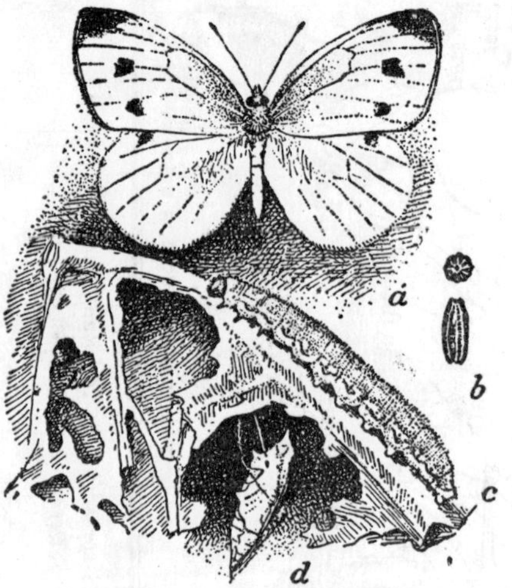

FIG. 176. The cabbage worm: *a*, female butterfly; *b*, egg, end and side views; *c*, larva on leaf; *d*, suspended chrysalis.

Courtesy of U. S. Department of Agriculture.

268. **Insects.** Insects are the most numerous of all forms of animal life visible to the naked eye. They vary greatly in appearance, but all have three pairs of legs and three distinct parts to their bodies, *head, thorax,* and *abdomen.* To the head are attached the *feelers,* or *antennæ* (ăn-tĕn′nē), the eyes, and the mouth parts. The thorax has three segments, to each of which is attached a pair of legs. In the adult stage one or two pairs of wings are also usually attached to the thorax. In nearly all cases insects hatch from eggs, and pass through several different forms before reach-

ing their final shape. The typical insect passes through four stages, the *egg,* the *larva* (lär′vä), the *pupa* (pū′pä), the *adult,* or *imago* (ĭ-mā′go). The larva may be entirely unlike the adult into which it will develop, as in the case of the caterpillar, which is the larva of a butterfly or a moth. During the larval stage the wormlike creature usually eats voraciously, does its great damage to crops, and grows rapidly until the skin hardens and refuses to grow further. Then it goes into a dormant-looking state and is called a pupa. The larva may spin a web case around itself, in which it lives as a pupa, or it may go into a cell in the ground or attach itself to a plant. While in its case the pupa [or *chrysalis* (krĭs′ȧ-lis) as it is called in the case of the butterfly] undergoes a wonderful change, and in due season comes out in the new form of the full-grown insect; as, for example, the ugly larval caterpillar pupates and comes out a beautiful butterfly or moth, and the cutworm becomes a moth. While the four stages—egg, larva, pupa, adult—are the usual stages, many insects omit one or two of these. For example, grasshoppers and several other insects are quite like the adult when hatched and have no pupa stage at all. These when young are called *nymphs* instead of larvæ.

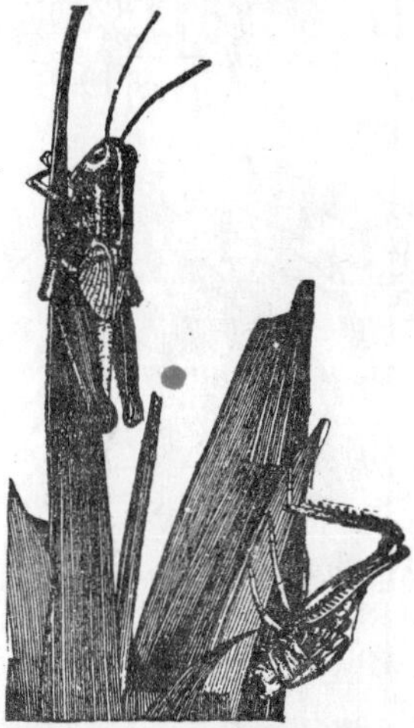

FIG. 177. Above, nymph of grasshopper in natural position; below, the empty pupa skin. *Courtesy of U. S. Department of Agriculture.*

Many kinds of insects after passing through their various stages of growth and becoming adults live only long enough to deposit eggs, not even living to see their own young hatched. Others may live through a season, producing several sets of offspring. The time required by insects to pass through all their various stages, or *life cycle* (sī′kl) as it is called, varies from a few days or few weeks in most cases to many years in a few cases. Insects live from one season to the next often only in the form of eggs or pupæ. In other cases the adults that may be still active at the approach of cold weather hide away under leaves or grass or trash or bark or in basements, or burrow into the ground and remain quiet until spring —occasionally even being frozen without causing death. This spending of the winter in an inactive state is called *hibernation* (hī-bĕr-nā′shŭn). In spring those hibernating insects that have not died become active again, lay eggs, and start a new generation.

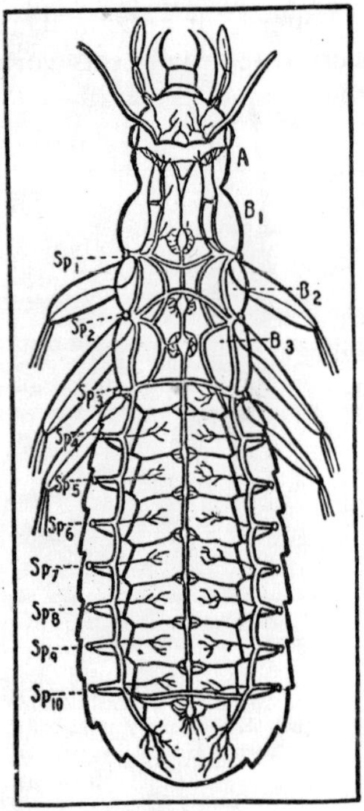

FIG. 178. A typical insect. *A*, head; *B*1 to *B*3, three segments of thorax, each with a pair of legs; *Sp*1 to *Sp*10, spiracles. —*After Kalbe.*

269. **How Insects Feed and Breathe.** Some insects feed by biting and chewing their food, as do the potato beetle, grasshopper, and cotton-leaf worm. Others, such as plant lice and San José scale, puncture the plant and feed by sucking out the juices. Insects do not breathe through a mouth or nose as we do, but through a number of delicate little slits called *spiracles* (spĭr′ȧ-klz) distributed along the sides of the thorax and abdomen. It is by a knowledge of the various stages through which insects pass, and of their eating and breathing characteristics, that successful methods of destroying them are devised.

270. **How Insects Are Destroyed.** If an insect gets its food by biting and chewing the plants, it is easy to destroy

it by putting poison on the part attacked, as we put arsenate of lead on the potato plants to destroy the potato-beetle larvæ which eat the leaves. If the insect gets its food by puncturing the plant and sucking out the juice, then such poisons will not reach it. The sucking insects must be killed by contact poisons or by sprays that get into their breathing pores or spiracles and choke them to death, such as the oil emulsions, and lime and sulphur sprays, with which plant lice and San José scale are destroyed. The formulas for making and directions for using the most important of these insect-destroying mixtures will be given later. Each insect usually has a very limited number of plants that it will eat. The earworm of corn will feed also on cotton, tomatoes, and tobacco; but many insects will feed on only one class of plants. By refraining for a year or two from planting the particular crops on which they feed, one may often starve to death insects which have a limited range of foods.

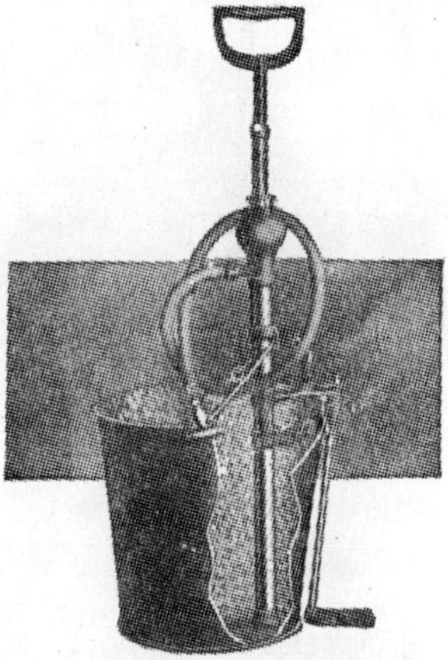

FIG. 179. A bucket spray pump.

271. **General Methods of Combating Insect Pests.** Besides being destroyed by poisons, insects may be very largely controlled by a wise general management of the farm through cleaning up the fields, destroying weeds and volunteer crops, deep fall plowing, trap crops, properly timing the crops, and the wise use of rotation.

When crops are harvested remnants are often left standing in the field on which insects continue to multiply until the end of the season, thus preparing fresh trouble for the following year. These crop remnants should either be used to poison the insects or should be plowed under completely so as to bury some of the insects and leave no food for others. The cornstalk borer winters in the stubble. The boll weevils continue to multiply on cotton until they hibernate. Many kinds of insects live in the grass, stubble, and rubbish left in parts of fields and along fence rows. Most of this should be turned under in the fall, and the fields and fence rows cleaned, only a few small piles of rubbish being left as traps. Hordes of insects will gather in these piles, which should then be burned. While usually all stubble should be turned under to enrich the soil, occasionally it is advisable to burn over stubble and grass land to destroy such pests as army worms, chinch bugs, and locusts.

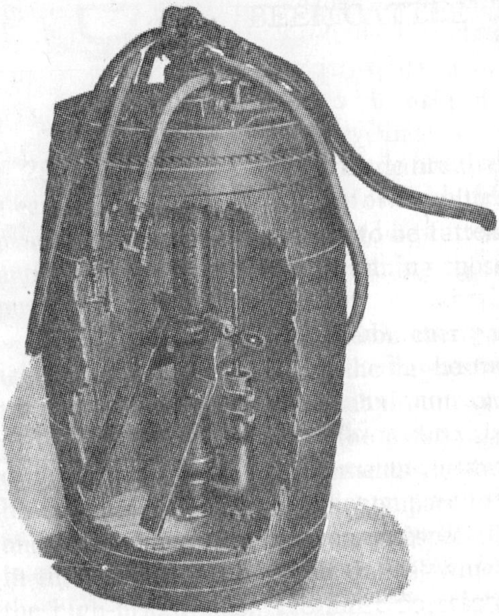

Fig. 180. A barrel sprayer.

Weeds and volunteer crops are another means of supporting pests when there is nothing else left for them to eat. The Hessian fly, for instance, feeds on volunteer wheat. Many larval forms also feed on weeds. The surrounding weeds are a great source of supply for garden cutworms.

In general, deep fall plowing is very helpful. In addition to turning under material on which the pests feed, or in which they hibernate, it helps to destroy many larval forms in the ground, such as the cornstalk borers, corn-ear worms, cutworms, locusts, and wireworms. Some of these are broken, some brought to the surface where they die, and some buried so deep that they can never get out. Grasshopper eggs and boll weevils are both said to be unable to do any harm if turned under as much as six inches.

At times trap crops, on which the insects feed, and by means of which they may be destroyed, are of great value. In this way the tomato, cotton, or tobacco crops, for example, may be protected from the worm which attacks all of them, and also the ears of corn, by planting a trap crop of corn. If

FIG. 181. A power sprayer.
Courtesy of "Farm and Ranch."

this crop is planted around the field that is to be protected, the moths deposit their eggs on the corn silks, as they seem to prefer these to the other plants. Before the eggs can develop into moths, which would produce a new crop of eggs, the corn is cut and fed to horses, thus destroying the insects. By the planting of a succession of trap crops of corn the other crops may be largely protected.

By learning the time at which the insect does the most damage and planting so as to avoid this season, often one can avoid, to a great extent, the injury from that insect. In this way, by planting early varieties of cotton which mature a large part of their bolls before the boll weevil becomes very plentiful, it has been possible to raise cotton profitably in spite of the weevil.

In a similar way, by forcing growth rapidly through the

FIG. 182. Dusting cotton with poison from an airplane.

use of fertilizers, the damage from insects is often reduced. In some cases kainit, lime, and nitrate of soda are thought to have a tendency to drive out certain insects.

Rotation is also a very effective method of handling insect pests. A crop that has been attacked by a certain pest should be followed by a different crop on which this insect cannot feed. In this way the pests developed one year find nothing which they can eat the next year. This is a matter that should always be carefully considered. For example, in a district in which corn-destroying insects are bad, one should see that land that has long been in pasture should not first be planted in corn, but in some other crop not related to the grasses, so that the insect pests that have accumulated in the grass will have nothing on which they can feed. Wireworms, Hessian flies, wheat plant lice, and many other serious pests may be largely controlled by rapid rota-

FIG. 183. Apples from a sprayed tree. On the left are the perfect apples (98.3 per cent), on the right are the wormy apples (1.7 per cent). In an orchard of 325 trees twelve to twenty years old, the cost of spraying was 23 cents per tree, materials being bought at wholesale prices. The sprayed trees averaged over 95 per cent sound and the unsprayed only 58 per cent sound apples. The yield was increased from three to seven bushels per tree by spraying.

Courtesy of Ohio Agricultural Experiment Station.

tion, while the lack of rotation causes their very disastrous increase.

272. **Natural Enemies of Insects.** Insects are destroyed in nature especially by animal and plant parasites, by other insects, and by birds, toads, lizards, and snakes. Man should make use of these in his fight with insects. This has been done in many remarkable cases. For example, the lady-bird beetle, the larval form of which destroys scale insects, was introduced into the orange and lemon groves of California, and practically exterminated a white cottony-cushion scale

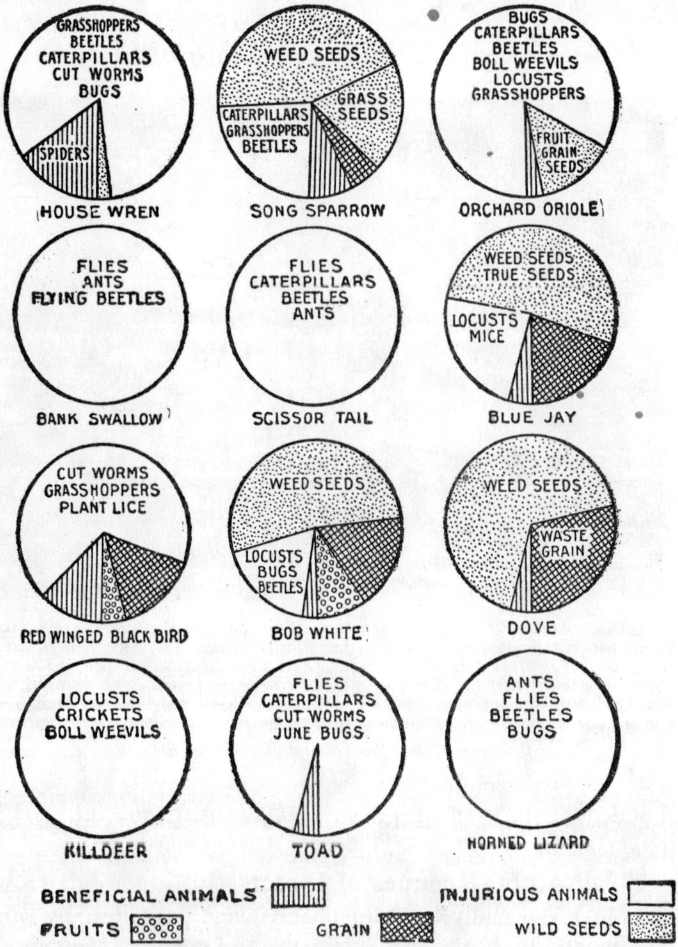

FIG. 184. The foods of some helpful birds and wild animals. *After Ferguson and Lewis.*

that had been causing losses of about five million dollars' worth of fruit a year. This little beetle saved also the cantaloupe-growing industry of California. Birds are among the farmer's most useful friends. In fact it is believed that the destruction of so many of our birds is one of the important causes of the disastrous increase of insect pests. All insect-destroying birds should be protected most carefully by farmers, as they are worth far more as insect destroyers than is the little that they eat or the food that they supply when killed. The interesting detailed facts about the work of parasites, helpful insects, birds, and other animals must be left for study in advanced classes.

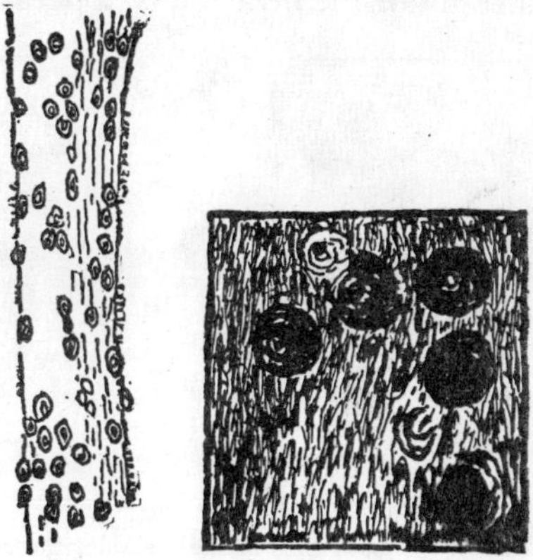

FIG. 185. San José scale. On the left in position on a small twig, on the right magnified.

273. The Causes of Plant Diseases. The causes of plant diseases are many but in the main may be divided into five groups, namely: those caused by bacteria, those due to fungi, the virus diseases, the physiological diseases, and the diseases and disturbances brought about by mechanical conditions. Bacteria and fungi are minute microscopic forms of plant life unable to manufacture their own food and hence must depend for their sustenance on living plants or decaying organic matter. The majority of bacteria and fungi are friends of the farmer, since they help in the decay of dead organic matter. Other bacteria and fungi are involved in liberating soluble plant food in the soil or in fixing in the soil free nitrogen from the air. Still others are important in the fermentation of various food products and beverages, and others are utilized in industry. Only a small percentage of the bacteria and fungi are harmful and able to cause diseases in plants. Most bacteria reproduce by a single cell dividing in two. Fungi reproduce by spores or seed-like bodies. A spore is capable of germination and the germ tube is able to pierce through the epidermis of the plant into the remainder of the tissue. A common mode of entrance is through the stomata, or breathing pores, of the plants. As a result of the invasion by a parasitic bacterium or fungus the

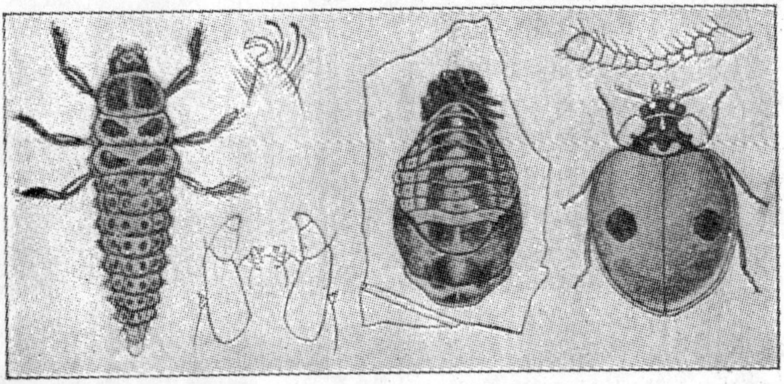

FIG. 186. Lady-bird beetle. Besides these with two spots there are many other kinds. One, about a quarter of an inch long with black head and body and orange wing covers on which are nine black spots, is very active against plant lice. Another has thirteen black dots; one has pink head, thorax and wing covers, with ten spots on the wings. Those feeding on scales are smaller and black, sometimes spotted with red or orange. One should watch these, learn to recognize the larval forms, and not destroy them.

Courtesy of U. S. Department of Agriculture.

plant cells usually break down, resulting in the wilting, and in the dying of the affected plant. In other cases, infection may result in the formation of abnormal growth, such as galls or warts. After it has invaded and destroyed certain plant tissue, the fungus usually develops a crop of minute summer spores or seed-like bodies which are easily blown about by the wind, or carried by insects to other parts of the plant and to other plants. In this way a disease is quickly scattered over a field or an entire territory. In addition to the ordinary summer spores, many fungi develop extra tough types of spores which help to carry over the fungus during the winter months or during unfavorable conditions of excessive moisture or drought. The red rust of wheat, for instance, is due to the countless red summer spores which are produced during the growing season, while the black stem rust stage is due to the tougher kind of spores which develop later and which, under cold climatic conditions, survive during the winter season.

As previously stated, bacteria are one-celled plants which

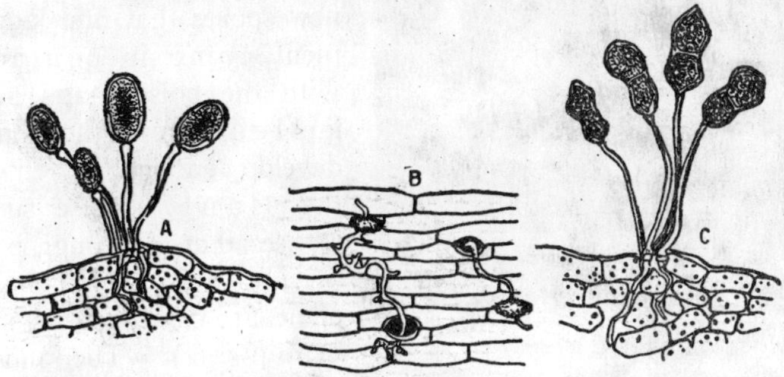

FIG. 187. Wheat rust. *A*, summer spores or "red-rust" stage; *B*, spores germinating and penetrating the plant; *C*, late spores or "black-rust" stage. *After Ferguson and Lewis.*

multiply rapidly by cell division, in which a single individual divides itself into two distinct individuals. The disease-producing bacteria when once successfully lodged on or within a plant multiply with enormous rapidity. Bacteria, as well as fungi, frequently find lodgment more easily when they fall upon a cut or bruised surface or upon tender buds or blossoms as is the case with the bacterium which causes fire blight of apples and pears, or the one causing brown rot on peaches.

274. **How to Control Plant Diseases.** Nowhere is it more true that "an ounce of prevention is worth a pound of cure" than in the handling of plant diseases. When once the fungus or bacterium is within the tissue of the plant, there is no successful way known of reaching it. Then all that can be done is to prevent the further spread by spraying, and by cutting off and burning diseased parts, or even

FIG. 188. Peach mummy caused by brown rot.

an entire tree or crop. While sprays will not destroy the fungi that are within the plant, they destroy those on the surface with which they come in contact. If all exposed surfaces are sprayed the new spores that find lodgment come in contact with the poison, and are killed before they can develop enough to enter the plant. These substances that kill fungi, but do not destroy the host plant, are called *fungicides* (fŭn′jĭ-sīdz). The most common fungicide is *Bordeaux* (bôr′dō) *mixture*, the formula for which you will find farther on in this chapter. If fungicides are to be of much value, the plants must be thoroughly covered, and the spraying must begin before the spores left from the past season have begun to germinate, and before the new crop of spores is formed. If fungicides are applied thoroughly and repeated at the right times, most plant diseases that affect the parts exposed to the air may be controlled.

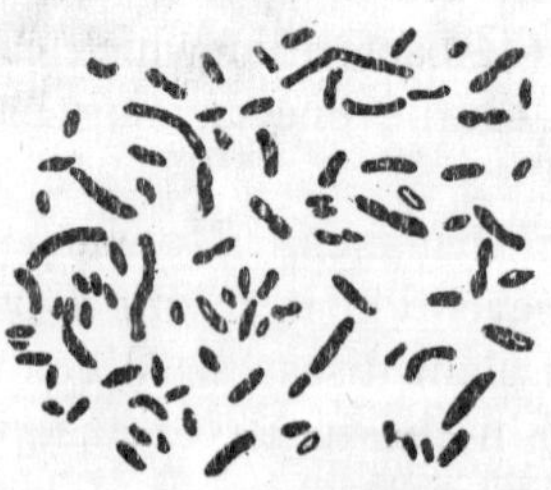

FIG. 189. The bacteria that cause pear blight. *After Warren.*

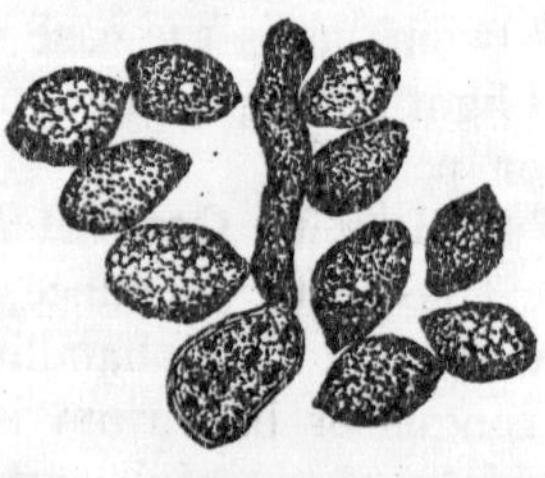

FIG. 190. Spores of brown rot of peach. *After Warren.*

For those diseases that affect the roots, or find entrance through the roots, such as cotton wilt and club root of cabbage, spraying will not avail. With the underground diseases the first precaution to take is to see that the fungi or bacteria that may be on the seeds are destroyed before planting the crop, as the potato scab is destroyed by soaking the potato in a solution of formalin. When once the land is infected, then rotation should be practised, and for several years no crop on which that fungus or bacterium can live should be grown on this land. In this way they may usually be destroyed. Cultural methods also are often a great aid. Damp soil favors the growth of most injurious soil fungi, and air and sunshine hurt them. Drainage and thorough tillage, opening up and getting air into the soil, help to purify it. Occasionally the application of lime will be helpful. In very limited areas, such as the germinating bed used for tobacco, sterilization of infected soil by "live" steam has been successfully employed.

FIG. 191. Potato infected with scab on left, sound potato on right.

275. **Disease-Resistant Varieties of Plants.** For reasons difficult to explain, certain plants are able to resist successfully the attacks of a disease that destroys other plants of the same variety growing around them. By selection of the resistant plant and multiplication, it is possible in many cases to develop a resistant variety which will be immune to this disease. This has been done in several cases. The rust-proof oats, the wilt-proof varieties of cotton, and the Iron peas, that resist wilt, are examples. Right here is one of the most valuable fields of work for the thoughtful farmer boy and girl. Wherever an insect or disease has attacked a field, a careful search should be made for individual plants

that have successfully withstood the attack. The seeds of these should be saved, and planted again where the plants will be exposed to attack. Each year the non-immune plants should be destroyed and seed saved only from those that are

FIG. 192. Ordinary cotton on the left and Dillon wilt-resistant cotton on right, grown in adjoining fields, the soil on the right being worse infected than that on the left, but making no impression on the resistant cotton.

Courtesy of U. S. Department of Agriculture.

immune. In this way a variety may be developed that is wholly or practically immune.

INSECTICIDES AND FUNGICIDES

276. **Formulas Vary with the Conditions.** Insects and fungus diseases are usually controlled by spraying or dusting the affected plants with poisons. Spraying as a rule is the most effective, because the affected plants can be more thoroughly covered than with a dust. On the other hand,

a greater area can be covered in a short time by the use of dust, and usually the expense of applying dust is much less than spraying. The amount of material to use in preparing sprays and dusts will vary with different times of the year and with climatic conditions. Hence, only general directions can be given. More detailed information may usually be obtained from the Agricultural and Mechanical College or the State Department of Agriculture or the County Agent or the United States Department of Agriculture.

Insecticides are generally considered under four main divisions:

(*a*) Stomach Poisons
(*b*) Contact Poisons
(*c*) Fumigants
(*d*) Repellents

Stomach poisons should be used when the insect has chewing or biting mouth parts. Contact poisons may be used when the insect has sucking or piercing mouth parts, and where it is impractical to use a stomach poison. Recently some new insecticides have become available that are both stomach and contact poisons. While deadly to some insects, they are not poisonous to man and should be used where the vegetables or fruits to be eaten might have a poisonous residue if the arsenicals were used.

Fumigants can be used most successfully in enclosed spaces where the gas which is produced can be confined for several hours at a time.

Repellents are substances, such as lime, which are used to drive away insects when it is impossible to use any of the other three classes of insecticides.

STOMACH POISONS

Arsenate of lead, calcium arsenate, Paris green, and sodium fluoride are the most common stomach poisons now used.

Arsenate of lead is used both as a dust and as a spray. When used as a spray the usual formula is:

ARSENATE OF LEAD

Arsenate of lead	1 to 3 lbs.
Water	50 gallons

Mix the arsenate of lead to be used with a gallon or two of water, so that all the particles are thoroughly wet, and then add it to the larger quantity.

Calcium arsenate is usually used in the dust form because of its fineness and the ease with which it may be used to cover large areas such as cotton fields. Five to seven pounds per acre are usually used. If it is used as a spray, it should be used at the rate of one to three pounds per fifty gallons of water.

Paris green was used formerly much more than it is today. It should never be used on trees or shrubs with tender foliage, as it is liable to burn the foliage. It is sometimes used in dusting by adding it to calcium arsenate at the rate of one part of Paris green to ten to fifteen parts of calcium arsenate. If Paris green is used as a spray, it is necessary to add stone lime or hydrated lime (builder's lime) to the mixture. The usual mixture is:

PARIS GREEN

Paris green	⅓ lb.
Stone lime or hydrated lime	3 lbs.
Water	50 gals.

For poisoning grasshoppers, cutworms, sow bugs, wire worms, etc., poisoned bran is good. The usual formula is:

POISONED BRAN MASH

Wheat bran	24 lbs.
Paris green, or sodium arsenite, or white arsenic	1 lb.
Black strap molasses	2 qts.
Lemons or oranges	3

Water enough to make a stiff mash, or about 3 gallons.

Mix the bran and poison thoroughly when dry. A part of the water is added to the molasses, the oranges or lemons are then chopped fine and added, and the whole added to the dry mixture which is shoveled over and more water added from time to time until the desired wetness is obtained. Such a bait is scattered thinly over the ground, five to ten pounds being necessary to cover an acre. For grasshoppers it is best to distribute this material early in the morning. For cutworms it is better to distribute it in the evening.

SODIUM FLUORIDE

Sodium fluoride is a good remedy for the control of lice on poultry. It is usually applied by the pinch method, which consists in placing a small pinch of the material on the head, a pinch on top of the back, one under each wing, another on the breast, and another near the vent of the fowl.

ANT POISON

Sugar	1 lb.
Water	1 pt.
Arsenate of soda	125 gr. or about ¼ oz.
Honey	1 tablespoonful

Boil the first three ingredients a few minutes but not long enough to decrease noticeably the amount of the liquid. Add the honey. If a stock solution is wanted, do not add the honey until ready to use the solution, as it is liable to make the material ferment if kept long. This is best put out in small tin cans with removable tops, containing a sponge filled with the poison. The sides of the can are crimped at the top and the lid replaced. Such a can will allow the ants to enter, and lessen the danger of dogs, cats, or other small animals getting the poison syrup. Ants may also be controlled by freely dusting sodium fluoride in their runways.

NON-TOXIC POISONS

These insecticides are not poisonous to man. They may consist of finely ground roots of plants such as derris and cubé (koo-bĭ′), sometimes called rotenone-bearing materials, or the ground flower heads of another plant known as pyrethrum. These insecticides deteriorate upon exposure for any length of time to light or air, and should be mixed only with sulphur, clay, talc, or similar material, but not with lime.

The following mixture is suggested for cabbage worms:

Derris or Cubé } containing 5 per cent rotenone	15
Sulphur, clay, or talc	85

For the bean leaf hopper and the beet webworm, use the mixture below:

Pyrethrum (containing 1 per cent pyrethrins)	10
Sulphur, clay, or talc	90

CONTACT POISONS

Perhaps the best contact insecticides for general use are the various compounds of nicotine, because they are easily applied and have high killing powers. Unfortunately, these are somewhat expensive. The manufactured nicotine products are sold under various trade names. Most of these contain 40 per cent by weight of nicotine.

The usual formula is:

NICOTINE SPRAY

Nicotine sulphate .... 1 pt.
Water .... 100 gals.

or

Nicotine sulphate .... 2 teaspoonfuls
Water .... 1 gal.

Unless an arsenic mixture is to be used at the same time, soap should be dissolved in hot water and added to the mixture at the rate of three pounds for fifty gallons of water.

Nicotine dusts are also extensively used. In this case nicotine sulphate is added to either air slaked or hydrated lime. Nicotine dusts are effective against plant lice. A two or three per cent nicotine mixture is the one usually used. The best results are obtained from the application of this mixture when the temperature is above 70° Fahrenheit.

NICOTINE DUSTS

STRENGTH OF DUST DESIRED	QUANTITY OF NICOTINE SULPHATE TO USE	QUANTITY OF HYDRATED LIME TO USE
1% of actual nicotine (or 2½% of nicotine sulphate)	1¼ lbs. or 1 pint	50 lbs.
2% of actual nicotine (or 5% of nicotine sulphate)	2½ lbs. or 2 pints	50 lbs.
3% of actual nicotine (or 7½% of nicotine sulphate)	3¾ lbs. or 3 pints	50 lbs.
4% of actual nicotine (or 10% of nicotine sulphate)	5 lbs. or 4 pints	50 lbs.

SULPHUR

Sulphur is toxic to all kinds of mites, such as chiggers and red spiders. It is excellent also in controlling the cotton-leaf hopper. Finely ground sulphur is used as a dust in such cases.

The most common compound of sulphur used is lime-sulphur, which may be obtained in either a liquid or dry form. It is made according to the following formula:

LIME-SULPHUR SPRAY

Stone lime	50 lbs.
Ground sulphur	100 lbs.
Water	50 gals.

Mix the sulphur with enough water to form a thick paste. Place the stone lime in an iron kettle and add enough water to slake the lime. Then add the sulphur and boil for one hour. The impurities will settle to the bottom, when the liquid should be allowed to cool, and should be strained and put into airtight containers. This spray is used only when trees are in a dormant condition. If homemade lime-sulphur is used, it is necessary to have a Baume hydrometer or tester to ascertain the strength.

If commercial lime-sulphur is used, it is only necessary to dilute it, using:

COMMERCIAL LIME-SULPHUR

Lime-sulphur	1 part
Water	8 parts

LUBRICATING OIL EMULSION

Lubricating oil	2 gals.
Water	1 gal.
Fish-oil soap	2 lbs.

Put the oil, water, and soap in a kettle, or some other receptacle that will stand heating, and heat to the boiling point. While the liquid is very hot, pump it back into itself by means of a bucket pump. It is a good idea to dissolve the soap in a little hot water before adding the oil.

If it is desirable to make the solution in large quantities, the materials can be heated in a large galvanized iron tank and *emulsified* (ė-mŭl′sĭ-fīd) by means of a spraying machine. First, place the suction hose in the tank and the discharge hose in the spray tank, and pump the liquid into the spray tank. Then reverse the procedure and pump it back into the heating tank. Several barrels of solution can be made in this manner in the course of twenty or thirty minutes.

Before any spray material is used, it should be tested to see if it is properly emulsified. To do this add a small amount to soft water. If no oil floats and no thick greasy scum forms, it is satisfactory. If free oil appears, it should not be used. Be sure to stir the contents of the barrel thoroughly before any tests are made.

These emulsions cannot be used in hard water unless the water is softened or the emulsions are stabilized.

In order to obtain the different strengths in diluting the stock emulsion, use the following table:

LUBRICATING OIL EMULSION SPRAYS

For	1%	strength	add	5/6	gal.	stock solution	to	49 1/6	gal. water.
"	1.5%	"	"	1¼	"	"	"	48¾	" "
"	2%	"	"	1⅔	"	"	"	48⅓	" "
"	3%	"	"	2½	"	"	"	47½	" "
"	4%	"	"	3⅓	"	"	"	46⅔	" "

Lubricating oil emulsion is used for about the same purposes as lime-sulphur, that is, the control of insects with sucking mouth parts, such as scale insects and plant lice.

A substitute for oil emulsion, although not as effective, is whale-oil soap solution.

WHALE-OIL SOAP SOLUTION

Whale-oil soap . . . . 1 lb.
Water . . . . 5 gals.

The whale-oil soap is first dissolved in a small quantity of hot water, and this is added to the larger amount of water.

FUMIGANTS

CARBON BISULPHIDE

Carbon bisulphide, or "highlife," is a good fumigant for grain weevils and similar pests. It is necessary to keep in mind its explosive qualities in using it. A lighted cigar, a match, or even a spark is sufficient to ignite it and cause an explosion. Carbon bisulphide *volatilizes* (vŏl′ȧ-tĭl-īz-ĕz) on exposure to air and forms a poisonous gas which is heavier than air. On that account the corn or other material to be fumigated should be placed in a solid box, or, if in very large quantities, in a room that can be made quite tight, and the carbon bisulphide placed on top of the material in open shallow vessels at the rate of ten to twenty pounds per thousand cubic feet of space in the box or room. The box or room should be closed at once and kept closed for twenty-four hours. If it is a room, it should be thoroughly aired for two or more hours before any one enters it.

CALCIUM CYANIDE

Calcium cyanide is a recent addition to our fumigants. It is valuable against bed bugs, roaches, fleas, and all kinds of household pests. It is a gray powder which is applied by spreading it out on newspapers to a depth of about ⅛ inch. Two pounds are used to each thousand cubic feet of air space. It is necessary to make the room or enclosure as nearly airtight as possible by stopping up all cracks with paper pulp or by pasting paper over them. This gas is very deadly to human beings; therefore, great care should be exercised not to breathe any of it. It may be used against fleas in barns and semi-tight enclosures effectively in warm weather.

FUNGICIDES

The most commonly used fungicides for the control of plant diseases are Bordeaux mixture and lime-sulphur.

The formula for making Bordeaux mixture is as follows:

BORDEAUX MIXTURE

Stone lime	4 lbs.
Copper sulphate	4 lbs.
Water	50 gals.

Several hours before the plants are to be sprayed, the copper sulphate should be placed in a muslin sack and the sack suspended in two gallons or more of water in a wooden bucket. This must be thoroughly dissolved some time before the mixture is to be made. The lime is slaked by adding enough water to slake it thoroughly but not enough to "drown" it. Then add about half of the remaining water to each vessel. Next, the solutions of copper sulphate and slaked lime are poured in two streams which meet in a barrel or large container. A blue precipitate will result which is called Bordeaux mixture. The strength of Bordeaux mixture may be increased when necessary by using more lime and more copper sulphate, six pounds copper sulphate and six pounds stone lime to each fifty gallons water.

Self-boiled lime-sulphur is a summer spray for the control of brown rot of plums and peaches. The formula for it is as follows:

SELF-BOILED LIME-SULPHUR

Flour of sulphur	8 lbs.
Stone lime	8 lbs.
Water	50 gals.

Place the lime in a receptacle and add sufficient water to cover it. With poor lime, hot water should be used to start the slaking. With good quick-acting lime, cold water should be used or else the mixture is liable to get so hot as to trouble the operator in mixing. While the lime is slaking, sift in the sulphur. Stir the mixture constantly and add water from time to time until a thin even paste is secured. Do not allow it to get dry. If the boiling tends to become too vigorous, add a little cold water.

With good lime, the boiling should be stopped when the lime is slaked by adding cold water. Poor grades of lime are allowed more time to cook. The mixture should not be allowed to remain hot for any great length of time, as this results in the formation of too high a concentration of sulphides which will injure the foliage. The mixture should be run through a 20-mesh strainer in order to remove the coarse particles of lime. However, the sulphur should be worked through the strainer, as it is the fungicidal agent. After diluting to the required amount, it is ready for application.

Self-boiled lime-sulphur is merelv a mechanical mixture

of lime and sulphur, with a small amount of sulphides in solution. The constant stirring, high temperature, and violent cooking result in a very uniform mixture of finely divided lime and sulphur particles. It is used in place of Bordeaux mixture and lime-sulphur solution on certain stone fruits. It is the standard summer spray for peaches to control brown rot and scab. The larger the quantity made up at one time, the more effective will be the heat of the lime and the better the results secured.

When good stone lime cannot be easily obtained, a good substitute for self-boiled lime-sulphur can be made as follows:

SELF-BOILED LIME-SULPHUR SUBSTITUTE

Flour of sulphur	8 lbs.
Hydrated lime	8 lbs.
Water	50 gals.

Take the eight pounds of sulphur and eight pounds of hydrated lime and mix together dry. Pour over the mixture eight gallons of boiling water and stir vigorously for five minutes. At the end of this time, cool with cold water. Then strain through a 20-mesh sieve and dilute to fifty gallons. This spray is practically the same as the ordinary self-boiled lime-sulphur, and is used for the same purposes. It should take the place of ordinary self-boiled lime-sulphur in localities where it is difficult to secure fresh stone lime, and also where it is impossible to keep the unslaked lime for any length of time.

To control plum curculio on plums and peaches, add two pounds of arsenate of lead to each fifty gallons of the above spray material.

To prevent oat smut, concealed smut of wheat, and scab of Irish potato, *formalin* is used. Grain is moistened in a solution of one ounce of formalin to three gallons of water

SPRAY CALENDAR

INSECT OR DISEASE	WHEN TO SPRAY	WITH WHAT TO SPRAY
All scale insects.	Early spring, before buds swell.	Lime-sulphur spray. (During growing season, use oil or soap spray.)
Striped melon beetles.	When the young plants appear above ground.	Tobacco dust or hydrated lime.
All leaf-eating insects.	When insects appear.	Arsenate of lead, or other arsenical poison, or one of the non-toxic poisons.
All sucking insects, such as plant lice.	When insects appear.	Nicotine spray or dust, or lubricating-oil spray.
Fruit rot.	Before blossoms open, and repeat till crop is safe.	Bordeaux mixture or self-boiled lime-sulphur spray.
Mildews and black rot of grapes.	Before blossoms open, when leaves are one-third grown, just after fruit sets, and every two weeks thereafter.	Bordeaux mixture.
Codling moth.	Just after blossoms fall, and again later.	Arsenate of lead.
Leaf curl.	Before buds swell, and later.	Bordeaux mixture, or lime-sulphur.
Mites, spiders, and the cotton leaf hopper.	When they appear.	Sulphur.
Potato scab.	Before planting.	Formalin, 1 oz. to 2 gals. water.
Grain smut.	Before planting.	Formalin, 1 oz. to 3 gals. water.

and kept moist for two hours, after which it is dried. Care must be taken that it is not allowed to come in contact with smut again before being planted. Potatoes used for seed should be soaked for two hours in a solution of one ounce of formalin to two gallons of water in order to kill the scab.

277. **Caution—Danger.** As arsenic, arsenate of lead, Paris green, carbon bisulphide, formalin, and most other insecticides and fungicides are poisons, they should be handled with care, always labelled, and never left in the pantry or kitchen or in reach of children, stock, poultry, or other animals.

QUESTIONS, PROBLEMS, AND EXERCISES

117. Make a list of the harmful insects in your neighborhood, and collect a set of bulletins that deal with these. Bulletins on insects and on scores of other interesting subjects may be secured free from the United States Department of Agriculture, the various state colleges of agriculture, state experiment stations, and state departments of agriculture.

118. Collect two varieties of biting insects. Draw and describe each.

119. Collect two varieties of sucking insects. Draw and describe each.

120. Place the eggs of some insect in such a cage as is shown in Figure 166, and make notes from day to day of the development. Watch some insect and, if possible, get eggs just as they are laid. Be sure to give the larvæ plenty of fresh food.

121. Make a list of the birds of your neighborhood. Find what each one lives on at each season of the year.

122. Find all of the kinds of helpful insects in your community. Bring some of each of these for the school garden.

123. If any insect or disease has afflicted your father's farm, study this pest, write out a practical plan for combating it. Show this to the teacher and, when it is approved, carry it out and report results.

124. What fungus plant diseases are in your community? How should each be treated?

125. What bacterial plant diseases are in your community? How should each be treated?

126. What plant diseases in your community are due to infected soil? How could this be remedied?

127. Keep a lookout for some immune plant in a crop that has been destroyed by some insect or disease. Save seeds and see if you can breed a resistant variety. Once in a while, in a field in which a crop has been ruined by some insect or disease, there is found a single plant that for some unknown reason has escaped. At times, this is due to an exceptional immunity that this plant alone has and can transmit to its offspring.

CHAPTER XII

ANIMAL HUSBANDRY

278. **The First Reason for Raising Livestock on the Farm.** We have already seen that heavy crops take out of the soil large quantities of the food materials necessary for plant growth, and that unless these are put back into the soil the land will soon become too poor to produce a good crop. We have also seen that when the crop is fed to animals and the manure properly saved and put back into the soil, between 70 and 80 per cent of the valuable plant-food materials are thus returned. On the other hand, if the crop is sold and carried off the farm, the farmer must constantly purchase large quantities of expensive fertilizers or his fields will soon not repay him for the labor of cultivating them. This is why the thoughtful farmer should always raise enough livestock to eat practically all the foodstuffs produced on his farm. By feeding his crops to stock and then selling the stock, he retains at home in the manure nearly four-fifths of the value of his crop, and sells the animals for as much as, often for more than, he could have sold the crop.

279. **Other Reasons for Raising Livestock.** Besides this there are seven other advantages that in most cases come from raising livestock on the farm instead of raising only cotton, grain, and other plant crops. First, the raising of some livestock necessitates the growing of hay, clover, alfalfa, peas, peanuts, and other cover crops and legumes which add humus, and in some cases nitrogen, to the soil. This diversifying the crops also makes successful financial results from farming more certain, as then no one failure due to unfavorable season or insect pest can affect all the crops of any one year. Second, this diversification and the

livestock feeding distribute the labor of the farm more evenly through the year, instead of causing great rushes at special seasons. A good part of the work of feeding stock for market comes in late fall and winter when the crops are out of the way. Third, there are always remnants of crops and a great deal of grass left in fields that can be gathered at no expense by animals and converted into salable meat. Fourth, a considerable part of the expense of harvesting and of hauling out fertilizer is saved with many crops by turning into the field the stock, which do their own harvesting and drop the manure in the field. Fifth, when several thousand pounds of crops are fed to stock, there are only a few animals to be driven to market instead of the several thousand pounds of produce to be hauled. This saves time and teams. Sixth, in most cases, except where markets are very near or the soil and

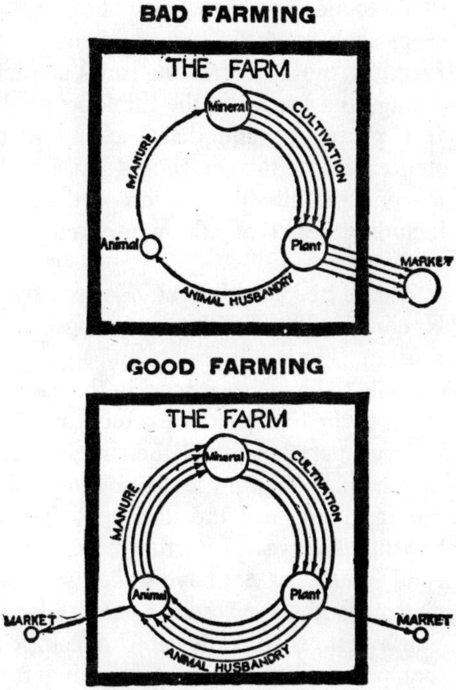

FIG. 193. The upper figure illustrates poor farm management: the minerals in the soil are converted into plant crops and four fifths of the products are carried off the farm to the market, while only one fifth is fed to stock and thus left on the farm in the form of manure. The lower figure illustrates good farm management: the same minerals in the soil are converted into plant crops, but only one tenth of these is taken off the farm to market, the other nine tenths are fed to stock. The stock leave on the farm in their manure seven tenths or more of the minerals that were taken from the soil and carry away only two tenths when they are sold in the market.

climate are especially adapted to some particular crop, more money can be made by devoting a considerable part of the farm to raising livestock and growing the crops suitable for feeding them economically than can be made by raising all market crops such as cotton. Seventh, the raising of livestock makes farming more interesting and attractive to both old and young, and broadens the thinking of the farmer. It is therefore perfectly plain that except under very special circumstances every farmer should raise at least enough livestock to consume all the food crops that a well-planned rotation, including legumes and winter cover crops, would demand on his farm.

280. **The Southwest Especially Adapted to Stock Raising.** The Southwest is especially adapted to livestock raising. The mild winters and long growing seasons make it possible to have green food in the field almost all the year, and to allow the livestock to exercise and to gather their own feed in large part nearly all the time. Such expensive barns and long winter feeding as are demanded in the North are not required, nor are the dangers of diseases caused by close housing so great. Furthermore, such a large part of the food eaten does not have to be used by the animal in keeping warm. With vast acres of pasture land and mild climate, the Southwest should develop livestock-raising rapidly, now that practical methods of handling the cattle tick and other animal pests and diseases have been learned.

281. **The Loss from Raising Scrub Livestock.** There are about 12,500,000 cattle in the States of Texas, Arkansas, Louisiana, Oklahoma, and New Mexico. This is between one fifth and one sixth of all the cattle in the United States. The value of these cattle is about $22 a head (based on 1936 figures), whereas the average valuation in the United States is over $34 a head. A part of this difference is due to location, but even allowing for this, tens of millions of dollars of wealth could be added to these states through improvement of cattle alone. If values of Southwestern cattle were compared with those in the states having the best cattle, the

FIG. 194. Above, an inferior feeder; in the center, a choice feeder; below, a fat steer of the correct type.

difference would be still greater. It takes nearly as much labor and feed to raise a scrub as it does to raise a purebred or high-grade animal. The raising of scrub stock is therefore very wasteful and unintelligent. In former years, when there were millions of acres of cheap land, it was possible to make a profit from scrub stock turned out to graze with very little oversight. Now, with higher-priced land and the country rapidly being broken up into small farms and ranches, the ranchman and farmer can no longer afford to waste time and food on scrub stock. Noticeable improvement has taken place in recent years in the cattle of the Southwest, but there is still room for improvement. When a pure-bred or high-grade calf at one year of age will weigh eight hundred to a thousand pounds, is it reasonable to raise scrubs that will weigh at the same age only three or four hundred?

282. **How to Improve the Quality of Stock.** While it is not practicable for all farmers at once to buy and raise only purebred stock, it is practicable to grade rapidly a herd at small expense by breeding only from purebred males. As you know, the parents of a scrub do not belong to any particular breed, but are a mixture of many inferior types, whereas both parents of a purebred belong to the same breed of selected stock. The result of this is that when a pure-bred male is crossed on a scrub female, the offspring, which is called a *grade,* is more likely to resemble the purebred male parent than the scrub female. For example, if a purebred Hereford bull is used, nearly all the calves will show the fine Hereford qualities. None of the males of these *half-bloods,* as the offspring of a full blood and a scrub are called, should be allowed to breed. The female half-bloods should be crossed again with a purebred and thus secure a three-quarter purebred. These similarly being crossed with a purebred bull will produce calves that are seven-eighths pure, or *high grade.* For practical beef and dairy purposes, such high grades are nearly as good as purebreds, but they would not bring high prices for breeding purposes. Grade bulls

should not be used for breeding, as with a mixed ancestry the calves from them would not be apt to come true. As long as the grade females are always bred to a purebred bull, however, the calves are very apt to possess the qualities of the good stock.

283. **What Must Be Known to Get Highest Profit from Livestock Raising.** In order to get the greatest profit from his stock raising the farmer must know two things: first, what kinds of animals and animal products—meat, milk, butter, wool, and eggs—the market demands and pays best for; second, how to produce these at the least cost. In order to produce at the least expense animals that will bring the highest prices, three things must be learned. These are: first, livestock judging; second, livestock breeding; third, livestock feeding. Let us now study each of these.

284. **Livestock Judging.** Livestock judging is the basis of all success in stock raising. If one does not know what are desirable points in an animal he will not know how much to pay for animals that he buys, nor what to charge for those that he sells, nor will he know which animals to select and breed from in his herd. One horse sells for $500, while another that looks very much like it to the untrained observer brings only $150. One bull sells for $50, while another that does not look very different to the average boy sells for $500. Let us take up the several farm animals in turn and find out what points are important and what relative value should be given to each different quality.

285. **Types of Cattle.** Cattle are divided into three classes: *beef cattle,* or those raised for beef; *dairy cattle,* or those raised for their milk and butter; *dual-purpose cattle,* or those raised both for beef and for milk and butter. Each of these types has its special points which have definite values in estimating the quality of the animal judged. These can be learned thoroughly only by study of actual cattle with the aid of a trained judge, but with the aid of pictures and the following descriptions any boy or girl may make a good start in learning to judge cattle.

CHAPTER XIII

BEEF CATTLE

286. **Beef Cattle.** Beef cattle are divided into: (1) "fat steers," meaning those ready for the butcher; (2) "feeders," meaning those that are ready to be fattened for the butcher; and (3) "breeding cattle," meaning those used for breeding purposes.

The *fat steer* for which the butcher pays the highest price is one that will dress out the highest per cent of salable meat and that carries the maximum amount of this meat in the regions from which the most valuable cuts come. In order to meet these requirements the fat steer must have a broad, deep, low-set, smooth, compact form with straight top and under lines. He must show especially high development in the ribs, loin, rump, and thighs, which are the regions of the high-priced cuts. He must possess good quality, as indicated by fine, soft hair; loose, pliable skin of medium thickness; even, firm, mellow flesh; and clean, medium-sized, dense bone. He must be in good condition, as indicated by a deep, even covering of firm flesh, especially in the region of choice cuts. The scrub steer with swayed back; high flanks; long, narrow, shallow body; long legs; large paunch; coarse bone; thick hide; and thin covering of flesh not only dresses out a low per cent of salable meat, but too large a proportion of this meat is located in the regions of the low-priced cuts. Figures 194, 195, 196 will make this description clear.

The score card found on page 287 presents the points in

detail to be considered in judging fat cattle and shows the relative value of those points. The score card is of great aid

FIG. 195. Points of the beef animal: 1, muzzle; 2, face; 3, eyes; 4, forehead; 5, ears; 6, poll; 7, jaw; 8, neck; 9, shoulder vein; 10, shoulder; 11, dewlap; 12, chest; 13, brisket; 14, breast; 15, arm; 16, knee; 17, shin; 18, hoof; 19, fore-flank; 20, crops; 21, ribs; 22, back; 23, loin; 24, rump; 25, hips, or hooks; 26, hind-flank; 27, purse, or cod; 28, tail-head; 29, pin bones; 30, thigh; 31, twists; 32, hocks; 33, shank; 34, tail.

to the beginner in stock judging, by familiarizing him with the ideal type, enabling him to distinguish clearly and fix in memory the points to be observed, and also to judge in a systematic way. As soon as these purposes are accomplished, further use of the card is not necessary. The student should then be able to judge and criticise an animal without referring to the card. After becoming proficient in judging a single animal, comparative judgments of two or more animals may be made.

The feeder steer is the one not yet fat but ready to be fattened for the market. The ideal feeder is one that will make the most economical gains in the feed lot and put the largest percentage of these gains on those regions which yield the high-priced cuts, so that when he is fat he becomes an ideal fat steer. The difference between the ideal feeder and the ideal fat steer is a matter of condition, or flesh covering. The most important points to be considered in feeders are the following. The body should be deep and wide, the top and bottom lines straight, legs short, and general appearance smooth and compact. The depth and thickness are not, of course, as great in the feeder as in the fat steer, but the more pronounced they are in the feeder the greater they are likely to be in the fat steer. A wide back; well-sprung ribs; wide, thick loin; level, long, wide rump, giving squareness to the hindquarters; thickly fleshed thighs; and deep twist are demanded to make sure of large valuable cuts when the animal is fat. The skin should be loose, pliable, and of medium thickness; the hair soft and glossy; the bone clean, dense, and of medium size. Medium-sized bone is preferred to small bone, because it has been found that animals possessing medium-sized bone have better constitutions and when fed make faster gains than do those with small bones. The loose, pliable skin and the glossy hair indicate good digestion, which is essential to economical gains in the feed lot. While not fat, the feeder must possess a large amount of flesh or lean muscular tissue, otherwise it will not dress out a large per cent of good quality of meat when fat. The feeder should have a strong constitution, as is indicated by deep, wide chest; large nostrils; large muzzle and mouth; bright, clear, quiet eyes; short, broad head; well-arched deep ribs and low flanks, giving large capacity for food. The butcher does not care for large head or large paunch, but in

SCORE CARD FOR BEEF CATTLE (FAT)

SCALE OF POINTS	PERFECT SCORE	STUDENT'S SCORE	CORRECTED SCORE
I. GENERAL APPEARANCE—38			
1. WEIGHT: Score according to age: at 12 mos. 850 lbs.; at 24 mos. 1,250 lbs.; at 30 mos. 1,500 lbs.	6		
2. FORM: Straight top and underline; deep, broad, compact, low set, blocky, proportioned, stylish	10		
3. QUALITY: Head medium size, clean cut; hair fine, silky; bone medium size, clean; hide of medium thickness, loose, pliable	10		
4. CONDITION: Deep, even covering of firm mellow flesh, especially in region of valuable cuts; free from patches, rolls, and ties; full purse and flanks indicating finish	12		
II. HEAD AND NECK—8			
5. MUZZLE: Broad; mouth and nostrils large	1		
6. FACE: Short, broad; jaw strong	1		
7. EYES: Large, clear, mild	1		
8. FOREHEAD: Broad	1		
9. HORNS: Medium size, even, oval, tapering	1		
10. EARS: Medium size, fine texture, well carried	1		
11. NECK: short, thick, blending smoothly with the shoulder; throat clean; dewlap light	2		
III. FOREQUARTERS—10			
12. SHOULDER VEIN: Full, smooth	2		
13. SHOULDER: Not prominent, compact on top, smoothly covered	4		
14. BRISKET: Advanced; breast wide, full; dewlap light	2		
15. LEGS: Short, wide apart, straight, arm full; shank fine, smooth	2		
IV. BODY—28			
16. CHEST: Deep, wide, full; girth large; crops full	4		
17. BACK: Straight, broad, thickly and smoothly fleshed	7		
18. RIBS: Long, close together, arched, thickly and smoothly covered with flesh	7		
19. LOIN: Broad, thickly fleshed	8		
20. FLANKS: Full, even with underline	2		
V. HINDQUARTERS—16			
21. HIPS: Smoothly covered, distance apart in proportion with other parts	2		
22. RUMP: Long, wide, level, not patchy; tail-head smooth	3		
23. THIGHS: Wide, deep, full	5		
24. TWIST: Deep, full, plump	4		
25. LEGS: Short, wide apart, straight; shanks fine, smooth	2		
Total	100		

Name of Scorer.................................Animal..................

Section..Date....................

the feeder they are desirable, as they indicate ability to make good use of food and make rapid gains in the feed lot.

Breeding cattle when thin should represent ideal feeders and when fat ideal fat cattle; but in addition to this they should possess qualities which indicate that they will breed regularly and that the offspring will resemble their parents.

FIG. 196. Wholesale cuts on a steer: 1, round; 2, loin; 3, flank; 4, rib; 5, plate; 6, chuck; 7, shank.

No matter how good the animals may be as individual specimens, they will not do as breeders unless they can reproduce their kind with regularity. The following points should be looked for in breeders.

1. The animal should be true to his type; that is, the Hereford should have the characteristics of the Hereford and the Jersey of the Jersey. The distinguishing features of each type have been fixed in it by long years of carefully selecting and breeding only animals of this type. Those that are good representatives of the type are therefore more apt to be able

to transmit this type to their offspring than would a specimen that had varied from the type. An animal capable of doing this is spoken of as *prepotent*.

2. The animal should possess the characteristics of the sex to which it belongs. Such animals are more apt to be prepotent. The bull should show the following masculine characteristics: bold expression in eyes; full forehead; thick neck, surmounted by heavy, well-developed crest; heavy, though not coarse, shoulders, giving him a strong, vigorous, burly appearance. The female should show the following feminine characteristics: mild expression in eyes; refinement of head and horns; neck slender and shoulders light as compared with the bull; more width and prominence of hips than the bull; and a generally gentle appearance.

3. The constitution must be strong, as only animals having such can stand the strain of producing offspring regularly and at the same time transmit to the offspring their strength and vigor. The signs of a strong constitution you have just learned in studying the feeders.

287. **Breeds of Beef Cattle.** There are seven breeds of beef cattle recognized in the United States: *Shorthorn, Polled Shorthorn, Hereford, Polled Hereford, Aberdeen-Angus, Galloway,* and *Brahman*. The Herefords, Shorthorns, and Aberdeen Angus are the principal breeds.

The Shorthorn. This breed originated in England, probably from the old Teeswater and Holderness stock, in the counties of York, Durham, and Northumberland. Shorthorns are sometimes improperly called Durhams. As early as 1780 the special selection and breeding were begun which produced this remarkable beef type, possessing easy-feeding qualities, early maturity, and thick flesh of good quality. The breed has long been prominent and steadily improved. In size the Shorthorn ranks first, bulls at maturity weighing two thousand to twenty-two hundred pounds. Mature cows weigh fourteen hundred to sixteen hundred pounds, some as high as two thousand pounds. The color may be pure red, pure white, red and white spotted, or roan, which is a mixture of red and white. The breed is sometimes called the "reds, whites, and roans." The horn, which is a well-marked characteristic of the breed, is usually short and small, preferably curved forward, with the tips bending inward and upward. The breed is noted for wide back, strong loin, and square, well-developed hindquarters. It is criticised because of long legs and lack of heart girth, as shown by insufficient

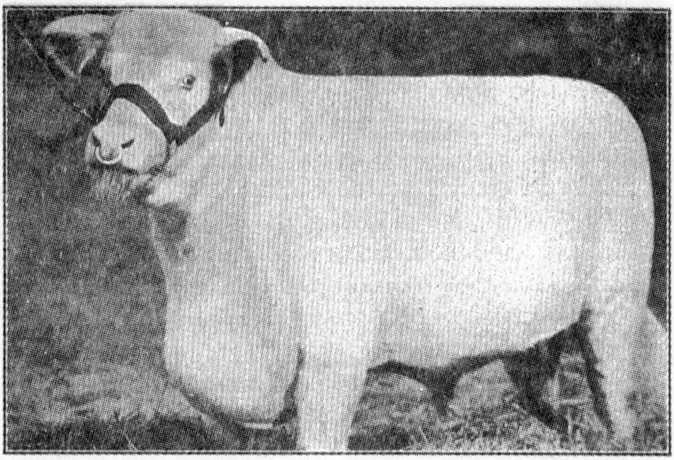

FIG. 197. Anoka White Frost, Grand Champion Shorthorn Bull.
Courtesy of the Singleton Farms.

fulness back of shoulders, in the crops and foreflanks. As milk producers they rank first among the beef breeds.

The Shorthorn is especially adapted to the farm, but not so well adapted to range conditions, particularly where exposed to severe winters, as the Hereford. Shorthorn bulls are used on the ranches, however, because of the marked improvement produced in the size of the offspring.

Polled Shorthorn cattle had their origin in the United States. About 1870 purebred Shorthorn bulls were bred to hornless cows and the offspring that inherited the hornlessness of the mother but the other qualities of the Shorthorn were selected. By continuous breeding and selection the polled breed of Shorthorns, called Polled Shorthorn, was produced. Those bred in this way are called "single standard." Another breed of polled cattle was developed by selecting a few pure-bred Shorthorn bulls and cows that varied from the normal in having no horns. These were bred to each other and the polled

FIG. 198. Publican Domino, Grand Champion Hereford Bull.
Courtesy C. M. Largent & Sons.

feature fixed in the offspring. Polled Shorthorns that originated in this way are called "double standard."

The Hereford is a native of Hereford County, England, the breed having originated early in the eighteenth century in efforts to produce a breed better suited to the production of fine beef by grazing. The Hereford is shorter of leg and somewhat more compact in appearance than the Shorthorn, but weighs practically as much. The color is remarkably uniform; the face, breast, top of neck, legs usually from slightly below the knee and hock down, the belly, and switch of tail are all white. The rest of the body is red. The head is shorter and broader than that of the Shorthorn, the horns longer and keener toward the tips. The horns are white or waxy yellow, and spring forward and usually down with a graceful curve. The Hereford is especially noted for its excellent constitution, thick middle, beautiful front, and early maturity. The most common defect in the form is light hindquarters, owing to a drooping, peaked rump and poorly developed thighs. The American breeders especially have in recent years greatly improved this breed in this respect. Hereford cows rank very low as milk producers.

Many Herefords have been imported and, because of their hardiness, excellent grazing qualities and adaptation to ranches, have been distributed widely over the entire range country. Hereford bulls are of immense value in grading up common herds because of the transmission of their fine beef qualities and ability to stand hard conditions. On account of hardiness and early maturity, Hereford steers stand in front rank as feeders.

The *Polled Hereford* was established in America in 1901 by Warren Gammon by mating Hereford cattle that were naturally polled. The breed has increased rapidly, as feeders prefer cattle without horns. If they can be bred polled, without sacrificing the desirability of type, it is obvious that many breeders would prefer avoiding the dehorning operation. Usually the mating of polled animals with horned ones will give a rather high percentage of polled calves. This breed has enjoyed an increasing popularity during recent years because of the great improvement in type among polled cattle. Much of the recent demand has come for bulls from range cattle breeders.

Double Standard Polled Herefords may be registered in both the American Hereford Herdbook and the American Polled Hereford Record.

Aberdeen Angus. This breed of hornless cattle originated in and around the shire of Aberdeen, in Scotland, taking its name from the shire and a near-by locality. While something had been done before, the real work of improving this breed began about 1815. Aberdeen Angus cattle are not as large as Shorthorns and Herefords, but are more cylindrical and compact in shape and are remarkably heavy for their size. Bulls weigh two thousand to twenty-two hundred pounds, cows fourteen hundred to fifteen hundred pounds, both sexes frequently passing these marks. The breed is noted for its smoothness, high percentage of dressed beef,

FIG. 199. Sir Perfection 45th Polled Hereford Bull.
Courtesy of L. A. Schreiner.

and the superior quality of the meat. The standard color is black, though occasionally solid reds appear. The *poll,* or top of the head, is a well-marked characteristic. It should be clearly defined and prominent, and there should be no traces of *rudimentary* (ru̧-dĭ-mĕn′tȧ-rў) horns. Any signs of scurs, horn buttons, or stub horns disqualify the animal for registration. The cows produce more milk than Herefords, but less than Shorthorns.

This breed was first brought to America in 1873, and has become quite widely spread and popular considering the short time it has been here. It has gained much favor in the upper Mississippi Valley and in the Western and Southwestern States. The bulls are excellent for grading up a herd, and the steers make excellent feeders. The absence of horns makes it possible to feed them in close quarters without danger of their injuring one another. While good

FIG. 200. Bluebonnet, a grand champion crossbred Shorthorn-Angus steer. *Courtesy Agricultural and Mechanical College of Texas.*

on the range the Aberdeen Angus is hardly the equal of the Hereford.

Brahman or *Zebu* cattle from Asia have attracted a great deal of attention along the Gulf Coast section, especially in Texas. There are several families of these cattle. The most important are the Nellore, Gir, Guzerat, and Krishna Valley.

This breed is especially adapted to regions where the climate is hot, cattle pests are troublesome, and a good deal of travel on the range is necessary on account of drought conditions or to secure drinking water.

Brahman cattle are very hardy and are little affected by heat. They are also more resistant to ticks than are our other cattle. This has in no small measure been responsible for their popularity in certain regions. Usually the desirable

FIG. 201. Grand Champion Guzerat Brahman Bull, weight 2200 pounds.
Courtesy of J. D. Hudgins, Hungerford, Texas.

characteristics of the breed are utilized by maintaining herds for commercial production with from one-eighth to one-half Brahman blood. Much of the production from the Southwest is marketed as heavy veal, as no large demand for feeder cattle of this breed has been developed. Brahman cattle of all grades are generally more alert and nervous than other breeds.

Brahman cattle are being used quite extensively for crossbreeding purposes in the South. Crossbreeding experiments by Robert J. Kleberg, Jr., manager of the King ranch in Texas, have established the Santa Gertrudis breed which has three-eighths Brahman blood and five-eighths Shorthorn blood. The United States Department of Agriculture

FIG. 202. Crossbred Brahman cow, three-fourths Brahman and one-fourth Hereford; six years old; weighs 1890 pounds; and is the mother of four calves.
Courtesy of J. D. Hudgins.

has produced at the experimental farm at Jeanerette, Louisiana, a crossbreed cattle containing one-quarter Brahman and three-quarters Angus blood. The crossing of the humped and horned Brahman cattle with our English breeds has produced crossbred cattle that are considered by many to be better adapted to conditions in the extreme South than are the beef breeds originating in Western Europe.

For QUESTIONS, PROBLEMS, AND EXERCISES, see the list at the end of the next chapter on Dairy Cattle.

CHAPTER XIV

DAIRY CATTLE

288. **Dairy Cattle.** A good dairy cow will return in milk and butter for a given amount of foodstuff a larger supply of human food than will the hog, sheep, chicken, or steer. This fact coupled with the ever-present demand for the products of the dairy makes dairying, when properly conducted, a most profitable business. No kind of livestock will as a rule yield a larger return from an acre of land than dairy cattle. In states that are thickly populated, and in which land is expensive, dairying is usually one of the chief occupations.

289. **The Southwest is Especially Suited to Dairying.** In many of the more thickly populated sections of the Southwest, dairying has advanced considerably in recent years; but, as a whole, this section is still behind in this important field. While Texas ranks fourth in number of dairy cows, it ranks only forty-second among all states in production per cow. This low production is partially due to the poor type of cow on many of our farms. More important, however, is the fact that too many farmers do not give their cows proper feed and care. The result is that the cows do not give the milk that they could produce with plenty of good feed and shelter. Many farmers also fail to handle the milk in a sanitary manner, with the result that much of it is of such poor quality it does not bring a fair price. Here in the Southwest where cows can stay in the open practically all of the year, and can have fresh green pasture, which is so important for dairy cows, nearly every day, it is time that more of our farmers realize the possibilities of this great industry. It was formerly believed that the Southwest could never com-

pete with the high-producing herds of the North and East, but fortunately this is not true. There are several herds in the Southwest that have made records of production that compare favorably with those of other parts of the country. Only recently, Magnolia Belle, a Jersey cow in southern Arkansas, produced over one thousand pounds of butterfat, or the equivalent of about one thousand two hundred and fifty pounds of butter in one year. While this is the first such record in the Southwest, it proves what can be done with the right kind of cows and proper feeding and care. Our farmers and farm boys and girls must learn more about judging, selecting, breeding, and feeding dairy cattle and about handling milk and cream, before the Southwest can take the high position in dairying that its natural advantages entitle it to hold. Let us begin the study now.

290. **Judging the Dairy Cow.** A dairy cow may be looked upon as a factory which takes in raw material in the shape of food and makes it into milk and butterfat. If this were all that had to be considered, the best dairy cow would be the one that yielded the largest amount of milk and butterfat from the smallest amount of food. By measuring the food given and the milk produced and testing the percentage of fat with the Babcock test[1] each day, one could keep records that would make it possible to judge the quality of the cow accurately. It is such constant testing and keeping of records on each cow, and then breeding from the best cows only, that has brought about the great improvement in recent years in dairy cattle. As the cream is the most valuable part of whole milk, it is especially important to test the percentage of cream.

But at times dairy cows must be judged when they are not giving milk, and when there are no records to go by.

[1]This is a test which was originated by Professor Babcock, of the University of Wisconsin, for finding the percentage of butterfat in milk. A little sulphuric acid is added to a bottle of milk, which causes the fat to be separated from the rest of the milk. The bottle is then rotated rapidly in a machine in such manner as to bring the fat to the top of the bottle. A scale is marked along the top part of the bottle by which the percentage of butterfat present can be seen at once.

Furthermore, there are other qualities besides capacity for milk production that must be considered, such as capacity to produce regularly offspring that will inherit the fine qualities of the parent, and capacity to maintain vigor for a number of years. For these reasons it is necessary to learn to

FIG. 203. Points of the dairy cow: 1, muzzle; 2, nostril; 3, jaw; 4, breadth between eyes; 5, forehead; 6, poll; 7, throat; 8, dewlap; 9, neck; 10, withers; 11, shoulder; 12, brisket; 13, chest; 14, heart-girt; 15, crops; 16, loin; 17, hips; 18, thurls; 19, tail-setting; 20, pin bones; 21, escutcheon; 22, thighs; 23, rear attachment of udder; 24, rear udder; 25, switch; 26, hocks; 27, stifle; 28, flank; 29, fore udder; 30, milk veins; 31, milk wells; 17–20, rump.

judge the qualities of a dairy cow by her physical make-up in a manner similar to that by which the qualities of beef cattle are judged.

291. **How Milk is Produced in the Cow.** In the beef type of cow the food consumed is in part turned into flesh and stored within the animal's body, but in the dairy type the food is turned into milk which is constantly being taken away from the body. We should therefore expect the two types to be very different in appearance. But before we can

know what the differences are and intelligently determine what is the best type for dairy purposes, we must know more about the means by which milk is produced in the cow. The parts most concerned in the production of milk are the digestive organs, the blood, the lungs, the heart, the *udder,* and the nervous system.

The digestive system must be strong enough to enable the

FIG. 204. The blood supply of the udder. Arteries (in white) lead from the heart to the udder, veins (in black) lead from the udder to the heart.

cow to consume and digest a large quantity of food in order to produce a great quantity of milk. She should therefore show a large middle, or "barrel," as it is called.

The blood, lungs, and *heart.* After the food has been digested or changed into a condition to be utilized by the animal it passes through the walls of the intestines into the blood. The material from which milk is to be formed thus becomes a part of the blood, which now goes through a large vein to the right side of the heart. From here it goes to the lungs to be purified by the air that is breathed in. It then returns to the heart, this time to the left side, and from there is pumped through the arteries to the various portions of the body. A part of it passes through a large artery under the backbone to the hindquarters. Here this artery sends out a large branch, which in turn throws out several smaller branches that distribute the blood through all the regions of the udder. After the blood has passed through the udder it appears on the outside of it in what are called the *milk veins*. These pass along the belly for some distance in front of the udder, enter the body walls through *milk wells*, and carry the blood back to the heart.

It is thus seen that the heart, lungs, arteries, and veins are of great importance in the manufacture of milk. The part played by the heart and lungs shows that it is very important for the cow to have a deep, wide, full chest, indicating that these organs are well developed and that she possesses a strong constitution. The size of the milk veins and milk wells is an indication of the amount of blood that passes through the udder to supply material for the manufacture of milk. On this account it is important that they be large.

The udder. It is in the udder that the process of making milk from the material supplied by the blood is carried on. The udder serves also as a reservoir for the milk after it has been made until withdrawn by the process of milking. It is especially important that the udder have a large capacity, and to this end it should be attached high behind and carried well forward, and be snugly attached to the body to prevent injury. The quarters should be even in size, free from fleshiness, and with the teats squarely placed and of convenient size for milking. When empty it should appear to consist of folds of soft, pliable, elastic skin.

The nervous system, represented by the brain and the spinal cord with its branches, controls the action of the various organs of the body. In the dairy cow it is very important that the nervous system be strong and well developed in order that the organs concerned in the manufacture of milk may carry on their work most effectively. The cow with a nervous system of this kind is spoken of as having a nervous temperament. This does not mean that she is irritable and excitable, as the term often implies, but that she possesses a strong set of nerves that has the various organs of the body under good control. The nervous temperament in the dairy cow is indicated by a lean yet vigorous condition, showing

FIG. 205. DeKol Plus Segis Dixie, No. 295,135, H. F. A. A., the Holstein-Friesian cow that gave in one year 33,464.7 pounds of milk, from which were made, 1,686.3 pounds of butter. Note the wedge shape, A B C.

that the feed she consumes is being used chiefly in the production of milk and not in the laying on of flesh. An animal of this temperament is active and wide awake. The temperament of the beef animal differs from that of the dairy animal, being what is called a *lymphatic* or lazy temperament, which is conducive to the laying on of flesh. Dairy cows that show a beefy tendency are not utilizing their food for milk production as they should.

292. **The Dairy Type.** Having learned the parts of the dairy cow that are chiefly involved in milk production, we are now in a position to understand the dairy type. We can

see that the digestive organs and the udder, on account of the important work they perform, should be highly developed. We can see also that the dairy cow should be lean in condition. A lean head; a rather long, thin neck; lean, thin withers; thinly fleshed back, ribs, loin, and rump; and thin, long thighs characterize the nervous temperament. The good dairy cow must also be wide of loin, hips, and rump. The high development of barrel and udder, the width of the loin, hips, and rump, together with the thin neck and lean condition throughout, give the dairy cow a wedge-shaped form. Three wedges may be seen, as indicated in Figures 205 and 206. This peculiar type which is so closely associated with high milk production has been intensified in each breed of dairy cattle by many years of careful breeding. The points in detail to be considered in judging dairy cows are given in the score card on pages 308, 309.

FIG. 206. Note the wedges *B A C* and *D A E*, characteristic of the dairy type. *Courtesy of the Agricultural and Mechanical College of Texas.*

293. **Breeder Type.** In addition to judging the dairy cow by the points indicated on the score card as a milk producer, she should be judged also as a breeder. The points to consider here are the same as those given for the breeder type when discussing beef cattle.

294. **The Dairy Bull.** The dairy bull may be judged by the records of his daughters as milk producers, but this method can be applied only to old bulls. The more common method is to judge by his agreement with a certain type proved to be valuable, and by the records of his ancestors.

A bull from good parents, grandparents, and great-grandparents is more likely to be a good breeder than one the ancestors of which are not of such merit. In judging the dairy bull the following points are especially important:

1. He should be typical of the breed he represents.

2. He should show in general the spare, angular form characteristic of the dairy cow.

3. He should show distinctly the nervous temperament, as indicated by an active, wide-awake appearance and lean condition.

4. He should possess good quality, as indicated by dense, clean bone, soft hair, and loose, pliable skin of medium thickness.

5. He should show a strong masculine character, as indicated by bold expression of eyes, burly head, strong horns, well-crested neck, and comparatively heavy though not coarse shoulders. The front of the dairy bull is necessarily much heavier than that of the dairy cow, but he should not show the same relative width of hips.

6. He should possess a strong constitution, as indicated by a deep, wide chest, large nostrils, bright, clear eyes, and a general appearance of health and vigor.

7. He should possess a large, capacious barrel, indicating plenty of room for food, for it is important that he be able to stamp this characteristic on his offspring.

8. He should possess a strong back, long, level rump and light, thin thighs, and should be cut up high in the twist. Thick, beefy thighs and deep, full twist are objectionable.

9. The rudimentary teats should be of good size and evenly placed, as they indicate to some extent the size and position of the teats in the female offspring.

295. **Breeds of Dairy Cattle.** The breeds of dairy cattle mentioned in order of popularity in the United States are: the *Holstein-Friesian* (Hōl′stīn-Frē′zhăn), the *Jersey,* the *Guernsey* (Gûrn′zŷ), the *Ayrshire* (Âr′shẽr), the *Brown Swiss,* the *Dutch Belted,* the *French Canadian,* and the *Kerry.*

Holstein-Friesian. The native home of this breed is Hol-

Fig. 207. Fern's Signal of Oaklands, grand champion Jersey bull, National Dairy Show.

Fig. 208. Stockwell's April Pogis: yielded 1,218.48 pounds of butterfat in 365 days, the highest yield ever made by a Jersey in the United States.

Courtesy of the American Jersey Cattle Club.

land. Little is known about its origin, but it is claimed that cattle of the Holstein-Friesian type have been kept by the people of Holland for the production of milk, butter, and cheese for over a thousand years. The size of the breed is greater than that of any other dairy breed. The average weight of mature cows is from twelve hundred and fifty

FIG. 209. Dunloggin Eleanor, Holstein-Friesian Grand Champion cow, National Dairy Show, 1936.

pounds to fourteen hundred, and of mature bulls from nineteen hundred to two thousand pounds. It is not uncommon for weights of both cows and bulls to exceed these figures. The color is black and white spotted, sometimes black predominating and sometimes the reverse. Black on the legs is considered objectionable. The Holstein-Friesian cow is famous for the large quantity of milk she produces. In this respect she is far ahead of all other breeds. The cow Carnation Ormsby Butter King holds the world's record over all breeds for milk and butterfat production, having produced 38,606.6 pounds of milk containing 1402 pounds of butterfat, in one year. The milk of the Holstein-Friesian is not rich in butterfat, but a large quantity of butter is generally produced on account of the large yield of milk. There are now more than two hundred and thirty Holstein-Friesian cows that have produced in excess of one thousand pounds of butterfat in one year.

Holstein-Friesian cattle were probably first brought to the United States by the early Dutch settlers of New York. During the latter half of the nineteenth century several importations were made, and from them the breed has been developed in this country. Since 1885, few importations have been made due to the prevalence of foot and mouth disease in Holland. While this is the most popular dairy breed in the United States, it has gained little attention in the Southwest due to the limited feed supply and poor pasture. With the greater attention now being given to the production of feed and pastures, we may expect to find more Holsteins in the future, especially around the city milk markets.

The *Jersey* came from a little island of that name in the English Channel, and is probably descended from two French types of cattle that had been taken to the island. As early as 1763, the interest in breeding a fine dairy type was strong enough to get a law passed forbidding the bringing to the island any cattle from France except for immediate slaughter. Soon similar laws were made against cattle from other countries. Since 1833 the most rigorous selection has been carried on, with the result that the Jersey excels all other breeds in richness of milk and in beauty and refinement. In size the Jersey ranks from medium to small. An average bull weighs about fifteen hundred pounds and an average cow about nine hundred and fifty pounds. There is, however, wide variation in weights of both bulls and cows. The color also varies considerably, a fawn-like color predominating. It may be a yellowish, reddish, grayish, brownish, or silvery fawn. Some are described as orange or lemon fawn, and others as squirrel gray or mulberry black. White markings often occur, but are not in favor. The Jersey is especially

SCORE CARD

Scale of Points for Jersey Cow

Approved by the Board of Directors of the American Jersey Cattle Club

HEAD, 7	*counts*
A—Medium size, lean; face dished; broad between eyes, horns medium size, incurving	3
B—Eyes full and placid; ears medium size, fine, carried alert; muzzle broad, with wide open nostrils and muscular lips; jaws strong	4
NECK, 4	
Thin, rather long, with clean throat, neatly joined to head and shoulders	4
BODY, 37	
A—Shoulders light, good distance through from point to point, but thin at withers; chest deep and full between and just back of fore legs	5
B—Ribs amply sprung and wide apart, giving wedge shape, with deep, large abdomen, firmly held up, with strong muscular development	10
C—Back straight and strong, with prominent spinal processes; loins broad and strong	5
D—Rump long to tail-setting, and level from hip-bones to rump-bones	6
E—Hip-bones high and wide apart	3
F—Thighs flat and wide apart, giving ample room for udder	3
G—Legs proportionate to size and of fine quality, well apart, with good feet, and not to weave or cross in walking	2
H—Hide loose and mellow	2
I—Tail thin, long, with good switch, not coarse at setting-on	1
UDDER, 26	
A—Large size, flexible and not fleshy	6
B—Broad, level or spherical, not deeply cut between teats	4

C—Fore udder full and well rounded, running well forward of front teats . . . . 10 . . .

D—Rear udder well rounded, and well out and up behind . . 6 . . .

TEATS, 8

Of good uniform length and size, regularly and squarely placed . . . . 8 . . .

MILK-VEINS, 4

Large, long, tortuous and elastic, entering large numerous orifices . . . . 4 . . .

SIZE, 4

Mature cows, 900 to 1100 pounds . . . . 4 . . .

GENERAL APPEARANCE, 10

A symmetrical balancing of all the parts, and a proportion of parts to each other, depending on size of animal, with the general appearance of a high-class animal, with capacity for food and productiveness at pail . . . . 10 . . .

TOTAL SCORE 100

noted as a producer of rich milk, that is, milk that contains a high percentage of butterfat. It is also noted for the comparatively large size of the fat globules in the milk, this being a great advantage in causing the cream to rise and separate more easily. Jersey milk is sometimes sold under the label "Jersey Cream Line" because of its deep rich layer of cream. The Jersey cow, Abigail of Hillside, No. 457241, A.J.C.C., holds (1937) the official test record of her breed for milk production. She produced in one year 23,677 pounds of milk, containing 1,197.51 pounds of butterfat. Another Jersey, Brampton Basilua, No. 48490, holds the world record for the breed in butterfat production, with a record of 1,312.8 pounds in one year.

The importation of Jersey cattle to the United States began early in the nineteenth century, but importations did not become frequent until 1850. The Jersey is one of the most popular breeds in the United States and is now found in every state. Jersey cattle are numerous in the Southwest, where they have been in strong favor for many years, almost to the exclusion of other breeds. They are more widely distributed over the world than any other dairy breed.

FIG. 210. Cathedral Rosalie, world record Guernsey cow, 23,714.5 pounds of milk and 1213.1 pounds of butterfat in one year.

The *Guernsey*. This breed, the native home of which is the islands of Guernsey and Alderney, in the English Channel, near the island of Jersey, has high merit. Several herds of Guernsey cattle are owned in the United States, chiefly in New England, New York, New Jersey, and Wisconsin. The breed, however, has not gained the prominence to which its merit entitles it. There are few Guernseys in the Southwest, though there is no reason why they should not do well here. The size of the Guernsey is generally larger than that of the Jersey, the average weight of mature cows being about a thousand and fifty pounds, and of mature bulls about fifteen hundred pounds. In color, animals of this breed may be either yellowish, brownish, or reddish fawn, with white markings frequently occurring on body or legs. The breed is noted for the rich golden color of the milk. Guernsey milk is frequently sold as "Golden Guernsey" milk.

FIG. 211. Grange Rosebud 2nd, world's record Ayrshire cow, 26,776 pounds of milk, 1303.66 pounds of butterfat in one year.

The *Ayrshire*. The native home of this breed is in the county of Ayr, in southwestern Scotland. In size the breed ranks as medium, the average weight for mature cows and bulls being about the same as stated for the Guernsey breed. The color is white, with red or brown markings. The breed ranks high in yield of milk, which, however, is comparatively low in butterfat content. Ayrshire cattle have been exported from Scotland to many different countries. In North America they are found chiefly in Quebec and Ontario, Canada, and in the New England and Eastern States of this country.

The *Brown Swiss* is a large, rather beefy breed of dairy cattle whose native home is in Switzerland. On account of its beefy tendency it is classed by some as a dual-purpose animal. This breed is gaining rapidly in popularity and is found principally in the corn-belt states. It is a heavy-producing breed and very rugged and hardy.

The *Dutch Belted* from Holland, the *French Canadian* from Canada, and the *Kerry* from Ireland are little known in this country and are of little importance to us.

296. **Dual-Purpose Cattle.** Dual-purpose cattle have been bred for both beef and milk production. From what you have learned of the beef and dairy types it should be clear to you that both beef production and milk production are not likely to attain the highest degree of development in the same animal. We therefore usually find the dual-purpose type first class neither for beef nor for milk, though there are some notable exceptions. Cattle of this type, however, meet the demand of many farmers for animals that will be superior to the dairy breeds for beef and superior to the beef breeds for milk. The two breeds of dual-purpose cattle of the most importance are the Red Polled and the Milking Shorthorn.

Red Polled. In the early part of the eighteenth century

FIG. 212. Red polled cow.
Courtesy of R. F. Hildebrand.

there existed a small, thin-fleshed, red-brindled, or dun-colored polled type of cattle in Suffolk, England, noted for its milk-producing qualities. About the same time in Norfolk there existed a type of cattle described as blood-red in color, with white or mottled face, having horns and possessing a strong tendency to fatten at an early age. These cattle were poor milkers, but of very good beef qualities. The red polled breed originated in a crossing of these two types. Careful selection was practised and the result was a polled dual-purpose breed, solid red in color. Mature males weigh from eighteen hundred to twenty-two hundred pounds and mature cows from eleven hundred to sixteen hundred pounds.

Red polled cattle were not imported into the United States to any extent until after 1873. They are now very well distributed throughout the Mississippi Valley States. They seem well adapted to Southwestern conditions and several prominent herds are owned in this section.

Milking Shorthorns. This breed originated in the shires of Durham, York, and Northumberland in northeastern England, and is sometimes known as the *Durham.* The Milking Shorthorn varies in type from beef to dairy conformation. The cows weigh usually about thirteen hundred pounds and are red, white, roan, or red and white in color. Like other dual-purpose breeds, the Shorthorn is not a persistent milker, frequently drying up after about seven or eight months, although some outstanding exceptions have been noted. The milk of the Shorthorn is richer than that of the Holstein but not as rich as that of the Jersey. There are many Shorthorns in the states of the Mississippi Valley. The breed seems well adapted to the grain sorghum section of the Southwest and a number of herds have been established there in recent years. In the British Commonwealth of Nations, and to some extent in the United States, the Milking Shorthorn is cultivated as a dairy breed. In Australia, the Milking Shorthorn cow, Melba 15th of Darbalara, is reported to have produced in one year 1614 pounds of butterfat.

THE CATTLE TICK

297. **Cause of Tick Fever.** One of the most expensive diseases the South and Southwest have ever known is the cattle "tick fever," as this disease is now called. For many years the losses through this fever from death, quarantine, and other effects have been estimated at over forty million dollars a year. The scientists of the United States Department of Agriculture, in co-operation with the Agricultural and Mechanical Colleges of Texas and other states, have now discovered the cause of this fever and devised methods of completely eradicating it. The fever was found to be caused by parasites which are taken in by ticks when biting infected cows. The parasites are then carried to other animals that are afterward bitten by these ticks, and are even transmitted to the eggs of the tick, and in this way to the next generation.

298. **Valuable Results of the Discovery.** When it was found that ticks caused the fever, and that they could be removed from cattle by oil and other dips, the rigid quarantine against Southern cattle was modified, and a considerable part of this expensive handicap was removed.

Perhaps the worst injury from the tick arose out of the fact that about four fifths of the fine-blooded cattle imported into the Southwest to breed up our scrub herds were given the fever. As they were less resistant to the fever than the native cattle, most of them died. This prevented the rapid improvement of our stock. The scientists next discovered that by injecting some of the blood of a native cow directly into a well one, the healthy animal would be given the fever. The fever properly transmitted in this way is not especially dangerous, as is shown by the fact that only 5 per cent of the animals infected die, whereas 80 per cent die from the fever caused directly by tick bite. The fever caused by direct inoculation, as this method is called, makes the animal immune to the disease thereafter in all forms. Now that this has been learned it is possible to import and inoculate the finest young bulls and heifers and breed up our low-grade herds economically. This great handicap being removed, the Southwest is now coming rapidly to the front in the raising of fine-blooded cattle.

299. **Exterminating Ticks.** By careful study of the habits and life cycle of the ticks, a method of entirely ridding the pastures of them has been devised, and the ticks have

FIG. 213. Cattle tick depositing eggs.
Courtesy of the U. S. Department of Agriculture.

been cleared out of a large number of states and will soon be wiped out of the United States. Investigators found that the grown female tick, when filled with blood, drops from the cow and lays about three thousand eggs. In warm weather tiny ticks soon hatch out and climb upon vegetation, where they are rubbed off by passing stock. As ticks can live only on blood, if no animal of the right kind is found, they finally starve to death. In summer they can live without food for about three months and in winter much longer.

The method of exterminating ticks is simple. The cattle are brought in from the pastures about once in two weeks and dipped in a solution that kills the ticks. In this way only the young ticks that can live to reproduce, namely, those upon the cattle, are killed before they have got their fill of blood and have dropped off and laid a new lot of eggs. This method does not necessarily kill all the ticks in a pasture, because there are other varieties of ticks that live on other animals. It does destroy the particular variety of tick that lives on cattle and transmits tick fever. In this way it clears the pasture of infected, fever-producing ticks.

QUESTIONS, PROBLEMS, AND EXERCISES

128. Make a list of all the reasons for and against raising stock on your farm.

129. How many beef cattle are on your farm? What is their value per head? How could their value be increased in a practical and economical manner?

130. How many dairy cattle are on your farm? How much milk and butter per year does each cow produce? How much more would these pay per year if each one produced one half as much as the Jersey Abigail?

131. Make what you consider a practical plan for livestock raising on your farm. Discuss this with the teacher and then at home.

132. How many of each of the following could be raised on your farm without interfering with the crops now grown there: 1, cattle; 2, horses; 3, sheep or goats; 4, hogs; 5, chickens and other fowls?

133. How many breeds of cattle are there in your community and what are they?

134. Find what kinds of purebred cattle are in your neighborhood and, together with the teacher and remainder of the class, make a visit, inspect, and score each variety.

135. Find and score one good specimen of each of these types: good feeder, poor feeder, fat steer, good breeder, poor breeder.

136. Study a dairy cow on your farm, locating all the points given in the text and figures.

If teacher and pupils can go together to a county or state fair and practise judging it will be very helpful.

CHAPTER XV

THE CARE OF MILK AND ITS PRODUCTS

300. **What Is Necessary in Dairying.** The first necessity in the economical production of milk, butter, and cheese is well-selected dairy cows. After securing cows of the right type one must then learn to handle the milk and butter properly and to feed economically before he can secure the largest return from his herd. Let us see what good milk and butter are and how they are produced. Later we shall study feeding.

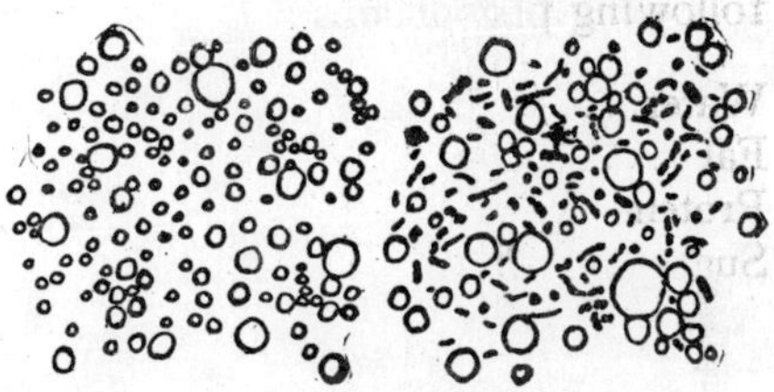

FIG. 214. On the left, pure freshly drawn milk as it looks under the microscope; on the right, impure milk.

301. **What Milk Is.** We have seen that in the good dairy cow a large supply of blood is carried to the udder, where there are organs which can utilize the materials brought by the blood in manufacturing milk. As the milk is made from the materials in the blood, the quality of the milk depends to a certain extent upon what food materials are in the blood, as well as upon the kind of milk-secreting organs there are in the udder. This is why milk from cows that are being fed on clover and peas has a different flavor from that produced by cows that are fed on cottonseed meal. When cows have been eating onions, for example, the flavor of the milk is directly affected.

The composition of milk varies with different breeds, and even with different individuals of the same breed. As a rule Holstein milk, for example, has about three and five-tenths per cent fat, while Jersey milk has about five and four-tenths per cent, yet some Holstein cow may have a much higher percentage of fat than the average and some Jersey much lower than the average. The ingredients of milk are usually in about the following proportions:

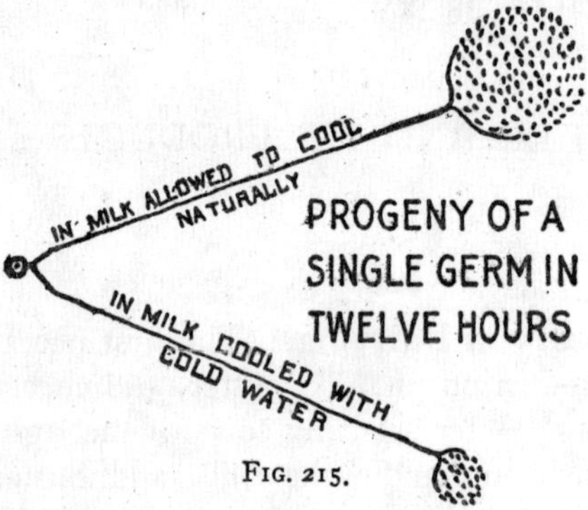

FIG. 215.

Water	87.0 per cent
Fat	4.0 " "
Protein	3.2 " "
Sugar	5.1 " "
Ash	.7 " "

The sweet taste of fresh milk is due to the milk sugar in it. Milk also contains bacteria which work on this sugar and produce an acid that gives the taste to sour milk.

302. **Why Milk Sours and How to Prevent It.** When milk is kept at a temperature of 75 to 100 degrees the bacteria in it multiply so rapidly that in ten to twelve hours the milk is sour. It is not practicable or desirable to keep these *lactic-acid* bacteria, as these are called, entirely out of milk, but if, as soon as it is drawn, the milk is cooled to a temperature of about 40 degrees and is kept cool the bacteria multiply very slowly and the milk will remain sweet much longer. Even at a temperature of 55 or 60 degrees the bacteria multiply slowly. Milk should therefore always be cooled as soon as drawn. Furthermore, every precaution should be

taken to prevent bacteria from getting into it. At milking time the cows and stables must be clean and free from dust, and before milking the cows must be brushed off and the udders washed and wiped clean. The hands of the person milking should be carefully washed before he begins to milk, and whenever soiled afterward. The milk pail should have a top that will admit a minimum of dust (see Figure 216). All milk vessels, no matter how well washed in warm or cold water, still have tiny particles of dried milk left in the cracks and angles. These tiny specks contain thousands of bacteria which will rapidly multiply in the next sweet milk that is put into the vessel and sour it. For this reason all milk vessels should be washed very carefully and sunned and should then be scalded before being used in order to kill all bacteria on them. If these precautions are taken, milk should keep sweet without ice for a day or two in even the hottest weather, provided it is cooled with spring or well water immediately after it is drawn and kept cool with running water or by the evaporation of water around it.

FIG. 216. A sanitary milk pail.

303. **Danger in Milk.** There are not only lactic-acid bacteria in milk, but many other bacteria are liable to get into it from the air, the vessels, and the persons handling it. Once in the milk, these bacteria thrive and multiply with wonderful rapidity. Fortunately many of them are harmless, but others are very dangerous. Typhoid fever, scarlet fever, tuberculosis, and many other serious diseases may be carried in milk. For this reason no one with a germ disease should work around a dairy or handle milk. The vessels and milk should always be protected from dust and especially from flies. The vessels should be washed only in water that is known to be pure. Two hundred and thirty-six people were given typhoid fever by one dairy at which the milk cans were washed in an infected stream. If milk is not kept scrupulously clean it is very dangerous. If it is not kept cool the bacteria in it multiply so rapidly that it is soon unfit

for use. When properly cared for it is one of the best and most wholesome of foods.

304. **Cream.** The cream which rises to the top when milk stands for several hours is composed mainly of butter-fat. This fat, which is composed of round globules so small that it takes 6000 of them side by side to measure an inch, is lighter than the rest of the milk, and hence it rises to the

FIG. 217. A sanitary and conveniently arranged dairy barn.

top when milk is allowed to stand. Cream may be separated immediately from the milk by a mechanical *separator,* in which the vessel of milk is rotated six thousand or more revolutions per minute. This rapid revolution tends to throw the milk off from the center of the vessel, as mud or water is thrown from a revolving wheel. The heavier part of the milk is thrown harder and hence is driven to the outer part of the vessel and the light cream is thus squeezed to the center. The milk escapes through an opening at the outer edge of the vessel and the cream through one at the center. In this way the immediate separation is brought about. Milk

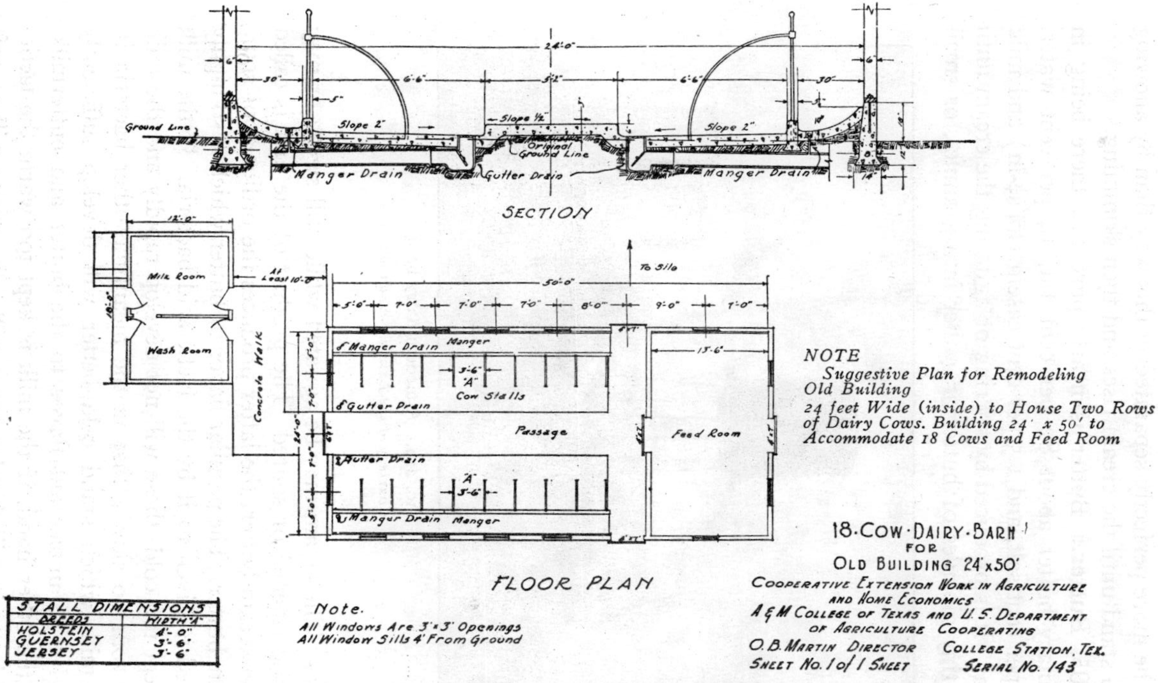

FIG. 218. Cross section and floor plan of dairy barn.
Courtesy of the Agricultural and Mechanical College of Texas.

can be more perfectly separated in this way than by allowing it to stand until the cream rises and then skimming.

305. **Butter.** Butter is nearly pure fat, there being in ordinary butter about 82 per cent fat, 14 per cent water, 2.5 per cent salt, and 1.5 per cent casein (kā'sē-ĭn) and milk sugar. It is produced by stirring or agitating the cream until the tiny globules of butterfat gather into granules, or small

FIG. 219. A modern dairy barn.
Courtesy of the Luling Foundation.

lumps. The cream may be agitated while still sweet, or may be first ripened or soured. The product of the first is called sweet-cream butter, the latter produces the ordinary or sour-cream butter. The peculiar flavor of butter which is so highly prized is given to it by the lactic-acid bacteria. If the milk is kept too cold, these will not develop rapidly and the milk will sour so slowly that in the meantime other bacteria in the milk which stand cold better will develop sufficiently to give an unpleasant flavor to the butter and buttermilk. On the other hand, if the milk is kept too warm the lactic-acid bacteria will develop too rapidly, the butter will be soft and of poor quality and the buttermilk too sour. Sixty to seventy degrees is a good temperature at which to ripen cream. Fifty degrees is the temperature used in some dairies. Sweet-cream butter has a milder flavor and much better keeping quality.

The best butter will result if the cream is at a temperature that will require thirty to forty minutes to churn. This will vary from 55 degrees in winter to 65 degrees in summer. It is a waste of time to churn cream that is at too low a temperature. If the temperature is too high the butter will be soft and mixed with the curds of the milk. A dairy thermometer costs very little and should always be used. The best churns are those that revolve. They should not be filled more than one-third to one-half full when churning. As soon as the grains of butter become as large as grains of wheat, draw off the buttermilk and rinse with water slightly cooler than the buttermilk, until the water runs clear. Then add as much water as there was buttermilk and churn until the butter gathers in a large lump. Draw off the water, add one-half ounce of salt per pound of butter, and work until the salt is thoroughly mixed. Care must be exercised not to overwork the butter, as too much working spoils the grain of the butter. When the butter comes slowly, it may be because the cream is not warm enough or the churn is too full. Butter, like milk, must be handled with the greatest of cleanliness, and should not be placed near anything having a strong odor, as it absorbs odors easily.

306. **Cheese.** Cheese is also made from milk. It consists principally of the protein part of the milk called casein, together with varying amounts of fat and water. In order to make the American cheese, such as is usually sold in the grocery stores, the casein is first precipitated by rennet, which is put into the milk for that purpose. When the casein and fat are precipitated the whey is drawn off and the fat and casein heated, drained, salted, compressed, and cured or ripened. The ripening requires from several weeks to several months, and usually demands cold storage and expert

handling. Bacteria also play an important part in the making of cheese and giving it flavor, but this must be left for later study, as must the making of all the other kinds of cheese except cottage cheese.

Cottage Cheese, or cream cheese, is very easy to make, and may be prepared in any home. The milk is allowed to clabber, then heated slightly to hasten the separation of the curd and whey. It is then hung in a cheesecloth bag or put into metal moulds which are made for the purpose and left till the whey is all drained out and the curd firm. It may be pressed into moulds and kept for several days if kept at a low temperature. When eaten it may be flavored with salt or served with sweet cream and sugar.

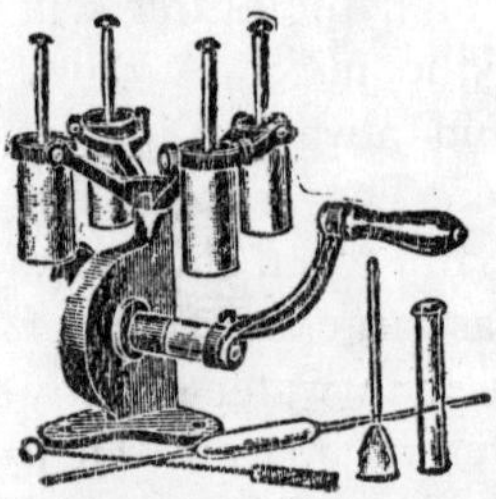

FIG. 220. A small, inexpensive Babock milk tester.

307. Sterilizing and Pasteurizing Milk. When milk is heated to the boiling point, 212 degrees, and boiled for a few minutes the germs in it are killed, so that it will keep sweet for a long time if protected from fresh infection. Such milk is said to be *sterilized* (stĕr′ĭl-īzd). Unfortunately, milk when boiled loses some of its food value. For that and other reasons milk is often heated to 145 degrees only and kept at this temperature for twenty minutes. This is called *pasteurizing* (păs′tŭr-īz-ĭng). If cooled promptly and kept cool, pasteurized milk will keep sweet for several days and will not have the cooked taste and other undesirable qualities of boiled milk. Pasteurizing kills bacteria, but does not kill spores, hence the milk is not sterile, and if it is allowed to get warm again the spores will develop. If milk, about the cleanliness of which there is any doubt, must be used it should be pasteurized. Pasteurization does not take the place of cleanliness at all, but where it is impossible to keep milk cool, pasteurization will help to delay the souring.

QUESTIONS, PROBLEMS, AND EXERCISES

137. Milk one quart of milk, using every precaution mentioned in the text, both in the milking and in the cleaning of the vessels. Place this beside a quart milked in the ordinary way and placed in vessels cleaned in the ordinary manner. Test both after ten, twelve, fifteen, eighteen, twenty-four hours to see which keeps best and has the best flavor.

138. Take three quarts of freshly drawn milk. Cool one at once to the temperature of spring or well water and keep it at that temperature. Wait an hour and then treat the second quart in the same way. Leave the third quart exposed to the summer heat all the time. Test these after six, ten, fifteen, and twenty-four hours and note the acidity and flavor of each.

139. Take several gallons of cream, churn one third while fresh, churn another third as soon as it clabbers, and the other third two days later. Note the difference in the butter from each third when churned and after it has been held two days.

140. Take several gallons of sour cream, churn one third of it at a temperature of fifty degrees, one third at sixty, and one third at seventy-five degrees. Note the effect of the different temperatures on the butter and buttermilk.

141. Take three quarts of fresh milk. Sterilize one, pasteurize another, do nothing to the third. Then place all three under the same conditions and taste each after twelve, eighteen, twenty-four, and forty-eight hours.

142. If your school has a Babcock tester, bring samples of milk from each cow and test for per cent of fat. Make test of milk drawn at the beginning of the milking and of that drawn when the udder is nearly empty.

CHAPTER XVI

HORSES

308. **The Modern Horse Not a Native of America.** The horse has probably been associated with man longer than any other domestic animal. A prehistoric horse existed in North America at one time, but the modern horse had its beginning here with the early settlement of the country by Europeans.

309. **Importance of Horse Raising in the Southwest.** Practically the whole Southwest is well adapted to successful horse raising, the western portion being especially so.

The number of horses in the states of Texas, Louisiana, Oklahoma, New Mexico, and Arkansas in January, 1936, was 1,507,000, which was about one eighth of the total number in the United States. The average value of horses in the South and Southwest is very low, being only $73 as compared to $97 for the entire United States. Massachusetts ranks first in value per head of horses, being $160. The low value per head of horses in the South and Southwest is due to there having been little attention paid to improvement through careful selection and breeding. Every farmer who raises horses should strive to produce animals of a definite recognized market type. He should decide on the type he wishes to produce and then work constantly toward that end. Only sound, purebred stallions should be used. It costs little more to produce a good horse, that will fill a definite market class, than an inferior one.

310. **Judging Horses.** In judging horses it is usual to consider them in two general classes, namely, *heavy* or *draft horses* and *light horses.* Light horses are of three types:

road, or light-harness horses; coach, or heavy-harness horses; saddle horses.

CLASSES OF HORSES

Road Horses. Horses of this class are used to draw light vehicles at a high rate of speed, which they must be able to maintain for a considerable distance without undue fatigue.

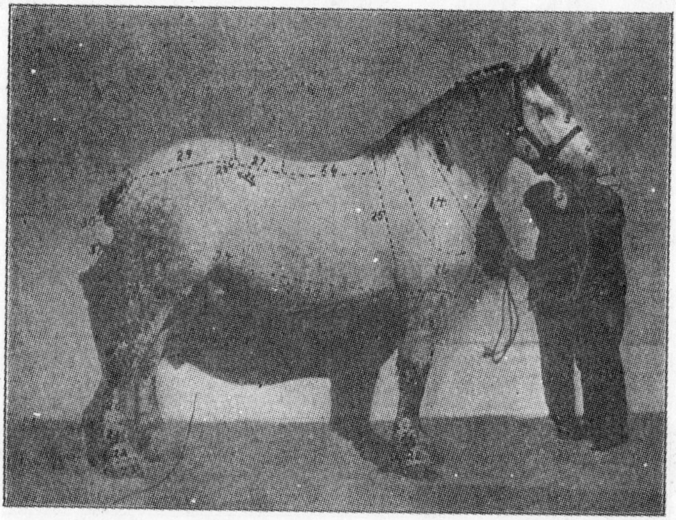

FIG. 221. Points of the horse: 1, mouth; 2, nostril; 3, chin; 4, nose; 5, face; 6, forehead; 7, ear; 8, eye; 9, lower jaw; 10, throatlatch; 11, windpipe; 12, crest; 13, withers; 14, shoulders; 15, breast; 16, arm; 17, elbow; 18, forearm; 19, knee; 20, cannon; 21, fetlock-joint; 22, pastern; 23, foot; 24, fore-flank; 25, heart girth; 26, back; 27, loin; 28, hip; 29, croup; 30, tail; 31, buttocks; 32, coupling; 33, belly; 34, rear-flank; 35, thigh; 36, stifle; 37, quarters; 38, gaskins, or lower thigh; 39, hock.

Courtesy of R. F. Hildebrand.

The chief characteristics are speed and stamina. The form is rather angular, the chest is very deep, and the loin and hindquarters, where the propelling power and chief strength of the horse reside, are very muscular. Good quality is indicated by clean bone, soft, pliable skin, silky hair. Horses of this type have been practically eliminated by motor vehicles. They are now used mainly for racing purposes and grades are used as farm animals and for the production of mules.

Heavy-Harness or Carriage Horses. Heavy-harness horses, used for drawing carriages, have been practically eliminated except for show and pleasure purposes. Symmetry of form and high, stylish action are their most marked qualities. Heavy-harness horses range in weight from nine hundred to twelve hundred pounds, and in height from four-

FIG. 222. Sir Gallahad III, one of the greatest Thoroughbred sires. The Thoroughbred is the foundation of the American Saddle horse and the Standardbred, and is responsible for much of the improvement in the polo pony and cattle pony.

teen hands to sixteen hands. The form of the coach horse is more compact and more smoothly turned than that of the roadster. It is characterized by fulness and symmetry throughout, owing to the plumpness of the muscle over all parts. A rather small, clean-cut head neatly attached to a well-arched neck is characteristic of the best type of coach horse and contributes much toward style. Such horses also possess quality in a high degree. The carriage horse in action is a "high-stepper." The action should be high, snappy, smooth, and graceful, the length of stride found in the roadster being sacrificed for high carriage of knees and hocks.

Saddle Horses. The name of this class indicates the purpose for which horses belonging to it are used. Their weight ranges from nine hundred to twelve hundred pounds, and their height from fourteen hands three inches to sixteen hands. The typical saddle horse, with its smoothness and symmetry of form, shows better quality and better manners than any other class of horses. In beauty of form, style, and graceful carriage the best saddle horses are unsurpassed. The action of the saddle horse is very important. Besides being able to move in a straight, true manner, horses of this class should have the following gaits: (1) walk, (2) trot, (3) single-foot, or rack, (4) canter, (5) slow pace, running walk, or fox trot. The market also recognizes a class of three-gaited saddle horses, the gaits required being the walk, trot, and canter.

In judging all classes great emphasis should be placed on sound, properly constructed feet and legs, and strength of constitution.

Draft Horses. Draft horses range in weight from fourteen hundred and fifty pounds for the lighter sorts to twenty-two hundred pounds or even more for the heavier kinds. In height they range from fifteen hands two inches to seventeen hands. Weight, made up of heavy bone and muscle, is absolutely essential to the drafter in order that great power may be exerted in the collar. Whereas long, slender bones and muscles are conducive to quick action, comparatively short, heavy bones and thick muscles are conducive to power. The form of the typical draft horse is therefore deep, wide, massive, and low set. Smoothness and symmetry, as in other classes of horses, are also highly desirable. The points indicating good quality and strong constitution must be present here as in all other types of livestock. The expression "no foot no horse" is very often heard, but certainly is not applicable to any class of horses more than to drafters. Their heavy bodies and the heavy work they have to do make large, sound, well-shaped feet and sound, properly constructed limbs of the utmost importance. The action of draft horses should be especially good at the walk, this being the gait at which they are generally required to perform

FIG. 223. Draft type. Linees, grand champion Percheron mare.

their work. However, good action at the trot is also highly valued. The walk should be straight, smooth, quick, and well balanced, with good length of stride. The trot should be free, straight, and regular. The points in detail to be considered in judging draft horses from the market standpoint are given in the card on pages 331 and 332.

311. **Judging for Breeding Purposes.** In judging horses for breeding purposes the market demands must be kept in mind, and in addition certain special breeding requirements. These are similar to those given under judging

SCORE CARD

DRAFT-HORSES **MARKET**

SCALE OF POINTS	STANDARD	POINTS DEFICIENT: STUDENT'S SCORE	POINTS DEFICIENT: CORRECTED
GENERAL APPEARANCE—19 PER CENT			
1. HEIGHT, estimated.... hands; actual....hands			
2. WEIGHT, over 1,600 lbs., estimated........lbs., actual......lbs., according to age..........	6	......	......
3. FORM, broad, massive, well proportioned, blocky, symmetrical........................	4	......	......
4. QUALITY, refined; bone clean, hard, large, strong; tendons clean, defined; skin and hair fine; feather, if present, silky..............	6	......	......
5. TEMPERAMENT, energetic; disposition good....	3	......	......
HEAD AND NECK—9 PER CENT			
6. HEAD, lean, proportionate size; profile straight.	1	......	......
7. EARS, medium size, well carried, alert.........	1	......	......
8. FOREHEAD, broad, full.......................	1	......	......
9. EYES, full, bright, clear, same color...........	2	......	......
10. LOWER JAW, angles wide, clean..............	1	......	......
11. MUZZLE, neat; nostrils large, open, free from discharge; lips thin, even, firm.............	1	......	......
12. NECK, well muscled, arched; throatlatch clean; windpipe large..........................	2	......	......
FOREQUARTERS—24 PER CENT			
13. SHOULDERS, moderately sloping, smooth, snug, extending into back......................	3	......	......
14. ARM, short, strongly muscled, thrown back, well set................................	1	......	......
15. FOREARM, strongly muscled, wide, clean.......	2	......	......
16. KNEES, deep, straight, wide, strongly supported	2	......	......
17. CANNONS, short, wide, clean; tendons defined, set back................................	2	......	......
18. FETLOCKS, wide, straight, strong, clean.......	1	......	......
19. PASTERNS, moderate length, sloping, strong, clean	2	......	......
20. FEET, large, even size, sound; horn dense, waxy; sole concave; bars strong; frog large, elastic; heel wide and one fourth to one half the lineal length of toe............................	8	......	......
21. LEGS, viewed in front, a perpendicular line from the point of the shoulder should fall upon the centre of the knee, cannon, pastern, and			

SCORE CARD
(Continued)

DRAFT-HORSES MARKET

SCALE OF POINTS	STANDARD	POINTS DEFICIENT: STUDENT'S SCORE	POINTS DEFICIENT: CORRECTED
foot. From the side a perpendicular line dropping from the centre of the elbow-joint should fall upon the centre of the knee and pastern joints and back of the hoof..........	3	......	......
BODY—9 PER CENT			
22. CHEST, deep, wide, large girth................	2	......	......
23. RIBS, long, well sprung, close; coupling strong..	2	......	......
24. BACK, straight, broad, strongly muscled.......	2	......	......
25. LOINS, wide, short, thickly muscled...........	2	......	......
26. UNDERLINE, low; flanks full.................	1	......	......
HINDQUARTERS—30 PER CENT			
27. HIPS, broad, smooth, level, well muscled......	2	......	......
28. CROUP, not markedly drooping, wide, heavily muscled..............................	2	......	......
29. TAIL, stylishly set and carried...............	1	......	......
30. QUARTERS, deep, broad, heavily muscled, thighs strong................................	3	......	......
31. GASKINS, long, wide, heavily muscled.........	2	......	......
32. HOCKS, large, clean, strong, wide, well set.....	6	......	......
33. CANNONS, short, wide, clean; tendons defined..	2	......	......
34. FETLOCKS, wide, straight, strong, clean.......	1	......	......
35. PASTERNS, moderately sloping, strong, clean...	2	......	......
36. FEET, large, even size, sound; horn dense, waxy; sole concave; bars strong; frog large, elastic; heel wide and one fourth to one half the lineal length of the toe	6	......	......
37. LEGS, viewed from behind, a perpendicular line from the point of the buttock should fall upon the centre of the hock, cannon, pastern, and foot. From side, a perpendicular line from the hip-joint should fall upon the centre of the foot and divide the gaskin in the middle, and a perpendicular line from the point of the buttock should run parallel with the line of the cannon................................	3	......	......
ACTION—9 PER CENT			
38. WALK, fast, elastic, regular, straight.........	6	......	......
39. TROT, free, springy, balanced, straight........	3	......	......
Total..............................	100	......	......

cattle for breeding purposes. The animal should be purebred, true to type, with a strong constitution and all the other marks of prepotency. The male should be distinctly masculine in his characteristics and the female distinctly feminine.

312. **Breeds of Light Horses.** The principal breeds of light horses are as follows:

Breeds of Horses

ROADSTER, OR LIGHT-HARNESS TYPE

Standardbred
Morgan

COACH, OR HEAVY-HARNESS TYPE

Hackney
French Coach
German Coach
Cleveland Bay

SADDLE TYPE

American Saddle Horse
Thoroughbred
Arabian

The Standardbred. The Standardbred is also called the American Trotter and Pacer. The Trotter and the Pacer are of the same breed, about the only difference being in the gaits. As the name indicates, this breed has been developed in the United States. It owes its origin chiefly to the Thoroughbred. The work of development began in the early part of the nineteenth century. The chief characteristics of the breed have already been set forth in the description of the roadster. The type varies considerably, however, and in addition to roadsters the breed furnishes many horses that are suited for carriage purposes. Practically all colors peculiar to horses are found among trotters and pacers, but brown and bay are the most common ones. The world's trotting record is held by Peter Manning, the time being one minute fifty-six and three-fourths seconds. The pacing stallion Dan Patch holds the world's pacing record, the time being one minute fifty-five and one-half seconds.

Hackney. The native home of the Hackney is England. The development of the breed began in the eighteenth century, Arabian, Barb, Turkish, and Thoroughbred stallions being crossed with the native mares of Norfolk and careful

FIG. 224. Red Oak, many times grand champion Morgan stallion. *Courtesy of the Agricultural and Mechanical College of Texas.*

selection practised. The result is a breed presenting the best traits of the coach horse. There is considerable variation in height, but fifteen and one-half to fifteen and three-quarters hands is the height desired by most breeders. The color varies a great deal, chestnut, bay, and brown being common. A great many animals have white markings on legs, feet, and face.

The French Coach and German Coach are similar to the Hackney, except that the German Coach is a little larger and not quite as flashy in action.

American Saddle Horse. Kentucky, Virginia, and Missouri have had most to do with the development of the American Saddle Horse, often called the Kentucky Saddle Horse. The breed had its beginning during the early days of Kentucky, when there were no railroads and horseback travel was common, a condition that caused a demand for

FIG. 225. Beau Peavine, Grand Champion Saddle Stallion, one of the most superb saddle stallions of all time.
Courtesy of Spindletop Farms, Lexington, Kentucky.

easy-gaited saddle horses. The foundation of the breed was laid in crossing Thoroughbred stallions of easy saddle gait on light types of mares possessing the same characteristic. This was followed by careful selection and resulted in a type of much uniformity. The chief characteristics of the breed were given under the description of market classes of horses. The color varies, but blacks, chestnuts, and bays are most common.

The Thoroughbred. The Thoroughbred horse had its

origin in England toward the end of the seventeenth century. During this period several strains of Oriental horses, among which was the Arab, were crossed on the lighter English horses, thus producing the Thoroughbred type. The breed is noted for its running speed, endurance, and quality, and for the influence it has had in the development of other

FIG. 226. Jupiter, grand champion Percheron stallion.

breeds, notably the Standardbred and American Saddle Horse. The height ranges from fourteen and one-half to sixteen and one-half hands, though from fifteen to fifteen and one-half hands is the most desirable height. The weight ranges from nine hundred to ten hundred and fifty pounds. The color varies considerably, blacks, bays, and chestnuts being most common. This breed has been used for racing, for polo playing, and for the production of cow ponies.

The Arabian. The Arabian is native to Arabia, where its

development began several hundred years before the Christian era. It is therefore the oldest of our present-day breeds. It has been noted for its beauty of form, its style, quality, endurance, and intelligence. The size is not as great as that of the average roadster, the height being from fourteen to fourteen and one-half hands. The color may be white, gray,

Fig. 227. Draft type, Clydesdale stallion.
Courtesy of R. F. Hildebrand.

bay, chestnut, or black. The breed holds a place of great importance on account of the influence its blood has had in the development of many other breeds of light horses.

313. **Breeds of Draft Horses.** The principal breeds of draft horses are the Percheron, the Clydesdale, the Shire, the Belgian, and the Suffolk.

The Percheron. The native home of the Percheron breed is France, chiefly in the district of La Perche. The best Percherons of today embody all the desirable features that were described in the discussion of the draft horse. The color is generally gray or black. Mature stallions usually weigh from seventeen hundred to twenty-one hundred pounds and mature females from fifteen hundred to nineteen hundred pounds. The breed is now widely distributed in this country and leads all other draft breeds in numbers. A number of Percherons are owned in the Southwest and do well there.

The Clydesdale. Scotland is the native home of this breed, where its development began in the early part of the eigh-

FIG. 228. Belgian stallion.
Courtesy of R. F. Hildebrand.

teenth century. It is noted for its good feet, sloping pasterns, quality of bone, and good action. A thick fringe of hair occurs on the back of the legs along the cannons, which is termed "feather." The color is generally bay or brown, with white in the face and on some part or all of the legs below the knees and hocks. Though Clydesdales have been owned in the United States for many years, they have never gained the popularity here that they have in Canada. Clean-limbed horses are preferred by farmers here.

The Shire. This breed of horses, which was developed in England, resembles somewhat the Clydesdale. The Shire is the largest of the draft breeds with the possible exception of the Belgian. Horses of this breed have been owned in the United States for many years, but the breed has never become popular.

The Belgian. The native home of this breed is Belgium, where it is bred under government supervision. Belgians are very compact and heavily built. Chestnut is the usual color, but roan, bay, and brown are common. This breed is increasing in popularity rapidly.

The Suffolk. This breed is native to Suffolk County, England. The color of the breed is always chestnut, varying from a light to a dark shade. The type is quite uniform. The weight is not generally as great as that of other draft horses. Suffolks have been imported into the United States to a limited extent. A few Suffolks are owned in the Southwest and seem well adapted to this section.

314. **Mules.** The mule is a hybrid, having for its dam a mare and for its sire a jack. Mules themselves usually do not have offspring, though there are records of a very few fertile mules. The mule has been for long years the principal draft animal used in the South and Southwest, and good mules are nearly always in demand at good prices. The quality of mules in the South and Southwest is very low, owing to the use of inferior jacks and small mares for breeding. As mules are used for draft animals they must have weight. Before our mules can take high rank it will be necessary first to breed our small mares to larger size by using purebred stallions of heavy type. Then by using the larger mares and selected jacks a better grade of mule will be produced.

QUESTIONS, PROBLEMS, AND EXERCISES

143. Find and score one horse of each type.

144. How many breeds of horses are there in your community and what are they?

145. Plan with the teacher and class a Saturday trip to visit and score each type of horse in your community.

CHAPTER XVII

SHEEP AND GOATS

315. **Sheep in the South and Southwest.** Sheep were probably among the earliest of domesticated animals, but there were no sheep in North America prior to its settlement by Europeans. The number of sheep in the United States, January 1, 1935, is given as 52,251,000. Texas, the leading state in number of sheep, was reported to have 7,092,000 sheep. The average fleece weight of the United States is 8 pounds as compared to 7.5 for the South Central States. The Texas average is 8.5. Texas produces about 16 per cent of the entire United States production of wool. Most of the Texas clip is graded as fine.

Sheep are well adapted to most sections of the South and Southwest where the land is well drained. None of our domestic animals is better adapted to the arid western regions. Many farmers make sheep raising a profitable part of their farming operations. It would be profitable for nearly every farmer favorably located for sheep production to make sheep raising a part of his farming operations.

Sheep afford two sources of income, namely, mutton and wool. They surpass all other farm animals in destroying weeds, thereby making more room for valuable grasses to grow and making use of plants that otherwise would be not only of no value but detrimental.

316. **Judging Sheep.** *Method of Examination.* In judging sheep it is necessary in examining them to use the hands as well as the eyes on account of the covering of wool hiding the shape. The use of the hands is important also in examining the wool. The best plan is to begin the examination by making a careful observation a short distance from

the rear, side, and front in turn before using the hands. Then use the hands. Start at the head, continue over the body to the hindquarters and then make a thorough examination of the fleece. In doing this it is very important that the

FIG. 229. Points of the sheep: 1, muzzle; 2, mouth; 3, lips; 4, nostrils; 5, nose; 6, face; 7, forehead; 8, eye; 9, ear; 10, neck; 11, shoulder vein; 12, brisket; 13, top of shoulder; 14, shoulder; 15, chest; 16, foreleg; 17, back; 18, loin; 19, rump; 20, crops; 21, ribs, or side; 22, hip; 23, fore-flank; 24, belly; 25, hind-flank; 26, leg of mutton, or thigh; 27, dock; 28, twist; 29, hind leg.

Courtesy of the Agricultural and Mechanical College of Texas.

hands be held flat with the fingers together in a sloping position, for in this way it is possible to feel the different parts of the sheep's body without breaking the fleece. It is objectionable for the fingers to be stuck into the fleece because they make holes in it, thereby giving access to rain and dirt and detracting from the appearance. The illustrations in Figures 231 and 232 show the correct method of handling sheep in making the examination.

317. **General Classification of Sheep.** Sheep are divided into two main classes, *mutton sheep* and *fine-wool sheep*. Mutton sheep have been developed primarily for mutton, with wool as a secondary consideration. Fine-wool sheep have been developed primarily for wool with little regard for mutton. We find that the two classes are represented by two distinctly different types.

318. **Mutton Sheep.** Mutton sheep are divided in the same way that beef cattle are divided, into three classes—fat sheep, feeders, and breeders. With the sheep, the hindquarter,

FIG. 230. Mutton cuts on the sheep: 1, leg; 2, loin; 3, short back, or rib; 4, breast; 5, chuck.

Courtesy of the Agricultural and Mechanical College of Texas.

loin, and ribs are the parts that produce the high-priced cuts.

319. **Breeds of Mutton Sheep.** The breeds of mutton sheep are divided into two classes based on the character of the wool. The two classes with the breeds included in each are as follows:

MEDIUM-WOOL BREEDS	LONG-WOOL BREEDS
Southdown	Leicester
Shropshire	Cotswold
Oxford Down	Lincoln
Hampshire Down	
Dorset Horn	
Cheviot	
Suffolk Down	
Tunis	

SCORE CARD

From Purdue University Circular No. 29

MUTTON SHEEP **FAT**

SCALE OF POINTS	STANDARD	POINTS DEFICIENT: STUDENT'S SCORE	POINTS DEFICIENT: CORRECTED
1. AGE	......	......	......
GENERAL APPEARANCE—38 PER CENT			
2. WEIGHT, score according to age	8	......	......
3. FORM, long, level, deep, broad, low set, stylish	10	......	......
4. QUALITY, clean bone; silky hair; fine, pink skin; light in offal; yielding high percentage of meat	10	......	......
5. CONDITION, deep, even covering of firm flesh, especially in regions of valuable cuts. Points indicating ripeness are: thick dock, back thickly covered with flesh, thick neck, full purse, full flank, plump breast	10	......	......
HEAD AND NECK—7 PER CENT			
6. MUZZLE, fine, mouth large; lips thin, nostrils large and open	1	......	......
7. EYES, large, clear, placid	1	......	......
8. FACE, short; features clean cut	1	......	......
9. FOREHEAD, broad, full	1	......	......
10. EARS, fine, alert	1	......	......
11. NECK, thick, short, free from folds	2	......	......
FOREQUARTERS—7 PER CENT			
12. SHOULDERS, covered with flesh, compact on top; snug	5	......	......
13. BRISKET, neat, proportionate; breast wide	1	......	......
14. LEGS, straight, short, wide apart, strong; forearm full; shank smooth, fine	1	......	......
BODY—20 PER CENT			
15. CHEST, wide, deep, full	4	......	......
16. RIBS, well sprung, long, close	4	......	......
17. BACK, broad, straight, long, thickly fleshed	6	......	......
18. LOIN, thick, broad, long	6	......	......
HINDQUARTERS—16 PER CENT			
19. HIPS, far apart, level, smooth	2	......	......
20. RUMP, long, level, wide to tail-head	4	......	......
21. THIGHS, full, deep, wide	4	......	......
22. TWIST, plump, deep	5	......	......
23. LEGS, straight, short, strong; shank fine, smooth	1	......	......
WOOL—12 PER CENT			
24. QUANTITY, long, dense, even	4	......	......
25. QUALITY, fine, pure; crimp close, regular, even	4	......	......
26. CONDITION, bright, sound, clean, soft, light	4	......	......
Total	100	......	......

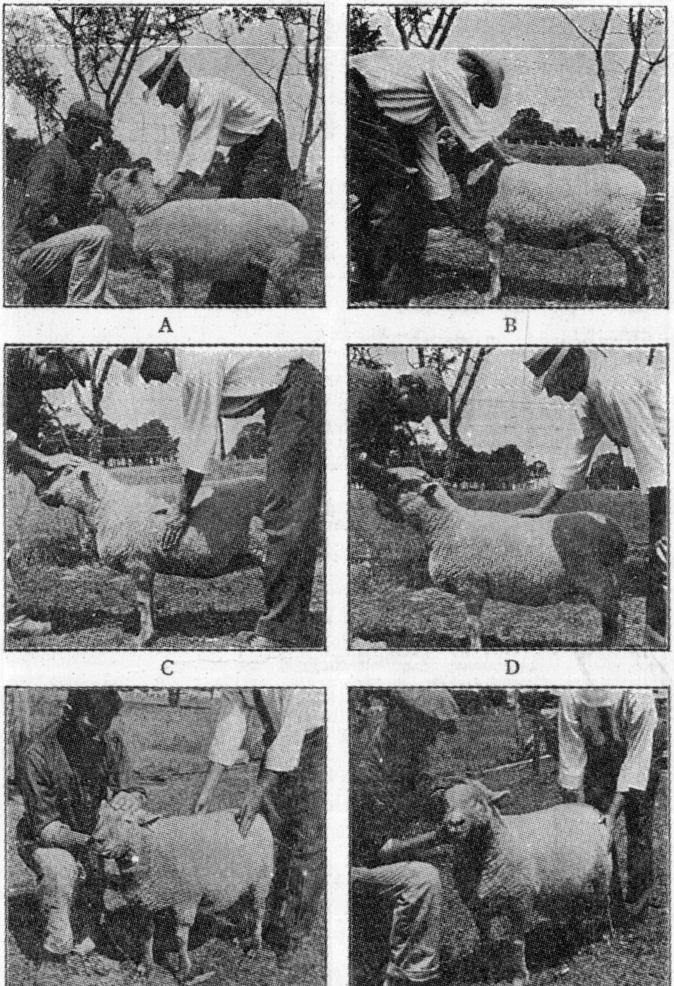

FIG. 231. Judging the sheep. Estimating: *A*, fulness of neck; *B*, depth of chest; *C*, width of chest and covering of ribs; *D*, firmness and covering of back; *E*, width of loin; *F*, width of rump.

Southdown. The native home of the Southdown breed is in the county of Sussex, England, the original stock being the native sheep of Sussex. The breed gets its name from the low range of chalky hills, known as the South Downs,

G H I J

FIG. 232. Judging the sheep. Estimating: *G*, length of rump; *H*, development of leg of mutton; *I*, first quality of wool and examining same; *J*, poorest quality of wool and examining same. *Figures 231 and 232, courtesy of the Agricultural and Mechanical College of Texas.*

which extends through the county. With its low-set, thick, smooth, plump form and high quality no breed of sheep excels the Southdown for mutton. It is also noted for its hardy character. Southdown sheep are not heavy wool producers and lack in size. Mature rams average about one hundred and seventy-five pounds and mature ewes one hundred and thirty-five pounds. The average clip of wool per year for ewes is about six or seven pounds. Rams average a little higher. The color of the face, ears, and legs of the Southdown is grayish or reddish brown. They have wide adaptability and are justly popular.

Shropshire. This breed of sheep originated in Shropshire and Staffordshire, England. The Shropshire is larger than

FIG. 233. Wether in fleece.
Courtesy of Professor W. C. Coffey.

the Southdown and produces a considerably heavier fleece. Mature rams average about two hundred and twenty-five pounds and ewes about one hundred and sixty pounds in weight. Ewes average from eight to ten pounds of wool per year and rams twelve to fifteen. The face, ears, and legs are usually dark brown or blackish brown. A distinguishing feature of the best specimens of the breed is the covering of wool over the head and face, leaving only a small space bare around the mouth and nostrils. The ears should be covered with fine wool instead of hair, and the legs should be well covered with wool down to the feet.

Oxford Down. The native home of this breed is Oxford County, England. It has a mixture of Southdown, Hampshire, and Cotswold blood. It is the largest of the medium-wool breeds. Mature rams weigh from two hundred and seventy-five to three hundred pounds, while ewes weigh about two hundred pounds. The Oxford Down is also in the first rank in the amount of wool produced. A good flock should average about twelve pounds of wool per year. Sheep of this breed are not so heavily wooled over the head as the Shropshire, and the fleece is longer and more open. The

FIG. 234. Wether shorn.
Courtesy of Professor W. C. Coffey.

color of the face and legs is a lighter brown than that of the Shropshire. On account of their large size and heavy fleece they have gained considerable favor in this country.

Hampshire Down. This is also an English breed, originated by crossing Southdown rams on native ewes. It is next to the Oxford Down in size, but only medium in production of wool. The wool is not of high quality. There are many Hampshires in the Southwest, and they seem to be well suited to this region. They have the highly desirable quality of dropping their lambs very early. The lambs grow very rapidly when under favorable conditions.

Dorset Horn. It is thought that this breed originated through the careful selection of breeding animals from native stock that existed in Dorset and surrounding counties in England. The other breeds of sheep that have been discussed are hornless, but this breed, in the case of both males and females, has horns. The color of the face, nostrils, legs, and hoofs is white. The head should have a short foretop of wool. The body should be well covered with wool, which should extend down to the knees and hocks. In size the Dorset Horn ranks as medium among the medium-wool breeds. Mature rams average about two hundred and twenty-five pounds, and mature ewes one hundred and sixty-five pounds. As a wool producer the breed does not rank high. Mature rams average about nine pounds and mature ewes about six pounds of unwashed wool. The breed is especially noted for the production of early lambs and for the good milk-producing qualities of the ewes, making them especially good mothers. Several high-class flocks are now owned in this country, chiefly in the Northeastern States. The fact that the ewes can be bred to drop lambs at practically any time of the year should do much to make the breed popular in this country, especially in the South.

Long-Wool Breeds. The long-wool breeds of mutton sheep are better suited to colder climates and hence are more popular in Canada than in the Southwest. The *Lincoln* is the largest of all breeds of sheep, the rams averaging three hundred pounds in weight and the ewes two hundred and seventy-five pounds. The fleece of rams weighs eighteen to twenty pounds and of ewes about fifteen pounds. The wool is noted for its length and lustre. The *Cotswold* is next in size, with a fleece eight inches long and weighing about ten pounds. The *Leicester* is the smallest of the long-wool sheep, but has a fleece of about the same weight as the Cotswold. The fleece of the long-wool sheep is more open than that of the medium-wool varieties.

320. **Fine-Wool Sheep.** The Merino or fine-wool type of sheep possesses a type of body closely resembling that of

the dairy cow. Instead of being full and square of form as is the mutton type, the Merino type is rather muscular and angular. The production of a heavy, dense fleece of fine quality, evenly distributed over the whole body, is the primary consideration in breeding Merino sheep. Spain is the native land of the Merino, but all the improved breeds are from other countries. Score cards for fine-wool sheep vary somewhat. The Texas show rules allow 50 per cent for body conformation and 50 per cent for wool. Most fine-wool sheep breeders strive to produce a mutton sheep conformation, because of the value of the lambs, and at the same time to produce as much fine wool as possible.

FIG. 235. A good mutton type. *Courtesy of Professor W. C. Coffey.*

321. **Breeds of Fine-Wool Sheep.** Fine-wool sheep in the United States are represented chiefly by three breeds, American Merino, Delaine Merino, and Rambouillet.

American Merino. This breed is simply an improved type of Spanish Merino, improved chiefly in this country. In size the American Merino varies considerably. Mature rams weigh from one hundred and thirty to over one hundred and fifty pounds, ewes average about one hundred pounds. Sheep of this breed stand in the front rank as to weight and quality of fleece and strength of fibre. Mature rams shear about twenty pounds and ewes from twelve to fifteen pounds of unwashed wool. One two-year-old ram in Vermont sheared forty-four pounds three ounces. The length of the fleece of one year's growth is about two and one-half inches. The American Merino is characterized by heavy folds or wrinkles over the whole body except the back. This feature is much more pronounced than in the other fine-wool breeds. The

Fig. 236. Mutton sheep: above, Shropshire ram lamb; center, Cotswold rams; below, Southdown ewe.
Courtesy of Professor W. C. Coffey.

fleece covers the entire body and legs, leaving only the nose and ears bare. All of the fine-wool breeds carry a much larger amount of oil or yolk in their fleeces than do the mutton breeds, but the American Merino carries more than any other breed. After a fleece of this breed has been scoured it may show a shrinkage in weight as high as 65 per cent. American Merino rams have large spirally twisted horns. The ewes are hornless. The large range flocks of Texas and other States of the West and Southwest were made up at one time largely of American Merinos. The breed is less popular now, partly on account of the increasing demand for mutton and partly because of the demand for a type of sheep with fewer folds on the body, so that shearing may be less difficult.

Fig. 237. Rambouillet ram above, Delaine ram in center, Hampshire ram below.
Courtesy of Professor W. C. Coffey.

Delaine Merino. The Delaine Merino is a branch of the American Merino that has been developed especially in Ohio.

The Delaine differs from the American Merino in the following ways. It is usually larger, is more thickly fleshed, thus making better mutton; is practically free of folds except about the neck; the fibre is longer, growing two and one-half to three inches a year, but the fleece weighs less, partly on account of having less oil in it; the rams may or may not have horns, the ewes are hornless. Delaine Merinos have become widely distributed in the United States. Rams of this breed have been used extensively in the range flocks of the West and Southwest.

Rambouillet. The native home of the Rambouillet is in France. In 1783 the French government purchased a large farm near the village of Rambouillet for the purpose of developing an improved type of fine-wool sheep. It is from this village that the breed gets its name. The Rambouillet is the largest of the fine-wool breeds. The average weight of mature rams is about one hundred and eighty-five pounds and of mature ewes one hundred and fifty to one hundred and sixty pounds. Mature rams shear on the average about ten pounds of wool per year. The fleece of the breed is not quite as fine as that of the other fine-wool breeds and does not contain as much oil. The length for one year's growth is about two and one-fourth to two and three-fourths inches. In mutton qualities the Rambouillet is the best of the fine-wool sheep. Folds occur usually only on the neck and breast. Rams usually have large spirally twisted horns, though some are hornless. The ewes are hornless. Wool covers the entire body and legs, leaving only the nose and ears bare. Numerous importations have been made into the United States and today the breed is extensively distributed throughout the country. On account of the ease with which it is sheared and its fairly good mutton qualities it has gained much favor in the West and Southwest, where the rams are used extensively.

GOATS

322. **Uses of the Goat.** Goats are valued chiefly for the production of fleece called *mohair,* and for the production of

The Bray and Hill

WOOL SCORE CARD

Breed.............................Student's Name...................

Grade.................................Date.............................

QUALITY or FINENESS—Fine fibre, breed or grade considered. Not a mixture of fine and coarse fibres. Not a wide difference between shoulder and breech. Fine areas large, coarser one small.	20
LENGTH—Should be clearly of combing length for the grade, *i. e.*, fine, 2½"; ½ blood, 3"; ⅜ blood, 3½"; ¼ blood, 4". Lengths more than ½ inch greater of no additional value except in wool coarser than ¼ blood. Fibres that lie together, all the same length. Little variation over main parts of the fleece. A minimum of short wool.	15
SOUNDNESS—Strong throughout, no weak spots. Fibres of uniform thickness from base to tip except for tapering of yearling fleeces.	15
PURITY—No hair, kemp, or black or dark brown fibres. Cut heavily for: beard-hairs on wrinkles of fine-wool sheep, dark fibres mixed through the fleece of downs, and coarse hair on the breech of crossbreds.	12
CHARACTER—(a) Fibre, evenly crimped throughout, crimp close and distinct, fibres parallel except for sufficient binders to hold the fleece together. Tips free from wastiness. No frowsy wool.	6
(b) Soft and springy to touch, elastic under pressure.	2
COLOR—White, bright. Main fleece free from stains. Minimum of stained areas around breech and on belly.	4
CONDITION—(a) Yolk, moderate in quantity, light color, evenly distributed.	3
(b) Free as possible from naturally adhering sand and dust, and from heavy tags and sweat locks.	3
(c) Free from burrs, chaff, seeds, or other foreign matter, not mouldy or mothy, free from excessive paint, not tied with sisal or rough jute twine or with excessive amounts of twine. Not cotted.	20
Total..	100

WEIGHT OF FLEECE (actual or estimated)..............................

ESTIMATED WEIGHT OF CLEAN WOOL..............................

FINAL SCORE (Total score, times estimated weight of clean wool)........

ESTIMATED PRICE PER CLEAN POUND..............................

TOTAL VALUE OF FLEECE..............................

VALUE PER GREASE POUND..............................

milk and *chevon*. The meat of goats is now called chevon. There have been no strains or breeds of goats developed primarily for chevon production. Though goat meat is used to some extent, the flesh of kid or young goat especially being of good quality and flavor, it has never become popular. Several breeds of milk goats have been developed in various countries, notably on the island of Malta, in Switzerland, Germany, Egypt, Abyssinia, and South Africa. In the United States milk goats have not come into much favor, though in recent years a number of importations have been made. The milk of goats is being used more and more for bottle-fed babies that are not digesting cow's milk well. The breed of goats most popular in the United States is the Angora, which has been developed primarily for its mohair.

FIG. 238. Angora Buck, showing approximately fifteen months' growth of mohair.

Courtesy of Texas Agricultural Experiment Station.

The Angora Goat. The native home of this goat is in the district of Angora, in Asia Minor. It is very probable that it has inhabited this region since before the Christian era. The Angora is adapted to a wide range of conditions, but seems to thrive best in a rather dry climate. Texas, New Mexico, and other Southwestern States are particularly well adapted to the raising of Angoras, and are noted for large flocks of both purebreds and grades.

Being browsers by nature, they are very effective in destroying tree sprouts from "cut-over" land and brush and undergrowth of all kinds. They are used extensively to keep down undergrowth. Goats also eat some of the weeds and coarser grasses that sheep refuse; but, especially during summer, they make about half of their ration on grass.

The Angora is smaller than the common goat, weighing usually about sixty pounds. The back should be straight, chest broad, head clean-cut, muzzle broad, and legs strong, with plenty of bone. The fleece should be creamy white, and should cover the entire body up to the ears and jaws. Sleek necks and bare bellies should be discriminated against. The mohair should grow at the rate of about one inch each month. Then, when the goats are shorn twice a year, each clip will have a five or six-inch staple. The locks should either be a ringlet or a flat lock. In either type a tight full lock is preferred. Short locks on the breech and belly are objectionable. The fleece should be uniform in fineness and length. A medium amount of oil is desired, in order to maintain lustre. Kemp is a serious objection.

Angora goats are very susceptible to heavy losses following cold rains after shearing if they are not protected by sheds. The average mohair fleece weight of the United States is 4.3 pounds, but many flocks produce considerably heavier fleeces.

QUESTIONS, PROBLEMS, AND EXERCISES

146. How many sheep are raised on your farm and what breeds are they?

147. How many sheep could be raised on your farm without interfering with the crops now raised?

148. Make a plan for a small beginning in sheep raising on your farm, finding exactly what this would cost and estimating the probable returns.

149. What other advantages not mentioned in the text are there in sheep raising? What are the difficulties in the way in your locality?

150. What difficulties are there on your farm in the way of raising goats?

CHAPTER XVIII

HOGS

323. Importance of the Hog Industry in the Southwest. Hogs should be one of the most important factors in diversified farming. Very few farms are complete without them, especially in the Southwest. As a rule the farmer who does not count them as one of his principal farm enterprises does not realize from his farm what it is capable of yielding him. Not only can the farmer through raising hogs often produce his own meat supply, by feeding what would otherwise be wasted, but he can market much of his grain and forage crops more profitably when converted into pork and lard than in any other form. The hog is excelled only by the dairy cow in economy in converting foodstuff into an animal product for use as food by man. The number of hogs in the United States, January 1, 1936, was given as 42,541,000. Iowa ranks first among the States in hog production, with 7,250,000 hogs. In some respects a large part of the Southwest is even better adapted to hog raising than is Iowa. Our farmers can grow successfully not only corn, grain sorghums, and other grains suited to hog production, but they can grow also different kinds of green forage crops practically throughout the year. These grains and green forage crops which hogs may graze and harvest themselves are among the most important factors in the economical production of pork. Hence, with her natural advantages, the Southwest could easily rank high as a hog-raising section, both in respect to numbers and value per head. In order to do this, however, our farmers must give more attention to the selection of good breeding stock and must learn to feed and care for their hogs better. In all cases only good, purebred males of the chosen breed should be used. The scrub and grade male should be sent to the butcher's block. The cost of a purebred boar is so small that there is no excuse for breeding scrub hogs. The purebreds and grades mature

FIG. 239. Champion Duroc Jersey Barrow, an excellent type of market barrow.

earlier and have more weight in those parts of the body that furnish the high-priced cuts. The demand for ham, bacon, lard, and other hog products is strong, and with proper management a good margin of profit exists in producing hogs for the market during most years. The meat that is used at home can usually be produced on the farm much more cheaply than it can be bought.

324. **The Care of Hogs.** Hogs are healthier and can be raised much more economically if they live partly on pasture and green crops than when kept in small pens and fed. They

should usually have a little grain to balance their ration properly, but by growing for hogs such crops as alfalfa, clover, peas, soybeans, vetch, sorghum, and peanuts the cost of pork is greatly reduced. Most of these crops may be harvested by the hogs themselves.

As the hog makes greater gain per hundred pounds of food consumed when it is young than when it is grown, it is usually more profitable to grow them rapidly and sell at six or eight months of age, when weighing 180 to 225 pounds.

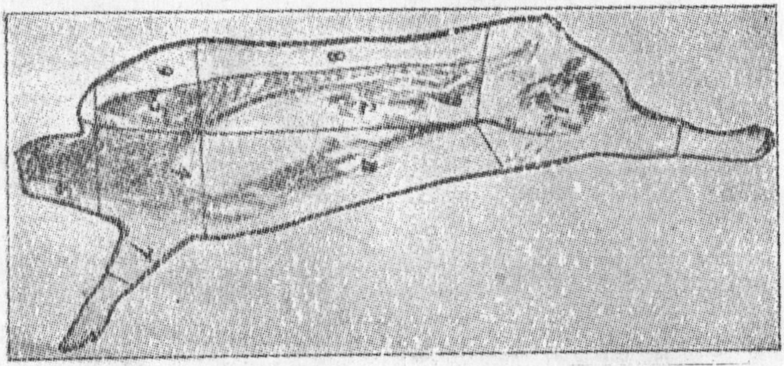

FIG. 240. Wholesale pork cuts: 1, short-cut ham; 2, loin; 3, belly; 4, picnic butt; 5, Boston butt; 6, jowl; 7, hock; 8, fat back; 9, clear plate; 2, 3, 8, side; 2, 8, back; 4, 7, picnic shoulder; 5, 9, shoulder butt; 8, 9, long fat back; 4, 5, 7, 9, rough shoulder.

From Purdue University Circular No. 29.

Sows need attention, especially at farrowing time, and should be protected against the weather, and the pigs should be protected against the awkwardness of the mother. The farrowing house shown in Figure 242 offers one easy and inexpensive means of meeting these needs.

Hog cholera, which for so many years was such a scourge, has now been conquered by the scientists, so that its ravages may be prevented by making the hogs immune through a form of vaccination. In order to prevent disease among hogs several precautions should be taken. First, any newly acquired hog should be kept to himself for several days before being put with the other hogs. Second, hogs should not be allowed to drink water that may be contaminated. Third,

FIG. 241. Points of the hog: 1, snout; 2, eye; 3, face; 4, ear; 5, jowl; 6, neck; 7, shoulder; 8, foreleg; 9, hind leg; 10, breast; 11, chest line; 12, back; 13, loin; 14, side; 15, tail; 16, fore flank; 17, hind flank; 18, hip; 19, rump; 20, belly; 21, ham; 22, stifle; 23, hock; 24, pasterns; 25, dew-claw; 26, foot.

From Purdue University Circular No. 29.

FIG. 242. A model movable hog house for farrowing.

Courtesy of the Agricultural and Mechanical College of Texas.

SCORE CARD

From Purdue University Circular No. 29

LARD-TYPE HOGS — **MARKET**

SCALE OF POINTS	STANDARD	POINTS DEFICIENT: STUDENT'S SCORE	POINTS DEFICIENT: CORRECTED
GENERAL APPEARANCE—30 PER CENT			
1. WEIGHT, score according to age	4	......	......
2. FORM, deep, broad, medium length; smooth, compact, symmetrical; standing squarely on medium short legs	10	......	......
3. QUALITY, hair smooth and fine; bone medium size, clean, strong; general appearance smooth and refined	6	......	......
4. COVERING, finished; deep, even, mellow, free from lumps and wrinkles	10	......	......
HEAD AND NECK—8 PER CENT			
5. SNOUT, medium length, not coarse	1	......	......
6. EYES, not sunken, clear, not obscured by wrinkles	1	......	......
7. FACE, short; cheeks full	1	......	......
8. EARS, fine, medium size, attached neatly	1	......	......
9. JOWL, full, firm, neat	2	......	......
10. NECK, thick, short, smooth to shoulder	2	......	......
FOREQUARTERS—12 PER CENT			
11. SHOULDERS, broad, deep, smooth, compact on top	8	......	......
12. BREAST, full, smooth, neat	2	......	......
13. LEGS, straight, short, strong; bone clean, hard; pasterns short, strong, upright; feet medium size	2	......	......
BODY—33 PER CENT			
14. CHEST, deep, wide, large girth	4	......	......
15. SIDES, deep, full, smooth, medium length	8	......	......
16. BACK, broad, strongly arched, thickly and evenly covered	9	......	......
17. LOIN, wide, thick, strong	9	......	......
18. BELLY, straight, smooth, firm	3	......	......
HINDQUARTERS—17 PER CENT			
19. HIPS, wide apart, smooth	3	......	......
20. RUMP, long, level, wide, evenly fleshed	3	......	......
21. HAM, heavily fleshed, full, firm, deep, wide	9	......	......
22. LEGS, straight, short, strong; bone clean, hard; pasterns short, strong, upright; feet medium sized	2	......	......
Total	100	......	......

SCORE CARD

From "Judging Live Stock," by J. A. Craig

BACON HOGS **MARKET**

SCALE OF POINTS	STANDARD	POINTS DEFICIENT: STUDENT'S SCORE	POINTS DEFICIENT: CORRECTED
GENERAL APPEARANCE			
WEIGHT, 170 to 200 lbs., largely the result of thick cover of firm flesh	6	......	......
FORM, long, level, smooth, deep	10	......	......
QUALITY, hair fine; skin thin; bone fine; firm, even covering of flesh without any soft bunches of fat or wrinkles	10	......	......
CONDITION, deep, uniform covering of flesh, especially in regions of valuable cuts	10	......	......
HEAD AND NECK			
SNOUT, fine	1	......	......
EYES, full, mild, bright	1	......	......
FACE, slim	1	......	......
EARS, trim, medium size	1	......	......
JOWL, light, trim	1	......	......
NECK, medium length, light	1	......	......
FOREQUARTERS			
SHOULDERS, free from roughness, smooth, compact, and same width as back and hindquarters	6	......	......
BREAST, moderately wide, full	2	......	......
LEGS, straight, short, strong; bone clean; pasterns upright; feet medium size	2	......	......
BODY			
CHEST, deep, full girth	4	......	......
BACK, medium and uniform in width, smooth	8	......	......
SIDES, long smooth, level from beginning of shoulders to end of hindquarters. The side at all points should touch a straight edge running from fore to hindquarter	10	......	......
RIBS, deep	2	......	......
BELLY, trim, firm, thick, without any flabbiness or shrinkage at flank	10	......	......
HINDQUARTERS			
HIPS, smooth, wide; proportionate to rest of body	2	......	......
RUMP, long, even, straight, rounded toward tail	2	......	......
GAMMON, firm, rounded, tapering, fleshed deep and low toward hocks	8	......	......
LEGS, straight, short, strong; feet medium size; bone clean; pasterns upright	2	......	......
Total	100	......	......

hogs should not be kept together in large herds, but in small separate herds.

325. **Judging Hogs.** Hogs are divided into two general classes: *fat* or *lard-type hogs* and *bacon hogs.*

Lard-type hogs supply the people of the United States with most of the lard, ham, bacon, shoulders, and other pork

FIG. 243. The bacon type. Champion Tamworth barrow.

products which they consume. The type demanded at present is a hog of moderate length; strong in the back; well developed in the hams; medium in thickness; smooth over the shoulder and along the side; and with straight legs that are of moderate length.

The score card on page 360 gives in detail the points to be considered in judging lard-type hogs.

Bacon Hogs. The bacon hog is comparatively narrow and upright in form, rather light in hams and shoulders, but long and deep in the sides. This type of hog supplies the market with bacon of the best quality. The points to be considered in judging bacon hogs are given in the score card on page 361.

326. **Hogs for Breeding Purposes.** In selecting hogs

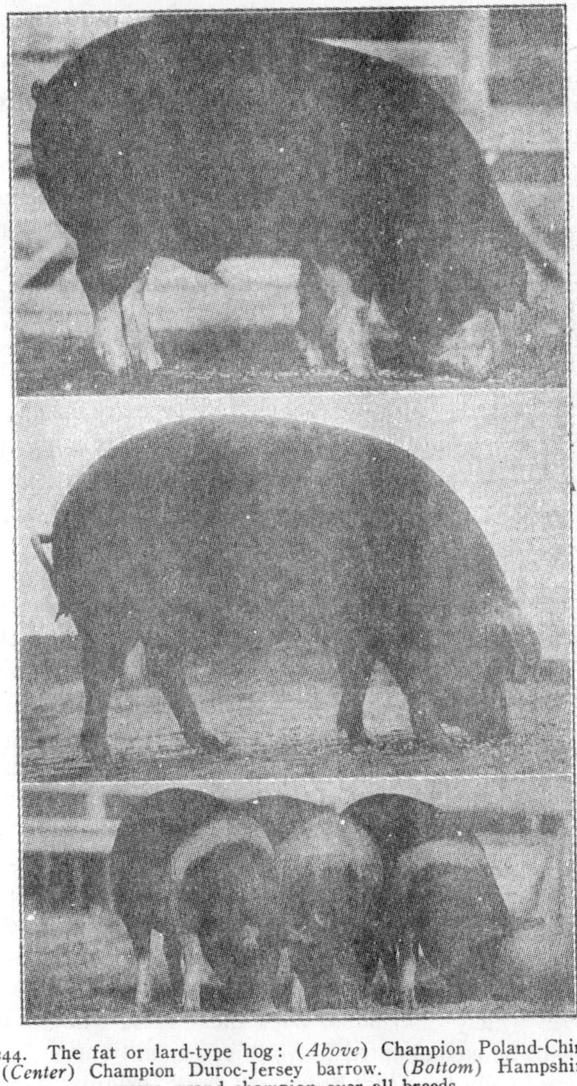

FIG. 244. The fat or lard-type hog: (*Above*) Champion Poland-China barrow. (*Center*) Champion Duroc-Jersey barrow. (*Bottom*) Hampshire barrows, grand champion over all breeds.

for breeding purposes special efforts should be made to get animals that are representative of their type, exhibiting a general appearance of strength and ruggedness without coarseness, well arched in their tops, and standing well up on their toes indicating strength of feet and pasterns. Particularly in selecting boars it is important to select one from a large healthy litter, of a good dam and good sire. The sow should be well developed in her underline, twelve well-developed teats being desired. It is best to select animals that are uniform in size, type, and color, so that their pigs will present a uniform and attractive appearance in the feed lot. Select only those animals that show good breeding qualities; short-middled sows of the chunky type are usually poor breeders; the long, deep-sided, stretchy sow is more apt to farrow large litters.

327. **Breeds of Lard-type Hogs.** The principal breeds of lard-type hogs are the Berkshire, Poland-China, Spotted Poland-China, Duroc-Jersey, Hampshire, and Chester White.

Berkshire. England is the native home of this breed of hogs. The Berkshire of today is characterized by a rather long body, short, dished face, and medium-sized, pointed, erect ears. The color is black, with white on face, feet, and tip of tail. White spots sometimes occur on the body, and though objectionable, they do not indicate impurity of breed. In size, hogs of this breed rank as medium. Mature boars in breeding condition average about seven hundred pounds and mature sows from four hundred and fifty to five hundred and fifty pounds. Many individuals weigh much more. The breed is adapted to a very wide range of conditions and today is widely distributed over the United States. It has proved to be well adapted to Southern conditions.

Poland-China. This breed of hogs originated in the United States, chiefly in Ohio during the period between 1825 and 1840. The modern type of Poland-China shows much quality and is of moderate length and width. The head is broad and of medium length and the face is practically straight. The ears should be of medium size and fine and the top third should droop. The color is generally black, with white on face, feet, and tip of tail. White spots on the body are not uncommon, however. This breed has become widely distributed in the United States, and has met with special favor in the corn-belt region on account of its good

FIG. 245. A cool and substantial hog house.

feeding qualities. It is well adapted to the Southwest and has long been very popular here.

Duroc-Jersey. This American breed of hogs had its origin in the combination of the large, coarse Jersey Red of New Jersey with the finer red Duroc breed of New York. The breeders of these two breeds decided on a definite standard for the Duroc-Jersey in 1877. The best breeders of other red hogs soon afterward adopted the Duroc-Jersey standard and a systematic development of the breed followed. The Duroc-Jersey of today resembles the Poland-China to a considerable degree except in color. The head is moderately wide and of medium length and the face is only slightly dished. The ears are of medium size, the upper third droops forward. The color varies from a light or yellowish red to a cherry red, the latter being in greatest favor. In size the

Duroc-Jersey ranks among the largest of the lard-type breeds. Mature boars in good condition should average from six hundred to eight hundred and fifty pounds, and mature sows from five hundred to six hundred and fifty pounds. The breed is widely distributed in the United States, and has gained much favor. It is well adapted to the South, and in Texas is one of the most popular breeds.

Hampshire. Hampshire hogs are of moderate size, being not so large as the Poland-China or the Durocs. They are black with a white belt from four to twelve inches wide encircling the body over the shoulders and foreribs. This breed has been increasing in popularity very rapidly during the last decade and now is found in large numbers in the South and Southwest.

Chester White. This breed originated in Chester County, Pennsylvania, about the beginning of the nineteenth century. It resulted from crossing a white hog common in that region known as the Big China with some white hogs of Yorkshire descent. This breed also resembles the Poland-China except in color, and is about the same weight. In the Eastern and corn-belt States the Chester White has long been held in high favor. It is not so well adapted to Southern conditions as the Berkshire, Poland-China, and Duroc-Jersey, on account of its color. The long, hot summers of the South cause the skin to sun-scald and become scurfy.

328. **Breeds of Bacon Hogs.** The breeds of bacon hogs are the Tamworth and the Large Yorkshire.

The *Tamworth* is one of the oldest English breeds of hogs. Since the early part of the nineteenth century it has been improved through careful selection and without any infusion of foreign blood. The result is that today representatives of the breed are very uniform in type and color. Tamworths are decidedly of the bacon type, being long and deep of body and lacking the width which characterizes the lard-type breeds. The head is long of snout, the face is slightly dished, and the ears are rather large and carried erect or tilted slightly forward without drooping. The color varies from a dark to a light shade of red, a cherry red being preferred. In size the breed is of the first rank. Mature boars average seven hundred pounds and mature sows five hundred and fifty pounds. Some boars weigh as much as one thousand pounds. Tamworths are very hardy and are industrious in seeking food. The sows are noted for being prolific breeders. The breed is popular in Canada, where there is a greater demand for the bacon type of hog than in the United States. The Tamworth is well adapted to the South and Southwest.

Large Yorkshire. This breed of hogs originated in England. It is quite uniform in type, with a long, deep body of medium width, making it well suited to bacon production. The head is of medium length, the face somewhat dished, and the ears, though often inclined to droop, should be carried erect. The hair is white and the skin pink, with an occasional bluish or black spot. Mature boars usually weigh from five hundred to eight hundred pounds and mature sows from four hundred to six hundred. In Canada this is the leading breed of hogs. It is popular also in a few of the Northern States. A few Yorkshires are owned in the Southwest, but they are not well adapted to that section because they are inclined to blister, due to their light color and pink skin.

QUESTIONS, PROBLEMS, AND EXERCISES

151. How much meat is bought per year on your farm, and how much does it cost?

152. How many hogs would it be necessary to raise per year to supply the meat needed on your farm, and how many acres of land would be required to feed the hogs?

153. Find out how much a young boar and young sows of the leading breeds cost in your community, and figure the expense of starting a small purebred or grade herd.

154. If it were true that a slow-growing scrub would give as large a carcass per hundred pounds of food consumed as a purebred hog, what would still be the advantages in raising purebred or high-grade hogs?

155. Get your father to let you feed and care for a litter of

pigs. Keep an exact account of all food used, and find what it costs per pound to produce this meat. Charge all home-raised food at its market value and all bought food at cost.

156. What foodstuff is wasted on your farm that could be eaten by hogs?

157. Is there idle land on your farm on which crops could be grown easily for feeding hogs?

CHAPTER XIX

POULTRY

329. **Poultry in the United States.** Poultry is raised on a very large per cent of the farms in the United States today. Formerly they received little care, but now a flock of 50 to 1000 purebred, well-culled laying hens is a profitable addition to the farm's livestock. It furnishes the farmer with a weekly cash income. Improved methods of feeding and caring for the poultry flock have enabled many farmers to rank the poultry project as the most profitable one on the farm.

The total production of poultry and poultry products amounts to $1,000,000,000 annually in the United States. The Southwest, with 41,000,000 laying birds, is becoming each year a more important factor in our national egg production.

The State of Texas leads in turkey production, raising about 4,500,000 turkeys each year, mostly for the Eastern holiday market. The annual shipment of dressed turkeys out of Texas is from 1200 to 1600 carloads.

330. **The South and Southwest Are Well Suited to Poultry Raising.** The mild winters and the comparatively dry climate of most of the Southwest are highly favorable to poultry raising. The fowls can exercise in the open and secure fresh, green foods almost every day in the year, which greatly promote both growth and egg laying, and lessen the cost of housing and feeding. Only a reasonable amount of intelligent care is needed to increase vastly the value of the poultry products of this section.

331. **Advantages of Poultry on the Farm.** (1) The fowls on the farm bring a material net profit above the cost of production if they are allowed to range widely and secure a part of their food from weed seeds, insects, and waste grains. This cheap feed needs to be supplemented with only a moderate supply of bought or home-grown feed, including a heavy protein feed, in order to furnish the hens all the necessary materials to make eggs. (2) Eggs and market

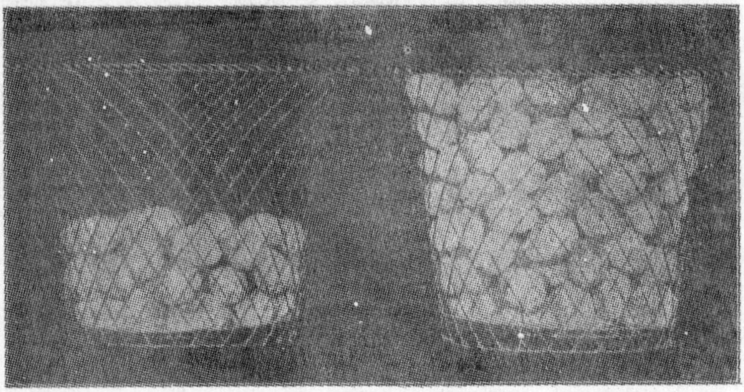

FIG. 246. On the left is the annual product from the average hen, 75 eggs. On the right is the product of one of the good hens at the experiment station, 220 eggs. Why waste food and labor on poor stock?

Courtesy of the University of Minnesota, Department of Extension.

poultry are sold for cash at all times of the year. (3) Eggs and fresh-killed poultry form a very cheap and palatable addition to the farm food supply. Any surplus stock may be canned, and in cases of unexpected company or other emergencies this canned meat saves many hurried trips to town and also provides a satisfying meal at lower cost than fresh meat bought at the market.

332. **How to Start and How to Improve the Flock.** One may start a profitable flock of poultry by purchasing: (1) hatching eggs, (2) day-old chicks, or (3) a number of mature birds from a reliable hatcheryman or from a neighbor. No one breed can be said to be better than all others under all circumstances. Usually it is best to select a breed

that is popular and well adapted to your locality. Get your stock from a flock that is strong, active, and healthy. Start with good stock or failure is certain, no matter how much care is given to the flock. Under ordinary conditions a hen that produces less than a hundred eggs per year is not profitable. It is a waste to keep such stock when the flock, by in-

FIG. 247. White Leghorn hen that produced 327 eggs in 365 days. Note the great length and depth of body, giving exceptional capacity.

telligent selection and care, may be brought to produce one hundred and forty or more eggs per hen during the year. When selecting the breeding stock, choose both males and females of good size with deep, broad bodies; clear sharp eyes; stocky well-developed legs and feet; broad and well-proportioned heads. Select those which have a good appetite, are alert and active, and lay large eggs. Even when one starts with good stock some of the offspring will be poor, hence constant watch should be kept and all hens culled out that by their size, shape, activity, and general appearance are found to be poor layers. Cull out excitable, flighty birds and those that lay small or poorly-shaped eggs. The male is a powerful factor in transmitting egg-laying qualities, so it is very necessary to keep only those that are the offspring of the best-laying hens. Males therefore must not be selected merely by their appearance. They should have good individual size, shape, and health, but they must also be the offspring of parents having desired body and egg-laying qualities.

333. **Feeding and Care of the Flock.** No matter how good the stock may be, there will be few eggs and poor chicks if the flock is not given proper food and care. First of all an egg is 65 per cent water, hence poultry must have a plentiful supply of clean, fresh water so placed that they cannot easily pollute it or get their feet into it. The fountain type of water supply is usually very satisfactory. An empty can makes an inexpensive one if a hole is made on the side about an inch from the top. When this can is filled with water and inverted in a pan about two inches deep and slightly larger in diameter it makes a satisfactory fountain. The water will flow into the pan until the hole is covered, then it will stop until the water again falls below the level of the hole.

Some desirable types of both homemade and ready-made feeding racks are shown in Figures 248 and 249.

Chickens and all other animals must have proteins, carbohydrates, mineral foods, and vitamins in order to thrive, hence the flock should have a varied and complete diet. Poultry seldom can find a plenty of bugs and worms to supply the needed protein, and when enough protein is not supplied in this way it must be provided in some other form, such as skim milk, meat scrap, fish meal, soybean meal, peas, clover, or alfalfa. If hens are fed only corn and milo they get too much carbohydrate and not enough protein, hence they become very fat and lay only a few eggs. Hens must have plenty of protein in order to lay eggs, but too much causes indigestion and kidney trouble. The ration must be complete in order to keep up the energy and animal heat of the hen and at the same time produce eggs. Oyster shell should be supplied freely to furnish the material from which the hen makes the eggshell. Poultry must also have green food in order to promote health and appetite. If this is not

FIG. 248. On the left, a convenient out-door feed hopper; in center and on right, convenient and safe feeding and watering devices.

Courtesy of the U. S. Department of Agriculture and the University of Minnesota.

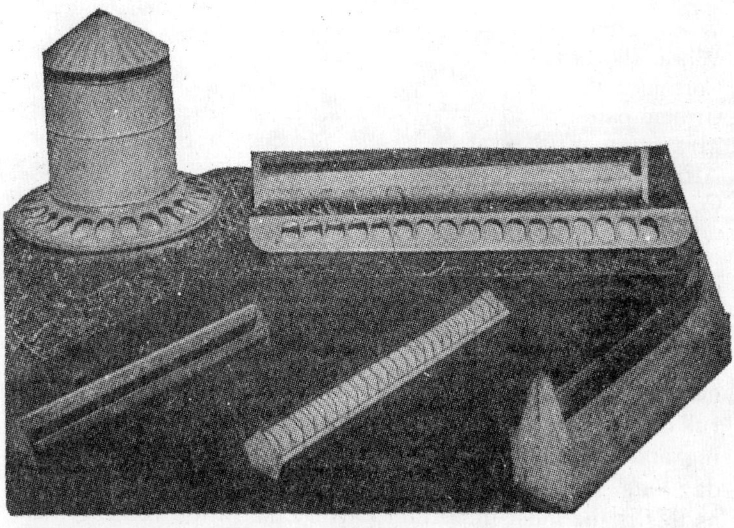

FIG. 249. Convenient and inexpensive ready-made feeders.

supplied plentifully by the natural range, then green food must be produced and fed to them. Good range for poultry can be provided in winter and early spring by growing oats, barley, wheat, clover, or alfalfa; in late spring and summer clover, peas, millet, and Sudan grass should be grown.

334. **Ration for Egg Production.** Many agricultural colleges use the following ration as the basis in feeding laying hens:

GRAIN	
Whole corn	50 lbs.
Whole milo	30 lbs.
Barley, wheat, or oats	20 lbs.

The percentage of each of these grains may be varied according to price. Feed most of the cheapest.

MASH	
Wheat bran	20 lbs.
Wheat shorts	20 lbs.
Corn meal	20 lbs.
Ground oats	15 lbs.
Meat scrap or fish meal	20 lbs.
Alfalfa leaf meal	5 lbs.
Common salt	½ lb.

The grain should be fed in a clean place. Give the birds a light feed in the morning (one-half ounce per bird is plenty) to sharpen their appetites and to start them to work; then allow them all they will eat (about one and one-half ounces per bird) late in the day in order to send them to roost with full crops. The dry mash should be put out fresh every morning and kept before them in hoppers or protected pans all day long. Encourage the birds to eat all the mash they will, as this is the most important part of the ration for egg production. In addition to the food ration, hens should be supplied with crushed limestone or oyster shells in order to produce good shells for their eggs. The chicken has no teeth. Therefore, the food is swallowed whole and must be crushed and ground fine by grit kept in the gizzard for this purpose.

Oyster shell is used to some extent as a substitute for grit, but a small amount of fine gravel or granite grit should be kept before the birds at all times.

335. **Food Needed Per Hen Per Year.** The Texas National Egg-Laying Contest, held at the Agricultural and Mechanical College, showed that a good Leghorn hen consumed 31 pounds of grain, 37 pounds of mash, and 25 pounds of miscellaneous feed in producing 211 eggs a year. Barred Plymouth Rocks ate on the average 37 pounds of grain, 35 pounds of mash, and 25 pounds of miscellaneous feed in producing 183 eggs a year. The 620 hens in the contest ate on the average 5.7 pounds of feed for each dozen eggs produced. The amount of grain and mash consumed by a flock would, of course, vary with the amount and quality of range provided.

336. **Incubation and Brooding of Baby Chicks.** Always choose a quiet, motherly hen for setting. Make sure that she is free of lice and vermin. Watch the hen, and if vermin appear, dust her thoroughly with a good louse powder; repeat in six days and again about three days before the eggs are due to hatch. Place the nest on the ground in a well-drained and protected place, if practicable. Never set a hen in an exposed box on the side of a wall, as the evaporation is so rapid that many dead chicks will result. If the nest must be in a box, then place fresh, well-moistened sod in the bottom and lay the nesting material on top of that. However, it is not usually economical to hatch eggs under hens. The Southwest has an egg capacity of more than twenty-two million in its hatcheries. These hatcheries are so widely distributed that practically everybody is within easy reach of one that can incubate his eggs at a nominal cost and a maximum efficiency. Under these conditions, it does not pay the average farmer to use hens for hatching or to run his own incubator.

Baby chicks received from the hatchery should be put in a dry, well-heated brooder and given fresh water and feed in clean dishes as soon as possible. Keep these chicks quiet and comfortable so that they will spend most of their time sleeping. Skim milk is the very best food they can receive for their first feed. This should be followed with some good chick starter, many brands of which are on the market today. Most of the baby chick starters are bulky enough so that they can be left before the chicks at all times.

A method of telling the sex of baby chicks a few hours after hatching has been perfected which enables the hatcheryman to sell baby pullets or baby cockerels. The baby pullets usually bring twice the price of baby chicks, the baby cockerels from one quarter to one third the price of ordinary baby chicks.

A good homemade baby chick starter can be made from the following:

Ground yellow corn	32 lbs.
Wheat shorts	20 lbs.
Ground oats	20 lbs.
Bone meal	2 lbs.
Ground limestone or oyster shells	2 lbs.
Salt	1 lb.
Dried milk	6 lbs.
Meat scrap	6 lbs.
Alfalfa leaf meal	5 lbs.
Cottonseed meal	6 lbs.

If fresh milk is given in addition to this starter do not put dried milk in the feed, but let the chicks drink all the skim milk or buttermilk they will. All feed should be clean, fresh, and free from mould, as mouldy feed will kill the chicks within a short time. The brooder house must be absolutely sanitary, as there are many baby chick diseases that are the result of faulty sanitation. It is practically impossible to raise chicks in a dirty place. Green feed and sunlight are necessary supplements to the ration. The brooder house should be comfortable, and not more than four chicks per square foot of floor space should be placed in any one house.

337. **The Poultry House.** The poultry house should be located on a slight rise that is well drained. A sandy loam

soil is the best. While expensive poultry houses are not necessary, there is real need for a house to protect chickens from sudden and often severe changes in temperature and weather. Allow at least three square feet of floor space per bird. The house needs especially good ventilation, yet the birds must not be exposed to drafts. The north and west

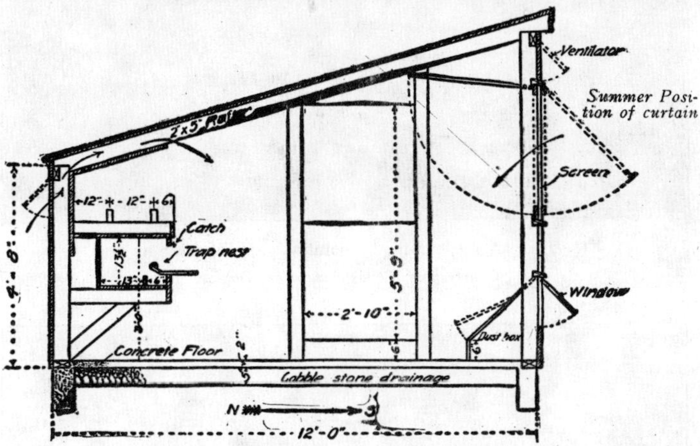

FIG. 250. A model chicken house.
Courtesy of the U. S. Department of Agriculture.

sides should have hinged windows which can be opened when the weather is hot, but these should be kept closed in cool weather. Some opening in the wall on the east side and a great deal on the south side will provide the necessary ventilation. These openings may be covered with netting tight enough to keep out birds and varmints, and should be provided with means of covering them temporarily during extremely bad weather. Arrange to have as much light and sunshine in the house as possible, but it must be so arranged as to shut out the sun during the hot summer months. Attach nests and feed hoppers to the walls where they can be easily reached, leaving the floor free. The floor may be of concrete, wood, or sand, raised above the level of the surrounding earth. Roosts should be smooth and movable and

FIG. 251. A modern model poultry house of medium size.

Courtesy of Texas Extension Service.

FIG. 252. A large commercial poultry laying house. The ventilators in the roof and the windows are highly desirable. The louvers in the gable ends are questionable. In winter all openings are closed except the lower covered windows.

Courtesy of the U. S. Department of Agriculture.

about two feet from the floor with screen in place for droppings below them. Allow ten inches roosting space for each hen. Figures 250, 251, and 252 show standard plans for poultry houses. The house must be kept clean and dry and free of vermin, if the chickens are to thrive.

338. **Protect Poultry from Insects.** Fowls must be protected from insect pests. A plentiful supply of dry road dust or coal ashes kept in a box in the corner of the poultry house will help the fowls in their natural method of suffocating insects. A pound of sodium fluoride mixed in with this dust will help to keep the birds free from lice. Whenever necessary, the fowls should be dusted with louse powder or commercial sodium fluoride, which is an excellent dusting material. Another good, inexpensive powder is made by moistening plaster of Paris with a mixture of three parts of gasoline to one part of crude carbolic acid. The resulting brown powder can be dusted into the feathers thoroughly. However, this is not always sufficient; so the roosts, the house, and everything in it should be sprayed with kerosene, stock dip, or wood preservatives as often as is necessary in order to kill off mites and other pests. Nests should be cleaned out, all litter removed and burned, and the whole inside of the house sprayed. Painting the roosts with nicotine sulphate, or Black Leaf 40 one-half hour before the birds go to roost, for two consecutive nights, will kill all lice on the birds. By the use of these and other means that can be learned in further study, the returns from poultry can be made much greater and more certain.

339. **Breeds of Poultry.** "The American Standard of Perfection," the accepted authority on poultry, recognizes forty standard breeds of chickens. Hundreds of others exist, but are not yet recognized as standard. Only a few of the most important breeds can be mentioned here.

There are three utility classes of chickens: the egg breeds, the dual-purpose breeds, and the meat breeds. The three most popular breeds in the Southwest are the Leghorns, Plymouth Rocks, and Rhode Island Reds. These three breeds compose probably 80 per cent of the chickens on Southwestern farms. Other less important breeds are the Minorcas, Wyandottes, Anconas, Orpingtons, and Cornish, the last named breed being kept largely for its excellence as a meat-producing fowl. The heavy Asiatic breeds, Brahma, Cochin, and Langshan, are not popular in this section. The ornamental class of chickens includes such breeds as the Bantams and Games.

340. **Turkeys.** About seven million turkeys are raised in the Southwest each year. Twelve to sixteen hundred car loads, each containing about two thousand dressed turkeys, are shipped from Texas each year. The popular breeds in order of their importance are as follows: Bronze, Narragansett, White Holland, Bourbon Red, Black, and Slate.

A good turkey breeding hen should have a broad, long back, a deep body, a broad breast, and the legs should be medium short and wide apart. She should also be fully matured before time for the Thanksgiving market. Early maturity in turkeys is inherited, just as early egg production is in chickens. The offspring will have a market sales value because of this inherited quality. There is a definite trend towards raising smaller and earlier-maturing turkeys for the Eastern markets. Several farmers' organizations in the Southwest are pushing turkey improvement along these lines in the hope of getting Southwest turkeys to compete on an even basis with the fancy market turkeys produced in the Northwest.

The sale of turkey hatching eggs has become a definite part of the turkey farm project in the Southwest. More than one million of these hatching eggs are sold to Northern farmers each year at prices ranging from fifteen to twenty cents each.

Turkey eggs are incubated exactly the same as chicken eggs, except that more attention is paid to having plenty of moisture in the incubator. The young poults are treated like baby chicks, except that they require at least 6 per cent more protein and more vitamins than do baby chicks, until they weigh from two to four pounds. Then they are turned out on range and secure most of their living from insects, weed seeds, and waste grains. A small feeding of growing mash once a day is advocated by the best turkey raisers. Raising turkeys in confinement has not been profitable in the Southwest. The turkeys are fattened in October and marketed in November. The minimum market weight should be nine pounds for a young turkey hen, and fourteen pounds for a young turkey tom.

341. **How to Market Eggs and Poultry, and How to Keep Eggs Fresh.** Eggs evaporate after being laid, are quickly affected by odors, and in all cases deteriorate with age. In warm weather the deterioration is very rapid. They should be gathered as soon as possible after being laid, and put in a cool place. If soiled, they should be wiped off at once with a wet cloth. Grade the eggs according to size and market them as soon as practicable. People pay more for fresh, uniformly graded, and clean eggs, especially if they are packed in neat egg cartons.

All male birds should be removed from the flock and either sold or confined by themselves at the close of the hatching season, as infertile eggs bring a premium in the market, and keep during the summer months very much better than do fertile ones.

QUESTIONS, PROBLEMS, AND EXERCISES

158. Make a plan for taking over the poultry at home and managing it scientifically. Discuss this plan with the teacher and then with your parents.

159. Make out complete plans for starting the raising of chickens at home yourself, giving complete bill of costs and probable returns for two years: (*a*) When starting with eggs for hatching; (*b*) when starting with baby chicks; (*c*) when starting with pullets or young hens. Give full cost of chickens, feed, housing, equipment, and other things needed, and estimate the probable returns from eggs and chickens sold and from chickens retained for the flock.

CHAPTER XX

THE CARE AND FEEDING OF ANIMALS

342. **How Animals Use Food.** The animal uses food in two general ways: first, to build up the body and repair used-up tissue; second, to furnish energy for the production of heat and motion. In the first, the animal acts very similarly to a plant, only instead of taking in crude food material from the soil and air, and first manufacturing foods and then making tissue from these, the animal takes in the ready-made foods provided by the plants and turns these into flesh, bone, milk, wool, or other products. In the production of energy the animal acts similarly to an engine, only instead of burning crude coal in a firebox by admitting the oxygen of the air, the animal burns the foods and its own tissue by means of the oxygen admitted to the body through the lungs. Just as the growth of the plant is dependent upon the amount and kinds of food materials supplied, and the energy of the engine is limited to the fuel and oxygen used in the firebox, so the growth and working capacity of the animal are fixed by the amount and kinds of food and air supplied. We can see then that in addition to securing stock of good blood it is necessary to feed and care for them intelligently if the highest returns are to be obtained. Let us then see of what the animal body is composed, how its energy is produced, and how the necessary materials may be most economically supplied.

343. **Composition of Animal Bodies.** You will recall that plants are composed mainly of water, proteins, fats, and

carbohydrates, together with a little mineral matter. The bodies of animals contain these same substances, which the animals secure by eating and digesting plants or the tissues of other animals that have lived on plants. These plant substances are altered somewhat by the process of digestion, particularly the carbohydrates, which are mostly converted into fats or are burned to supply energy to the body.

344. **Water in the Animal Body.** The ordinary animal body is from 43 to 67 per cent water, which is in the blood, other fluids, and all tissues, even the bones. Water cannot be combined with carbon dioxide in the animal body to form carbohydrates as it can in plants, but it is very important in many ways, helping to dissolve and carry around the foods and to carry out the waste materials. A plentiful supply of clean, palatable water is therefore the first essential of good feeding. A considerable part of this water is supplied in the green foods eaten. Green foods not only supply water and other food materials, but, also, being succulent, tend to increase the amount of other foods eaten by the animal.

345. **Protein in the Animal Body.** Protein is used by the animal chiefly in making lean meat, blood, tendons, skin, hair, hoofs, feathers, eggs, and milk curds. Nothing else can take the place of protein for these purposes. It may also be used to some extent to supply energy and heat, which are, however, mainly and more economically supplied by the fats and carbohydrates. From this it is plain that all growing animals, working animals, milk cows, laying hens, geese that are growing feathers, sheep or goats growing wool, must have a plentiful supply of protein in their food, whereas resting adult animals or fattening animals do not need so much protein or *nitrogenous* (nī-trŏj′ĕ-nŭs) food. This protein is found to some extent in practically all parts of all plants fed to stock, but in some, such as corn, kafir corn, sorghum, and prairie hay, the proportion of protein is so small that these should be supplemented by some other food containing a higher percentage of protein. Wheat bran, shorts, cottonseed meal, beans, peas, and other legumes and alfalfa leaves

are rich in protein and make good nitrogenous material to mix with the *carbonaceous* (kär-bô-nā'shŭs) foods, or those containing mainly carbohydrates and fats.

346. **The Use of Carbohydrates and Fat by Animals.** Animals neither manufacture carbohydrates in the body nor store them as do plants, nor is the fat which is stored by the animal exactly the same as that in the plant. The animal, however, uses the plant fats and carbohydrates in the pro-

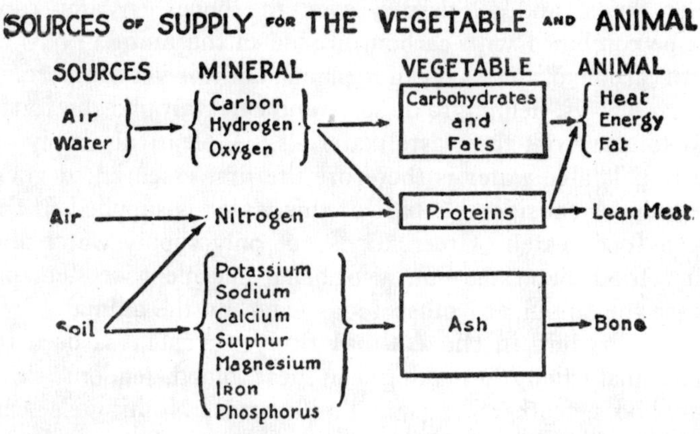

FIG. 253. For the sake of clearness certain details are omitted in the above diagram. For example, bone has some other matter in it besides ash, and protein has in it some of the elements in the lower group of minerals. The diagram is in general correct and affords a good summary to keep in mind.

duction both of animal fat and of heat and energy. The carbohydrates, being rich in carbon, furnish a much more economical material than protein for producing heat and energy by being burned in the body. The fats are even greater energy producers, one pound of fat being equal to two and one-fourth pounds of carbohydrates. The animal fat may accumulate in the body far beyond the present needs and be stored for future use. The amount of fat stored within the body depends upon the animal, the age, the work being done, and the food supplied. Animals hard at work or exposed to excessive cold have to use up the fat to supply energy and heat. The amount of fat in the body varies from 5 per cent to 30 per cent of the body weight.

Carbohydrates are present in all plants in the form of starch and sugar and of crude fibre. Sugar and starch are easy to digest and high in nutritive value; the fibre is hard to digest and low in nutritive value. Corn, kafir corn, milo, rice polish, and molasses are especially rich in digestible carbohydrates. Fat is not so widely and plentifully distributed as the carbohydrates, though all plants contain a little. Peanuts, cottonseed, and soybeans are especially rich in fat.

347. **Mineral Matter in Animals.** Mineral matter is found in all parts of the animal. The bones, which are composed chiefly of calcium and phosphorus, form the largest per cent of the body's mineral content. Mineral matter is also found in the blood, the digestive fluids, and protoplasm. Approximately 2 to 5 per cent of the animal's body is mineral matter. In addition to the calcium and phosphorus found in the bones, several other minerals are essential for the normal growth and development of animals. These are iron, sulphur, iodine, potassium, magnesium, and copper. Although these last named minerals are found in relatively small quantities, they are essential and have specific functions. Minerals are also found in all plants and this is nature's way of supplying animals with minerals. In the wild state, the animals usually found this source of minerals to be adequate for their needs. However, when man removed the animals from their natural surroundings and forced them to grow faster, work harder; and produce greater quantities of milk, he also made this natural supply of mineral matter become inadequate to meet the animal's requirement. Under such conditions man has resorted to the feeding of a mineral mixture to his livestock in addition to the regular feeds. In this way he has been able to get his rapid growth, his harder work, and greater milk flow without injuring the animals.

348. **Air, Shelter, Exercise, Rest, and Kind Treatment.** We have seen that all energy, even that by which

the heart beats, the lungs expand and contract, the digestive system works, and other internal bodily activities are carried on, comes from the combination of oxygen with the compounds in the body. For this and other reasons a plentiful supply of fresh air through well-ventilated stables is essential to the highest success in stock raising. On the other hand, cold draughts are dangerous, while standing out in the open through cold and stormy weather is injurious and uses up food for heat that should go toward flesh and energy production. Properly constructed stables and sheds, therefore, should be provided, having clean, dry beds so that animals may lie down and rest in comfort. It has been proved that a steer gives off from 30 to 50 per cent more heat when standing than when lying down, showing the increased amount of energy consumed in maintaining a standing position. A well-ventilated, comfortable shelter for stock, therefore, quickly pays for itself.

Animals differ from the engine in having a digestive system and assimilating powers by means of which they are able to repair the wear and tear of their own parts. They differ also in having minds that influence the activity of their digestive systems. Therefore all animals must be given exercise to improve appetite and digestion and to stir up the circulation of the blood, which helps to build new tissue and to carry off waste material from the body. They must likewise be given rest always before wear of the tissues is too great to be easily replaced. They must have kind treatment, as the digestive system and other bodily organs do not work so well when animals are irritated and abused.

349. **Proportion of Concentrates to Roughage in Rations.** A food, such as wheat or corn or cottonseed meal, that contains a large percentage of nutrient is spoken of as a *concentrate,* whereas a coarse, rough food, such as hay or sorghum or fodder, that contains a comparatively small percentage of nutrient is called a *roughage.* The proportion of concentrates to roughage in the rations of animals varies greatly, and depends upon the class of animals fed, the pur-

pose in view, and the nature of the feed. Roughage, being generally cheaper than concentrates, should be utilized as much as the demands of the particular animal will allow. Generally speaking, growing stock, stock kept for breeding purposes, and idle horses may be given much the greater portion of their food in roughage. Fattening cattle usually give the best returns when the amount of concentrates in the ration is almost double the amount of roughage. On the other hand, dairy cattle generally produce milk most economically when the amount of roughage is about twice the amount of concentrates. Fattening sheep do best usually when roughage constitutes a little less than half of the ration and concentrates the remaining portion. Horses doing hard work require a ration of more than half concentrates, whereas horses doing light work may get along well on a ration made up chiefly of roughage of good quality. Due to the nature of the hog's digestive system this animal cannot utilize much coarse, bulky material, and therefore its ration must be made up practically altogether of concentrates. However, hogs may utilize advantageously tender green forage plants.

350. **Diet Should Be Varied and Mixed.** It is always best to vary the diet from time to time, and to feed a mixed ration, as experience has shown that good flavor and variety improve the appetite and digestion of stock as well as of man. The daily ration should contain part roughage, part concentrates, and part green succulent food. It is highly desirable that a portion of succulent or juicy food, either grass, fresh green crops, silage, turnips, or other root crops, be used all the time.

351. **Vitamins.** All animals, including man, require certain organic substances, termed vitamins, in very small quantities. Although fed sufficient quantities of digestible protein, carbohydrates, fats, and minerals, the animals will become sick if the proper vitamins are not present. Vitamins are generally present in ordinary feeds and will be sufficiently supplied if the feeds are properly selected. Vitamins are usually named by the letters of the alphabet, on account of

our lack of knowledge as to their exact nature. But they are now being given other names. Vitamin A is found in green plants, and yellow feeds such as yellow corn and sweet potatoes. It may be needed by animals that do not secure enough good green hay pasture. Vitamin B is found in wheat, corn, rice polish or rice bran, and many other feeds. Vitamin C is found in fruits and fresh vegetables. Cows do not seem to need vitamin C. Vitamin D is found in hay or cod liver oil, but its chief source is sunshine. Animals that receive even a moderate amount of sunshine have all needed vitamin D. Vitamin E is extensively distributed and is very abundant in cottonseed oil or cottonseed meal. Animals probably receive plenty of it in their natural foods. Our knowledge of vitamins is increasing every day and you will learn more about them in advanced courses.

352. **The Basis of Calculating Animal Rations.** By careful chemical analyses the dry matter content, the digestible protein content, and the productive value of each of the common feedstuffs have been determined. Rigidly controlled experiments have also shown the daily requirements of the different kinds and sizes of animals for dry matter, digestible protein, and productive energy. Tables I and II are summaries of the results of these analyses and experiments. It is upon such data that the calculation of animal rations is based. Through the application of the information given in these two tables one is able to calculate a daily ration for an animal, which will fulfill his nutritive requirements. Such a ration is called a *balanced ration.* If the proper growth, development, and functioning of an animal are to be obtained it is absolutely essential for him to be fed a balanced ration. If the ration fed him fails to supply him with an adequate amount of each of the essentials, undernourishment will result. This may be shown by retarded growth, a lowered resistence to disease, or a lack of ability to perform the amount of work expected of him. If, on the other hand, the ration supplies a surplus of one essential, an unnecessary expense is incurred. Such a ration may also place an undue strain

TABLE I

AVERAGE PERCENTAGE COMPOSITION AND PRODUCTIVE ENERGY IN COMMON FEEDS

FEED	DRY MATTER PER 100 LBS.	DIGESTIBLE CRUDE PROTEIN	PRODUCTIVE ENERGY IN THERMS
Alfalfa hay	91.7	11.0	37.4
Barley	90.7	9.6	74.4
Bermuda hay	92.2	3.0	31.3
Clover, sweet, hay	90.8	13.5	42.8
Corn, dent	90.9	6.4	84.8
Corn fodder, field cured	57.8	2.3	24.3
Corn silage	29.5	1.1	15.6
Cottonseed	93.0	15.4	76.6
Cottonseed, whole pressed, 28% protein	94.3	22.8	61.0
Cottonseed meal, 43% protein	93.2	35.9	74.9
Cottonseed hulls	90.6	0.4	17.9
Cowpeas	88.4	20.6	80.5
Cowpea hay	90.4	9.0	32.6
Darso heads	90.2	6.6	79.9
Darso (seed) or chops	90.3	7.8	84.9
Ear corn chops	88.7	5.5	74.7
Ear corn chops with shuck	90.4	5.2	72.8
Hegari grain	88.2	8.8	83.7
Hegari silage	32.8	0.7	13.8
Hominy feed	90.5	7.6	85.5
Johnson grass hay	92.6	2.7	34.7
Kafir grain	89.5	8.6	85.8
Linseed meal, 34% protein	99.9	29.4	77.6
Milo grain	89.3	8.1	84.8
Milo heads	90.0	7.6	77.5
Molasses, feeding	72.7	0.5	61.6
Oats, Red No. 2	91.8	8.7	70.8
Oat hay	91.6	4.5	34.7
Prairie hay (upland)	92.6	2.7	35.3
Rice bran	91.0	8.9	69.9
Rice polish	90.3	9.0	90.7
Sorghum hay	81.4	1.8	35.7
Sorghum silage	30.9	0.3	13.6
Soybeans	91.8	35.8	96.3
Soybean hay	89.8	12.7	39.2
Tankage (digester 60%)	92.5	42.3	64.4
Wheat	90.0	11.3	78.8
Wheat bran	90.3	13.3	56.8
Wheat gray shorts	90.0	14.9	75.7

Tables from Texas Agricultural Experiment Station Bulletin No. 461.

on his digestive system. It is evident, therefore, that for the successful and economical feeding of livestock a balanced ration must be fed. Let us now see how to calculate a balanced ration.

353. **How to Calculate a Balanced Ration.** Let us suppose that we have a 1300-pound horse doing medium work for which we want to calculate a balanced ration. Let us also assume that we have oats, ear corn chops, and prairie hay available as feeds. Our first step is to determine the requirements for our horse. Table II tells us that for every

TABLE II

TENTATIVE FEEDING STANDARDS PER DAY PER 1000 POUNDS LIVE WEIGHT

	DRY MATTER	DIGESTIBLE CRUDE PROTEIN	PRODUCTIVE VALUE IN THERMS
Growing and fattening steers:			
Weight 400–500 lbs.	24.0–26.0	2.0–2.2	14.0–15.9
" 600–700 "	22.0–24.0	1.8–2.0	13.3–15.2
" 800–900 "	20.5–22.5	1.6–1.8	12.6–14.5
Fattening 2-year old steers on full feed	20.0–30.0	1.6–1.9	13.2–15.7
Horses:			
Medium work	16.0–22.0	1.2–1.5	10.0–13.0
Fattening lambs:			
Weight 70 to 90 lbs.	28.0–31.0	2.4–2.7	18.0–21.4
Dairy cows:			
For maintenance of 1000 lb. cow	....	.6	6.00
For each lb. of 3% milk	....	.045–.055	.25
" " " " 4% "	....	.053–.065	.30
" " " " 5% "	....	.060–.070	.35
150-lb. fattening pig	33.0–36.0	4.5–5.0	30.0–35.0

NOTE: Hog standard is estimate only and is not taken from Texas Experiment Station Bulletin.

Tables from Texas Agricultural Experiment Station Bulletin No. 461.

1000 pounds liveweight a horse doing medium work should receive daily between 16 and 22 pounds of dry matter, 1.2 to 1.5 pounds of digestible protein, and 10 to 13 therms of productive energy. Since our horse weighs 1300 pounds, his requirement will be 13/10ths of the above figures, or between 20.8 and 28.6 pounds of dry matter, 1.56 to 1.95 pounds of digestible protein, and 13 to 17 therms of productive energy. This is the requirement which we must fulfill. In our first attempt to formulate the ration let us take 8 pounds of oats, 6 pounds of ear corn chops, and 12 pounds of prairie hay. Referring to Table I we note the percentages of the dry matter, the digestible protein, and the productive energy for the three feeds. By multiplying the number of pounds of each of the feeds we plan to use by the percentage composition of each we arrive at the composition of a ration made up of the given amounts of these feeds. By making our calculations in the form given on the next page we will be able to follow the procedure more clearly.

	DRY MATTER	DIGESTIBLE PROTEIN	PRODUCTIVE ENERGY
8 pounds of oats			
8 × .918 dry matter	7.34		
8 × .087 digestible protein		.696	
8 × .708 productive energy			5.66
6 pounds of ear corn chops			
6 × .887 dry matter	5.32		
6 × .055 digestible protein		.330	
6 × .747 productive energy			4.48
12 pounds of prairie hay			
12 × .926 dry matter	11.11		
12 × .027 digestible protein		.324	
12 × .353 productive energy			4.24
Total	23.77	1.35	14.38
Standard requirement	20.8–28.6	1.56–1.95	13–17

After comparing our totals with the standard requirements we find that the dry matter and the productive energy are adequately fulfilled. The digestible protein, however, is somewhat below the requirement. This means that we are going to have to make some adjustment which will bring the amount of digestible protein up to the requirement. Let us now look at the composition of the three feeds. We find that oats have the highest digestible protein content of the three feeds. By increasing the amount of oats and decreasing the amount of ear corn chops we should be able to increase the total digestible protein content of the ration. Let us see if thirteen pounds of oats and three pounds of ear corn chops will bring about the desired changes. The same form may be used as in the first trial. (See table on page 392.)

By comparing the standard requirements with our totals we find that our ration is adequate in all respects. The total digestible protein content has been raised so that it meets the requirements. Thus we have balanced our ration.

At first it will take a number of trials to arrive at the correct amounts of the feeds to meet the standard requirements. It is through the calculating of a number of rations for vari-

	DRY MATTER	DIGESTIBLE PROTEIN	PRODUCTIVE ENERGY
13 pounds of oats			
13 × .918 dry matter..........	11.93		
13 × .087 digestible protein.....		1.13	
13 × .708 productive energy.....			9.20
3 pounds of ear corn chops			
3 × .887 dry matter............	2.66		
3 × .055 digestible protein......		.165	
3 × .747 productive energy.....			2.24
12 pounds of prairie hay			
12 × .926 dry matter..........	11.11		
12 × .027 digestible protein.....		.325	
12 × .353 productive energy....			4.24
Total..............................	25.70	1.62	15.68
Standard requirement...........	20.8–28.6	1.56–1.95	13–17

ous animals that one becomes adept in the computing of rations.

Because of the differences in the composition of feeds grown under different conditions and the variation in the digestive and assimilative power of different animals, the requirements given in the standard are rather wide in their range. In reality the standard requirements are averages which may be varied slightly to fit individual cases.

354. **Concentrated Manufactured Feeds.** Feeds sold in Texas must carry a printed tag showing the correct name of the feed, the names of ingredients, the quantities of the ingredients, and the guaranteed percentages of protein, fat, crude fibre and nitrogen-free extract. This information enables the purchaser to get a good idea of what the feeding value is, if he has sufficient knowledge to utilize the information given. Analyses are made to insure that the feed comes up to the guarantee and is correctly labelled. These results are published in bulletins which can be secured from the Feed Control Service, A. & M. College, College Station, Texas.

QUESTIONS, PROBLEMS, AND EXERCISES

160. Draw a plan of the barn lot, barn, and stock shed on your place, give a description of them and tell in what respects they are right and in what wrong.

161. Make a plan for a barn lot, barn, and stock shed for your farm that meets the requirements indicated in this chapter.

162. Weigh the rations given two different kinds of stock on your farm. Calculate the nutrients in these, and if they are not nearly in accord with the standards, prepare rations out of the foodstuffs used that are properly balanced.

163. Try to plan another practical ration for these animals that will accomplish the same result at less expense.

CHAPTER XXI

PLANNING THE FARM BUSINESS

355. **The Farmer's Economic Problem.** There is no longer among farmers such a thing as complete self-sufficiency. The farmer is dependent upon other peoples for a large part of the goods he enjoys and the tools he uses in farming. He may produce some of the goods he consumes on the farm but most of his energies are devoted to the task of creating goods that he hopes to exchange for money, which he must have to buy the many things necessary for the comfort of his family. The farmer therefore has to plan his farm operations to meet the desires of other people as well as to satisfy his own wishes. It is not enough to plan the farm for production of agricultural products. Production must be planned so as to produce the things that will sell in the markets.

356. **How to Obtain Greatest Possible Returns.** The chief problem of the farmer in planning the farm business is to obtain the greatest possible returns from his resources. How can he do this? The solution to this problem has two phases: (1) planning the arrangements of his fields, pastures, fences, woodlots, barn, farmstead, etc., in such manner as to give the most economical use of his labor and his other factors of production; and (2) the selection and combination of crops to cultivate and livestock to raise that will bring him

the most satisfactory income. This means that he must plan his production according to market demands.

357. **The Farm Layout.** An examination of the farms in any community reveals the fact that but few of them have any well marked out plan along which to develop. Fields are irregular in size and shape, often poorly arranged and located, making it necessary to lose much time in going to and from them. Fields of irregular shape and size cause numerous corners and nooks difficult to cultivate and thus make it impossible to cultivate all the land. Money is tied up in land, just as it is in work animals, and it is not wise to keep a work animal or piece of land idle when it could help to increase the production of crops on the farm. Close observation of farms will reveal poor judgment also in locating buildings, fences, roads, lanes, pastures, orchards, and gardens, and will give an impression that they were located by chance rather than by any well-developed plan.

358. **Why Have a Farm Plan?** Good plans will save time and labor and allow the best and most economical use of equipment and the most complete and profitable use of all the land. Plan to avoid unnecessary fences and field divisions. The dividing of tillable land into small fields is extravagant of fencing, wasteful of land, labor and time in cultivating. To fence a square field of two and a half acres requires eighty rods of fencing. Allowing a strip six feet wide around the field immediately inside the fence for turning uses up 7776 square feet or slightly over 7 per cent of the field. A square field of ten acres would take only twice as much fencing, and only about 3½ per cent of the ten acres would be needed for turning uses. The time consumed in turning in cultivating the two-and-one-half-acre field is twice as great in proportion to the area worked in the ten-acre field. A forty-acre field would have a proportionately greater advantage over the ten-acre field. It takes about twice the time to plow an acre in a field of less than two acres that it does to plow an acre in a field containing fifteen or more acres.

Shapes and Sizes of Fields. Long fields permit of better use of machinery, teams, and labor in tillage than short ones, especially when rows are run only one way. There is just half as much time consumed in turning when rows are doubled in length. On a fairly smooth-lying farm where the fields are made rectangular instead of square, each may have an entrance comparatively close to the barn and house.

Uniformity in size of fields is desirable, especially when a regular system of farming is followed, and when a system of rotation of crops is desired. The rotation or cropping system should be considered before deciding upon the number and size of cultivated fields. Another factor that should be given serious consideration in planning the layout of fields is the conservation of the soil. In addition to the rotation of crops, it is necessary to give proper attention to the prevention of soil erosion. Proper arrangement, shape, and size of fields are important to conserve time, labor, and expense of cultivating, but it is not profitable to have these savings at the expense of the soil. If terracing and contouring of the land are necessary they should not be omitted for the sole purpose of having a convenient arrangement of fields as to size and shape. A careful inspection of farms and their history in any community will reveal the fact that some farms have grown old as rapidly as the owner and some have become furrowed with gullies in the same space of time it has taken for the face of the owner to become wrinkled with age. A good farm manager should guard against soil erosion just as much as he should guard against plant and animal diseases. Erosion is an earth disease and it spreads and spreads until it destroys the ability of farm land to produce profitable crops.

359. **The Pasture.** Work stock render better service and last longer if they have a good pasture in which to graze while not at work. A good pasture affords the very best and cheapest food for livestock. Work animals, for example, on farms in Texas seldom work more than sixty to ninety days of the year. If good pastures are provided these animals

would not require much feed from the barn when not at work. The pasture should be in easy reach of the barn and should be so arranged that it can be divided into two or more fields to avoid the necessity of different kinds of stock being together at times when one is likely to interfere with the welfare of the other. A mean mule, for example, seems to delight in annoying hogs, cows, and calves and may seriously injure them. If possible, the rugged land that is apt to erode badly should be used for permanent pasture. Enough trees should be in the pasture to give shade and protection to the stock, but beyond this limit trees are a disadvantage because of their effect upon the growth of grass. The pasture should be large enough to provide grazing for all the stock on the farm and should be so arranged and managed as to furnish grazing over as long a period as possible.

360. **The Wood Lot.** It is advisable in areas in which trees grow to reserve enough land for a wood lot to give the annual fuel supply and to furnish at least part of the posts needed on the farm. Lands unsuited to cultivation, such as rough areas of doubtful value in producing regular crops, may be devoted to timber growing. Land set apart for woods should be made to grow trees of value. The timber lot should be so managed as to give some harvest each year. Stock should not be pastured in a wood lot if there is danger of damage to the young trees.

361. **The Homestead.** The farmstead should be convenient to the main part of the farm. It should be on a well-elevated site, convenient to roads. It should be so laid out that the work can be done with a minimum loss of time. The barns and water should be conveniently located. The garden should be near the house. Barns should not be too close to the house or any other building in which fire is kept. If there is a highway running through the farm and it is advisable to have the farmstead on this highway, barns should be placed behind the house and not across the road in front. Too much danger to livestock and the family would result where crossing the highway is frequent each day. The

dwelling house itself should be far enough back from the road to avoid dust, annoyance, and noise caused by travel. The farm is a place in which to live as well as a place to carry on a business, and should be made attractive. Where practicable, the farm should be connected with the rural electric line. These lines are now being installed in many sections by the agencies of the national government and by some private agencies. Where it is not yet possible to connect with a rural electric line, it is often practicable to install one's own home power and light plant. The addition of electric light and power to the farm home adds great comforts to farm living, saves much labor, and in many localities is no longer an expensive luxury.

362. **The Farmyard** should be neat, attractive and beautiful by having a good lawn and well placed shrubs and flowers. A simple home can be made attractive by proper planning of lawns and shrubs and flowers.

363. **The Garden** should not be overlooked in planning a farmstead or a farm business. This, however, has been well illustrated and discussed in Chapters VIII and IX. The farmer who fails to give proper attention to the garden in planning his farm business will not only fail to get the greatest amount of income but will also fail to realize the maximum enjoyment of living on the farm.

364. **The Barn,** the building second in importance on the farm, should be built in accordance with the needs. It should be at a convenient distance in the rear of the house. There should be a number of trees between the house and the barn in order to cut off objectionable views and to serve as a protection of one building from the other in case of fire. The barn should be large enough to house the farm produce and furnish quarters for the stock. It can and should be of artistic design and good, durable construction without being excessively expensive.

365. **The Workshop and Implement Shed.** There should be a workshop combined with a tool and implement shed or house. This building should be placed at a distance

of at least one hundred feet from the barn and at a point easily accessible.

366. **Calf Shed.** Near the pasture for calves and other young stock there should be located quarters for their protection. This building calls for nothing expensive in structure but should be substantially built.

367. **Poultry House.** There should be one or more well-designed poultry houses, located some distance from the barn and tool-house, planned with a view to sanitation. The buildings should be portable so that they can be moved at any time it is found advisable.

368. **Hog Houses.** The hog houses should be portable and located near to, or within, the hog pasture. At times they may be placed in the fields and lots in which special crops are being grown for the hogs.

369. **Tenant Houses.** Where farms are of sufficient size to require tenant houses, these should be located with a view to convenience of both landlord and tenant. The method of renting and the size of the farm are important factors in locating tenant houses. On farms where there are share-croppers their houses should be located near the barns, so that the tenant will not lose much time in going to and coming from the barns where livestock are kept. It is usually advisable to have one central feed lot, corn barn, and hay barn on such farms, so that the owner may keep a watchful eye on the stock and the feed supplies. If, however, the tenant furnishes his own teams, the tenant houses should be conveniently located near the fields he cultivates.

370. **Save Labor.** The planning of the physical layout of the farm should always be done with a view of saving as much labor and time in farm operations as possible. The fact that cash wages are not paid for family labor is no reason for not using it efficiently. There is no more reason for a farmer to be wasteful with labor by having a poor farm layout than there is for a manufacturing concern to waste the labor of its workers by having machines and equipment poorly arranged.

IMPORTANT FACTORS IN DETERMINING THE COMBINATION OF CROPS AND LIVESTOCK

371. **The Outlook for Prices and Markets.** After the farmer determines what crops and livestock are adapted to his farm, his next problem to solve is what combination of crops or crops and livestock will give him the greatest reward for his efforts. If a farmer cannot sell or use his crops profitably, he will not be able to realize any profit on his production, regardless of his efficiency. He must select and produce crops or animals for which people are willing to pay satisfactory prices. It is just as necessary to observe and study price trends as it is to study more efficient methods of production.

372. **Crop Rotations.** Most crops are more productive when they follow another crop. Very few crops can be grown continuously on the same soil without decreasing yields. The advantages of crop rotation, however, have already been discussed in Chapter V. It is the problem of the farm manager to give consideration to crop rotation when choosing the combinations of crops and livestock.

373. **Distribution of Risks.** There is always a great deal of uncertainty in farming, because the farmer has no control over weather and frequently little or no control over insects, diseases, weeds, and other troublesome things that may reduce yields of crops or livestock. Furthermore, there is never any guarantee that there will be good prices for a particular farm product, so that even a good yield may bring small profits to the producer. A failure of some particular crop, from one cause or another, is something that is liable to happen. On the other hand, a good combination of crops or crops and livestock gives some insurance against total loss. We seldom have a year in which all crops fail or the prices for all are extremely low.

374. **Distribution of Income.** A given amount of income coming in at different times of the year is for many farmers more valuable than the same amount would be if it

were received all at one time. There is a greater tendency to spend too freely when we have a large amount of money at one time. It is also less difficult to finance the operations of the farm when money is coming in regularly. Less credit is necessary, and therefore more of the income is available to spend for living. It is perfectly proper to make use of credit and pay interest, if the money borrowed will increase the production on the farm enough to repay the loan, pay the interest, and have something left in addition. One-crop farmers in the Cotton Belt, however, often get entangled in a vicious credit system that necessitates the paying of high interest rates and high prices for the goods they buy. One way to increase the amount of money available for living and to avoid excessive use of credit is to select a combination of crops and livestock that will give the farmer something to sell at different times of the year so that he can meet his normal operating expenses as they occur.

375. **Labor Distribution.** In selecting the combination of crops or crops and livestock, farmers should not lose sight of the need to develop a combination that will keep labor profitably employed throughout the year. A one-crop system, such as cotton farming, does not profitably employ labor the year round. There are two peak labor loads on a cotton farm, in the spring and in the fall months. The cotton farmer needs to select other farm enterprises that will give profitable employment during the slack periods of the summer and winter months. In selecting other enterprises, care must be taken not to choose additional crops or livestock that will demand the attention of the farmer at the same time as his main crop. If, for example, cotton is clearly the one most profitable crop, the additional crops or farm enterprises should be chosen so as not to interfere with the production and harvesting of cotton, but to give employment at times when labor is not needed in the cotton field.

376. **Use of By-Products.** On many farms there are portions of crops and livestock products that are unmarketable unless converted into some other form. For example,

if there is damaged grain, hay, or other feeds that are unmarketable, these might be converted into meat, butter, eggs, or milk which are easily marketed. The farmer, like the manufacturer, should convert all by-products into salable forms where possible.

377. **Efficient Use of Equipment.** Equipment purchased for the cultivating and care of one crop can, as a rule, be used to care for others. Plows used for cultivating cotton can also be used for other crops if the crops selected do not need the equipment at the same time. A proper combination of crops will, therefore, give a more profitable utilization of farm equipment.

378. **Conclusions.** We can now summarize by stating general principles to govern the selection and combination of farm enterprises.

1. If there are several crops that will grow well on the farm, but compete with each other for the farmer's labor, equipment, horses, etc., at the same time of year, select the crop that promises the greatest net return for the farm as a whole.

2. Combine with the main crop other crops that will aid in the production of the main crop; such, for example, as legumes or other soil-building or soil-conserving crops that will help to maintain or build up the fertility of the soil. Add any livestock that could utilize any of the crops that otherwise might be unsalable, or add livestock when any of the crops might be more profitably marketed through livestock.

3. Select supplementary enterprises, crops, or livestock that will contribute to the total income of the farm by giving labor profitable employment when it might otherwise remain idle. On cotton farms, the supplementary enterprises should give employment to labor during the winter and summer months.

The farmer should try to select a combination of crops and livestock which will yield the highest net return for the farm as a whole. Some one crop or type of animal might give better returns than any other for the farmer's labor, equipment, etc., but, so long as the addition of other enterprises will add something to the total income, they should be added.

QUESTIONS, PROBLEMS, AND EXERCISES

164. Draw a map of your father's farm, showing shape and size of fields, pastures, and lots, and the locations of dwelling, barns, and other houses. Write a description and criticism of the farm layout, pointing out what is good and what is bad, and suggesting improvements in the plan.

165. You have forty-eight acres, which you can put into either one field six acres wide and eight acres in length, or into three fields, each two acres wide and eight acres long. How much more fencing will have to be built to make the three fields? How much more land will be used up by the fences for the three fields, allowing a strip six feet wide around each field for turning?

166. A farmer finds it necessary to arrange for credit to make a crop and makes a trade with a local credit merchant to give him $200 credit at the store. He makes a note and mortgage for $200, discounted at 10 per cent. This leaves the farmer $180 with which to buy supplies. Assume that the prices of goods at this credit store are 10 per cent higher than at the cash stores with the best prices. How many more dollars would the farmer have to spend if he did not have to use credit but had the needed $200 cash?

167. If this same farmer arranged to get his credit from one of the Production Credit Associations organized by the Farm Credit Administration, and had to pay at the rate of 5 per cent a year for the $200, and if the actual period during which he used the borrowed money were only six months, how much would the farmer save by securing credit from this institution instead of using store credit?

APPENDIX I

WILD LIFE

379. Extent and Size of Wild Life Industry. The value of wild life throughout the United States is being realized as never before. By wild life we mean not only the game birds, but the song birds and insect-eating varieties, the game and non-game mammals, the reptiles, the amphibians, and even some of the native plants that furnish food and shelter for animals of various kinds. The wild flowers are sometimes regarded as wild life, too, and quite properly so. On a farm that has kept some of its original wild land, and that has avoided overgrazing by livestock and other destructive forces, there are likely to be some bobwhite quail, cottontails, and other game. If the farm is so large that it is in reality a ranch, there may be deer and wild turkey also. All these things have not only a value in themselves, which can hardly be put in terms of dollars and cents, but they also possess a definite money value to the farmer, the State, and the Nation.

The annual meat and fur value of Texas wild life has been estimated by Mr. William J. Tucker, executive secretary of the Texas Game, Fish, and Oyster Commission, at no less than $20,000,000. There are more than 500,000 hunters and fishermen in Texas, who pay close to $300,000 annually for hunting licenses. It costs them about $25,000,000 to harvest the crop of wild life, and they do it for fun. The money goes to the farmer and ranchman, the storekeeper, the clothing dealer, the hardware man, the hotel man, the restaurant keeper, the automobile dealer, the gasoline service station proprietor, and to the other people who sell supplies to these men. About 25,000 Texas hunters last year paid more than half a million dollars to farmers and ranchmen for the privilege of hunting on 3,500,000 acres of land where a legal charge was made under the Texas hunting preserve law.

There are said to be 36,000 clubs and organizations in the United States interested in the conservation of wild life. Until recently these local groups accomplished little, but now, through the general wild life federation, many of them are

FIG. 254. White-tailed deer, introduced into the Navasota River bottom in Leon County, Texas, as a part of the game restoration program of the Texas Game, Fish, and Oyster Commission.

getting together to exert the weight of their united influence on public opinion in behalf of wild life conservation.

Included in the list of native wild life of the Southern and Southwestern States are several varieties of deer, wild turkey, bobwhite, scaled and Gambel quail, gray squirrel, fox squirrel, peccary or javelina, mourning dove, white-winged dove, band-tailed pigeon, cottontail and swamp rabbit, beaver, bighorn sheep, antelope, cougar, black bear, coyote, fox, furbearers, waterfowl, prairie chickens, and many shore birds. Unfortunately, the inevitable changes brought about by civilization, coupled all too often with insufficient protection, have brought many of these species to the verge of extinction.

380. **Game, Fish, and Oyster Commission.** During the last fifty years the State of Texas has endeavored to care for and increase its fish and game resources through the setting up of a definite conservation agency for the purpose. The first Texas Fish Commissioner was appointed in 1879. The Commission was abolished in 1885 and for ten years there was no special organized agency charged with the enforcement of game laws and care of game and wild life in the State. In 1895 the Commission was recreated and a Fish and Oyster Commissioner was appointed to take charge of the work. In 1907 the title of the executive head of the Commission was changed to Game, Fish, and Oyster Commissioner. On September 1, 1929, the form of organization of the Commission was changed and six commissioners took the place of a single official. The Texas Game, Fish, and Oyster Commission functions through an executive secretary, and exercises broad powers in the conservation of Texas wild life, including fish, game, and non-game species. The Commission also regulates the use and sale of sand, shell, and gravel. Its full-time personnel includes more than 150 persons, its total income for the year ending August 31, 1936, was $380,636.37, and its assets, as shown by its balance sheet as of that date, aggregated $744,225.97.

While these sums are very substantial, it should be remembered that the area under the jurisdiction of the Commission is far greater than in any other State in the Union, reaching the enormous figure of 170,000,000 acres, approximately one twelfth the surface area of the entire United States. While Texas stands first in extent of lands and wild life resources, Mr. William J. Tucker, executive secretary of the Texas Game, Fish, and Oyster Commission, reports that she stands nineteenth in expenditures for conservation of wild life.

381. **The Farmer and the Wild Life.** The landowner is the key man in the production of wild life. The wild birds,

animals, and plants are as much a product of the soil as the trees or the field crops. If the farmer is interested in wild life production and takes the necessary steps to insure a plentiful game supply on his place, there will be an abundance of game. To a great degree the game supply in a farm or ranch area under private ownership depends upon the way in which the farmer handles his land. As Herbert L. Stoddard, the world authority on bobwhite quail, has said, quail can be eliminated from an area by unwise agricultural practices without the firing of a shot. They likewise can be increased on an area with very little expenditure by following wise farm practices, usually in the South without any restocking at all.

A plentiful supply of insect-eating and song birds, of valuable fur-bearing animals, of deer, rabbits, and wild turkey, and even of squirrels may add much to the value of a farm. The value of the insect-eating species is in the protection of the crops. The song birds increase the attractiveness of the home. The fur bearers may be trapped in winter, when they constitute a substantial addition to the income of the farm boy or the farmer himself.

382. **Natural Methods in Producing Wild Life Superior to Artificial.** Except in the case of certain fishes, which have been found to be easily and efficiently propagated in fish hatcheries, natural methods of producing wild life have been found to be considerably more effective and economical than artificial. By natural methods, we refer to production of game in the field under natural conditions. By artificial, we mean production by means of incubators and other means used by poultrymen. It has been found that propagating quail, for example, by keeping the birds in captivity, hatching the eggs in incubators, raising them in brooders or with bantam hens, and then liberating them is a difficult matter. Out of their natural home, the birds are more subject to disease than when wild. Breeding stock, pens, other equipment, and supervision are expensive. The fertility of the eggs of birds kept in captivity may be reduced. Most of

these difficulties are avoided if the farmer leaves sufficient food and cover about his place to accommodate the game in which he is interested. The expense is much less; the stock

FIG. 255. The Opossum is the most important fur bearer in number of skins produced, in the State of Texas. The illustration is of a very young animal.

is healthier and happier; the birds shift for themselves and only rarely need to be fed; and they doubtless make better objects of sport.

Not only game birds, but insectivorous and song birds also, are favored by the farm practices that afford the wild species adequate food and cover. Better shelter is provided for the useful species, and, of course, fur bearers tend to be

FIG. 256. The grazed area at the left of the fence is barren of quail. The ungrazed pasture to the right supports a covey of 17 birds.

FIG. 257. Partridge pea, a valuable quail food.

more abundant, provided other conditions are suitable. The soil is protected from erosion, there is more forage left on the ground, the farm wood lot is in better condition, and the farm is more nearly a balanced enterprise.

383. **Fur-bearing Animals.** Until very recently Texas

FIG. 258. Such beauty spots are suitable for such game as squirrel, wild turkey, and deer, as well as for fish and flowers.

furs paid the farm boys and trappers of the State approximately a million dollars a year. The number of skins taken within the State in the last few years has dropped off markedly. This fact indicates a disturbing decline in the amount of fur available. There is serious need for better care and culture of the fur bearers.

The leading fur bearers are the opossum, which leads all the others in number of skins produced in Texas, the skunk, the raccoon, mink, fox, civet cat (a kind of skunk), ringtail cat, muskrat, and badger.

384. **Flesh Eaters.** No hard-and-fast line can be drawn between the fur-bearing and the so-called predatory animals. When a fur bearer becomes a nuisance, it is usually regarded as a predator. The predator may possess appreciable fur value, and some of the flesh eaters, especially the mountain lion and the coyote, have real value as game.

385. **How to Improve Conditions for Wild Life.** It is obvious that wild birds and animals, including game, insectivorous kinds, and fur bearers, must have food and shelter in order to exist. A general principle to follow, therefore, is to increase food plants and needed shelter for the species desired about the place. Different species require different treatment. Just what to do depends on local conditions and on the kinds of wild life present.

If pastures are solidly turfed, there will be a minimum of food for birds and game. By plowing a strip from five to thirty feet wide along the margin of the pasture, preferably adjoining a wooded or bushy tract, you can stimulate the growth of a good crop of game food plants without any planting at all.

The conservation of soil and water will tend to increase the wild life. Terracing, contour furrowing, the planting of strip crops of legumes or small grains in fields of corn or cotton—all will favor wild life. Brush and trees should be left along creeks and branches, thus helping to prevent erosion and affording additional shelter for birds and game. In mowing pastures, leave strips unmowed five to thirty feet wide along the fences. If practicable, leave standing about field borders a number of rows of small grain or corn when the crop is harvested. This will furnish both food and cover for wild life.

Little winter feeding is called for in our mild southern climate. In most instances expensive planting for wild life is unnecessary. Just give the native plants a chance to grow.

QUESTIONS, PROBLEMS, AND EXERCISES

168. Make a list of the game birds, game animals, and game fishes on or near your father's farm. Are they scarce or common; increasing or decreasing from year to year?

169. What fur-bearing animals are found in your county? In what places would you look for them?

170. Make a list of things the farmers in your community can do to increase the wild life on their farms.

171. Make a rough map of a farm area you know and show where fence-rows should be allowed to grow up to native plants for game food; also where strip crops should be left and pasture borders plowed to encourage valuable wild life.

172. Count the quail coveys on your farm or some friend's farm, and mark where you find them on an outline map.

173. Why are natural methods better than artificial methods of producing quail and other wild life?

174. Make a list of the principal song birds about your home.

175. List some of the money values of wild life.

176. What is the name of the official State agency that helps to take care of wild life?

APPENDIX II

SILOS

386. **What Is a Silo?** A silo is an air-tight structure for the preserving of green forage crops, such as corn, sorghum, cowpeas, and kafir corn in their original succulent state. The material preserved in a silo is called *silage.* It fills the same place in the diet of livestock that canned fruits and vegetables do in the diet of people. We are all familiar with the value of green vegetables as a means of keeping our bodies healthy. Grass is just as necessary to livestock, but since we cannot always have green grass in the winter time or during seasons of drought, we build silos to preserve forage crops in a succulent state. When the silage is placed in the air-tight silo in the green state it ferments and becomes very hot. This fermentation consists largely of lactic-acid bacteria acting upon the sugar in the ensilage, converting it into acid. Yeasts likewise have a part in these changes.

387. **Uses of the Silo.** Silos are valuable in several ways. First, they furnish succulent food for the livestock all the year round. Second, they preserve the entire stalk in such a form that it can all be eaten by animals, while if it were cured dry the stock would waste a large percentage of it simply because they cannot eat the hard dried stalk. Third, when the season turns out so dry that corn or a similar crop would not produce any grain, it may be harvested while still green and preserved in the silo; whereas, if it remained in the field all the fodder would dry up and be destroyed by sun, wind, and rain. Fourth, more feed can be stored in a given space in the form of silage than in the form of fodder or hay. Fifth, an acre of corn or sorghum can be placed in a

silo at less cost than the same area can be headed or husked and shredded. Sixth, converting a sorghum or a corn crop into silage clears the land early and leaves it ready for another crop, preferably a soil-building crop.

388. **Kinds of Silos.** There are two general types or classes of silos, namely, above-ground silos and underground silos. The first type of silo used was the underground type, consisting of a square pit six to eight feet in depth and about the same in width and length. This type of silo was first built near Stuttgart, Germany, and was later modified to our present trench-type silo. The overhead or above-ground silo was an American idea, the first being constructed in square form, something on the order of a small granary. This type was not successful, due to the difficulty of packing the silage tight enough in the corners to exclude the air. It was modified, therefore, to our present cylindrical type silo. The above-ground silo, until recent years, was the only type used extensively in this country. The principal disadvantage of this type of silo is the cost of construction, which makes it prohibitive to the small farmer with only a few head of livestock. For the farmer with a large number of livestock, the above-ground silo has the advantage of reduced labor cost in feeding and possibly less waste of silage. The underground trench silo is more widely adapted to the use of the average farmer and is gaining rapidly in favor because of the fact that such a silo can be built at a cost of from thirty to fifty cents per ton capacity. The same quality of silage can be made in a trench silo as in any type of above-ground silo, and this silage can be had with much less investment. It may be had by any farmer, including the tenant farmer.

389. **Animals That Eat Silage.** Silage is more necessary probably for dairy cattle than for any other class of livestock, as it is important for them to have green or succulent food in order to produce large amounts of milk economically. The dairy cow will eat from thirty to seventy-five pounds of silage per day, according to her size, capacity, and amount of other feed consumed. The silo is also very

FIG. 259. Trench silo being filled. Note that one side is smooth.
Courtesy of the U. S. Department of Agriculture.

FIG. 260. Trench silo being emptied. Note the covering of soil that protected the ensilage.
Courtesy of the U. S. Department of Agriculture.

important in the feeding of beef cattle, as it helps to keep them in good condition and induces them to eat a large amount of foodstuffs that can be raised cheaply. A fine quality of silage is often fed to horses and mules to great advantage. It has too much juice in it to be used advantageously as a food for hard-worked horses or mules. The effect is very much the same as that of fresh grass when fed to such horses or mules. Silage is not a satisfactory food for hogs or poultry. They eat the grain in it, but will not eat anything else except a few of the tenderest blades.

390. **Crops Used for Silage.** The best crops for use as silage are corn, sorghum, kafir corn, and milo-maize. Sometimes such crops as alfalfa, clover, Johnson grass, sudan grass, oats, and cowpeas are also used. Alfalfa and clover usually contain too little of carbohydrates to make a good quality of silage; however, this difficulty may be remedied by adding sugar in some form to these feeds as they are put into the silo. This may be done by mixing with sweet sorghums or by adding blackstrap molasses.

391. **Time to Harvest Silage Crops.** The crop should be fairly mature before it is cut for the silo, since the feeding value increases rapidly as the crop approaches maturity. Corn should be placed in silos just as the ear begins to harden and the kernels begin to dent. Kafir corn, milo-maize, and sorghum should be placed in the silo when the grain is in the late dough stage.

APPENDIX III

4-H CLUBS AND FUTURE FARMERS OF AMERICA

392. **What Are 4-H Clubs?** Four-H Clubs are the junior organizations in which rural boys and girls are improving rural life and farm practices. These clubs are organized by Extension agents who are employed co-operatively by the United States and the State governments, co-operating with a county agency. The United States Department of Agriculture, the State Land Grant College, and the local agency designated by the Land Grant College furnish supervision and help in many ways.

393. **The Emblem.** The emblem of the 4-H Club is a four-leaf clover with an "H" on each leaf, representing the head, the heart, the hand, and health of the club members. The interpretation of the emblem is that a trained and alert brain, a good heart, a skilled hand, and good health will make for better citizenship.

394. **What 4-H Clubs Do.** The boys and girls who enroll in 4-H Clubs carry out some agricultural or home making activity under the supervision and guidance of the extension agent in charge. They are expected to keep records of what they do, what they expend, and the value of what they develop in their club work. Public exhibits of club work are held and serve to stimulate and train the club members.

The undertaking of a club member is called a "demonstration." It is used by the club as a whole as an object lesson

FIG. 261. A 12-year-old 4-H Club boy, feeding out his six Hampshire pigs. The boy made the pen, trough, and self-feeder, and took all the care of his pigs.

Courtesy of the U. S. Department of Agriculture.

FIG. 262. In Harris County, Texas, 2594 fruit trees were planted by 4-H Club girls in one year. The tree above was set out in February and the picture taken in July.

FIG. 263. One Southwestern school's boys carrying out baby beef projects.

or an illustration of the results that can be obtained by following certain practices.

Contests, too, are recognized agencies for promoting this work. These contests are local, state, and national, and embrace practically every type of work that the club members undertake. The prizes involved range from small local prizes to national trips and scholarships.

In most states the training of demonstration teams is a definite part of 4–H Club work. The program of every club includes the training of at least one team. A demonstration

FIG. 264. Members of canning club at work in their garden.

FIG. 265. A Southwestern Club girl, who built this poultry house her first year of club work and made a net profit of $107.20 on ten hens, thirty pullets, and four cocks.

team demonstrates a practice to the audience, actually showing them how it is done. A few of the subjects for team demonstrations are: fitting and handling livestock for shows; controlling insects in garden and orchard; grading and candling eggs for market; repairing and refinishing furniture.

A SUGGESTED OUTLINE FOR A TEAM DEMONSTRATION

395. **Equipment Needed.** Sweet potatoes; corrosive sublimate; barrel, water, and sack; charts showing relative yields of treated and untreated seed; chart showing contagious diseases.

396. The Treating and Growing of Sweet Potatoes for Seed: Program.

TEAMMATE A	TEAMMATE B
Tell what club is represented. Introduce "B" and self. State object of demonstration. Give local applications.	Acknowledge introduction. Get material in place.
Discuss contagious diseases to be controlled.	Show chart of diseases of potatoes. Point out losses due to disease.
Prepare mixture of corrosive sublimate in barrel.	Discuss relative yields of treated and untreated seed. Personal experience.
Dip sack of potatoes in solution.	Tell the method of treatment, why we treat seed.
Discuss use of sweet potato beds.	Use chart for illustrating preparation of hotbeds.
Chart showing advantage of slips for growing disease-free seed.	Discuss growing sweet potatoes from slips for seed.
Summarize the points. Ask for questions.	Assist in answering questions.

397. **Judging.** Judging done by 4–H Club members is a club activity which has great value. It establishes definite standards of perfection in the minds of club members and creates the desire to attain these high standards in their demonstrations. It also teaches closer observation and develops self-assurance and good sportsmanship. Crops and livestock may be judged at various times throughout the year, as available. In order that the club members may become thoroughly acquainted with the principal points on which to base judgment, score cards are used in the training in judging. As the members become more experienced, the score cards

may be discarded. Agricultural teams judge dairy cattle, beef cattle, sheep, swine, and poultry.

Definite details for arranging demonstrations and for planning and carrying out shows and contests may be secured from the United States Department of Agriculture, the State Land Grant College, or the county demonstration agent.

398. **Girls' 4–H Club Work.** Girls' 4–H Club work, like boys' 4–H Club work, is a pārt of a great national plan for the improvement of agriculture and home life. County home demonstration agents, representing the Extension Service, have direct supervision of the work in the counties. Girls between ten and twenty years of age are eligible to do 4–H Club work in counties having home demonstration agents. Usually a club of ten or more girls is organized. A woman in the community agrees to be a sponsor for the club, helping in whatever ways seem necessary to promote its purposes.

399. **Program.** The program of work in the county is decided upon by the County Agent, with the advice of the sponsors of the clubs. Each such county program must include two demonstrations per year. One of these must have to do with production, either garden, poultry, or orchard; the other may be concerned with the yard, wardrobe, bedroom, or some part of the farm food supply. Each club must have a demonstrator in each demonstration. The agent helps the groups who are carrying on their demonstrations to plan and develop their work to such a point that their accomplishments afford good examples for the other girls in the club to follow. The other members of the club are called co-operators. The county program sets up definite goals for achievement for all co-operators.

400. **Learn to Do by Doing and Teaching.** Like all co-operative Extension work, girls' 4–H Club work is a means of learning to do by doing. This learning by doing, followed in turn by teaching others, using what you have done as illustrative material, is a most effective means of education and development. The combination of this kind

of work with the social phases of club membership promotes rapid growth of intelligent citizenship.

SCORE CARDS FOR CORN

SMITH–HUGHES CONTEST SCORE CARD

Type (40%). Ears should be fairly large, be uniform, and have the general characteristics of a single variety. The kernels should be uniform in color and shape. The cobs of all ears should be the same color. Kernels should be of medium depth and possess large germs. Well filled butts, tips, and straight rows are show card characters which have little connection with yields.

Soundness (60%). Sound ears are indicated by good weight, tight-fitting kernels, sound cobs, freedom from weevil damage, ear rots, moulds, rodent damage, smut, etc. Kernels should not be starchy or shrunken at the tips. Indentation should be medium smooth to smooth. A golden yellow or pearly white color is more desirable than a dull lemon color or a chalky white. Ears should not have bleached areas or look old or dead. Any indication of unsoundness is undesirable.

1. Type and purity		40 points
(*a*) Shape of ear and kernel	(5)	
(*b*) Color of kernel and cob	(20)	
(*c*) Indentation and texture	(10)	
(*d*) Uniformity	(5)	
2. Quality of sample		60 points
(*a*) Maturity of sample	(20)	
(*b*) Freedom from disease	(20)	
(*c*) Size and condition of germ	(10)	
(*d*) Size of ear and kernel	(10)	
Total		100 points

POINTS TO CONSIDER IN SCORING SAMPLES

1. One off-colored cob in sample.
2. Several off-colored endosperm kernels.
3. Dead ears due to disease.
4. Sample infested with live weevils.
5. Evidence of defect as to true quality.

SCORE CARD FOR JUDGING SHELLED CORN

Because there are so many varying opinions as to what constitutes a good type ear, some states are judging corn on a kernel basis entirely.

The following score card was adopted for the International Hay and Grain Show in Chicago.

1. Freedom from disease	20 points
2. Plumpness of kernel tips	10 points
3. Freedom from white starch	20 points
4. Condition of germ face	15 points
5. Freedom from damage	10 points
6. Size and shape of kernels	10 points
7. Uniformity of sample	15 points
Total	100 points

Freedom from Disease. Indications of disease are mouldiness, yellow or brown staining near kernel tips, lack of luster. A flinty, lustrous appearance is associated with freedom from disease.

Plumpness of Kernel Tips. Shrunken kernel tips are considered to indicate immaturity.

Freedom from Excessive Starchiness. Excessive white starch may be found either at the tip, on the back, or near the crown of the kernel. In general, a starchy condition of the kernel tips is considered to be the most objectionable.

Condition of the Germ Face. Refers to freedom from blistering or discoloration. A badly blistered or dark colored germ face indicates that the germ has probably been injured by freezing.

Freedom from Damage. This refers to the presence of cracked, broken kernels, or mouse damaged kernels. Kernels with tip caps broken off are included under this heading.

Size and Shape of Kernels. Distinctly narrow, thin, or pointed kernels are considered objectionable.

Uniformity. Uniformity in size, shape, and color of kernels indicates freedom from mixtures of varieties or types.

FUTURE FARMERS OF AMERICA

401. **What Is the Future Farmers of America?** The Future Farmers of America is a national organization of farm boys studying vocational agriculture in the public schools throughout the United States. It is a non-profit cor-

FIG. 266. Members of the Future Farmers' Club, exhibiting their sheep in a contest in San Angelo, Texas.

poration designed to recognize and encourage the natural instincts and tendencies to organize into "gangs" or groups, and to put enthusiasm and pleasure into work and recreation. In nine years following the founding of the organization at Kansas City, 47 States, the Territory of Hawaii, and Puerto Rico have affiliated with the national organization. There are now (1937) 135,000 members in 5000 local chapters. Texas has 450 local chapters with 16,000 members.

402. **Purposes.** Among the purposes of the F. F. A. are the following:

1. To develop competent, aggressive, rural, and agricultural leadership, by providing participation in conducting meetings, and by public speaking, debating, one-act plays, extemporaneous speaking, news writing, and panel discussions.

2. To strengthen the confidence of the farm boy in himself

and his work. This is done by helping him to find out that he can do things by his carrying out an outstanding farming program, and by providing opportunity for him to match his knowledge and skills with fellow members in exhibits and contests.

3. To create more interest in the intelligent choice of farming occupations. This is accomplished by teaching him, under the supervision of his teacher of vocational agriculture, how to handle the various farm enterprises adapted to his area. During this period of instruction he studies the various types of farming by visiting farms; talking to managers and farm owners; participating in fairs, shows, and exhibits; and studying and discussing farming occupations in the classroom.

4. To create and nurture a love for country life, by making studies of the beauties of nature and the response which nature makes when given an opportunity. These studies are made during trips to fields, woods, canyons, gulfs, rivers, lakes, and farms.

5. To improve the home and its surroundings, by studying landscaping made by both nature and man, by beautifying home grounds through planting native trees, shrubs, flowers, and other plants which will increase the beauty of the home grounds. Members are also taught to improve the home by repairing and constructing home conveniences.

6. To encourage co-operative effort among students of vocational education in agriculture, by buying and selling farm products and supplies used co-operatively in farm practice work; and by participating in groups in carrying out improved farm practices, class feeding projects, egg-laying contests, the projects of cow-testing associations, and in grafting and budding pecan groves.

7. To promote thrift among students of vocational agriculture, by insisting that boys use high quality varieties, breeds, and individuals in their farming activities, and by making special effort to induce boys to save their money and invest it in progressive farm business.

8. To promote and improve scholarship, by rewarding members who make outstanding grades in all school subjects and do outstanding work on their farm practice work.

9. To encourage organized recreational activities among students of vocational agriculture, by leading them to develop and carry out as chapter activities Father and Son Banquets, parties for homemaking girls, educational tours, camps, and athletic contests.

10. To supplement the regular systematic instruction offered to students of vocational education in agriculture, by providing boy-initiated and boy-directed activities for members, which will give them practice in the knowledge and skills taught in class, such as: culling hens for egg production; putting into effect a soil conserving and building program on the home farm; improving home farm pasture; planting home gardens and orchards; selecting, caring for, feeding, butchering, curing, and canning animals and birds for home use.

403. **Degrees.** Four degrees are offered by the organization to its members. They are Green Hand, Future Farmer, State Farmer (Lone Star Farmer in Texas), and American Farmer. Members are raised to Green Hands by local chapters, and to Future Farmers by federation, district, and area organizations. The State Farmer degree is awarded only by State Associations and the American Farmer degree is awarded by the national organization of F. F. A. Members are raised to these degrees at annual conventions each year. Achievement of the individual is made the basis for advancement from rank to rank in the Future Farmers of America. Class work of a high order, development of a farming program, evidence of thrift, and a practical demonstration of leadership are some of the passports to higher degrees of the organization.

404. **Information.** For more complete information about the Future Farmers of America, write to the United States Office of Education, Washington, D. C., or to the Vocational Division of the State Department of Education.

APPENDIX IV

LENGTH OF TIME SEEDS MAINTAIN THEIR VITALITY

	AVERAGE YEARS
Barley	3
Bean	3
Beet	6
Buckwheat	2
Cabbage	5
Carrot	4
Celery	8
Clover	3
Corn	2
Cucumber, common	6
Eggplant	6
Flax	2
Hop	2
Lettuce, common	5
Millet	2
Muskmelon	5
Mustard	3
Oats	3
Onion	2
Orchard grass	2
Parsnip	2
Peanut	1
Peas	3
Pumpkin	5
Radish	5
Rape	5
Rye	2
Salsify	2
Soybean	2
Squash	6
Timothy	2
Turnip	5
Watermelon	6
Wheat	2

QUANTITY OF SEED SOWN PER ACRE

Alfalfa (broadcast)	20–30 lbs.
Alfalfa (drilled)	15–20 lbs.
Barley	6– 8 pks.
Beans (field)	2– 6 pks.
Blue grass (sown alone)	25 lbs.
Brome grass (sown alone)	12–20 lbs.
Buckwheat	3– 5 pks.
Cabbage	¾– 1 lb.
Carrot	4– 6 lbs.
Clover (alsike alone)	8–15 lbs.
Clover (red alone)	10–18 lbs.
Corn	6– 8 qts.
Corn (for silage)	9–11 qts.
Cotton	1– 2 bu.
Cowpeas	1–1½ bu.
Flax	2– 4 pks.
Mangels	5– 8 lbs.
Millet	2– 4 pks.
Oats	2– 3 bu.
Potato	10–14 bu.
Potato (recommended)	15–18 bu.
Pumpkin	4 lbs.
Rape (broadcast)	4– 8 lbs.
Red-top (recleaned)	12–15 lbs.
Rice	1– 3 bu.
Rye	4– 7 pks.
Sugar beets	15–20 lbs.
Sweet potato	1½–4 bu.
Timothy	10–20 lbs.
Timothy and clover:	
Timothy	10–15 lbs.
Clover	4–10 lbs.
Turnip (broadcast)	2– 4 lbs.
Vetch (hairy) 1 bu. plus 1 bu. small grain.	
Wheat	6– 9 pks.

WEIGHT AND MEASURE OF FEEDSTUFFS

FEED	ONE QUART WEIGHS	ONE POUND MEASURES
Cottonseed meal	1.5 lbs.	0.7 qt.
Wheat middlings (flour)	1.2 lbs.	0.8 qt.
Wheat middlings (standard)	0.8 lb.	1.3 qts.
Wheat mixed feed	0.6 lb.	1.7 qts.
Wheat bran	0.5 lb.	2.0 qts.
Whole oats	1.0 lb.	1.0 qt.
Ground oats	0.7 lb.	1.4 qts
Whole wheat	1.9 lbs.	0.5 qt.
Ground wheat	1.7 lbs.	0.6 qt.
Whole corn	1.7 lbs.	0.6 qt.
Corn meal	1.5 lbs.	0.7 qt.
Corn and cob meal	1.4 lbs.	0.7 qt.
Corn bran	0.5 lb.	2.0 qts.
Hominy meal	1.1 lbs.	0.9 qt.
Corn and oat feed	0.7 lb.	1.4 qts.
Whole barley	1.5 lbs.	0.7 qt.
Barley meal	1.1 lbs.	0.9 qt.
Whole rye	1.7 lbs.	0.6 qt.
Rye meal	1.5 lbs.	0.7 qt.
Rice bran	0.8 lb.	1.3 qts.
Rice polish	1.2 lbs.	0.8 qt.
Cottonseed hulls	0.26 lb.	3.8 qts.
Alfalfa meal	0.98 lb.	1.3 qts.
Molasses (blackstrap)	3.0 lbs.	0.3 qt.

THE USUAL LEGAL WEIGHTS PER BUSHEL OF SOME FARM PRODUCTS

NAME OF MATERIAL	WEIGHT IN POUNDS
Apples	48
Apples (dried)	24
Barley	48
Beans	60
Buckwheat	52
Carrots	50
Clover seed	60
Corn (ear)	70
Corn (shelled)	56
Cottonseed	32
Flaxseed	56
Kentucky bluegrass (seed)	14
Millet	50
Oats	32
Onions	57
Peas	60
Potatoes (Irish)	60
Potatoes (sweet)	55
Rye	56
Timothy seed	45
Turnips	55
Wheat	60

GLOSSARY

(The number indicates the paragraph in which the word is defined and the pronunciation given. When the word occurs in the Appendix, the reference gives the number of the Appendix.)

INDEX

(Numbers refer to paragraphs, except when chapter or figure is specified.)